AF361807

Pb-Zn Deposits and Associated Critical Metals

Pb-Zn Deposits and Associated Critical Metals

Editors

Jia-Xi Zhou
Changqing Zhang
Tao Ren
Yue Wu

Basel • Beijing • Wuhan • Barcelona • Belgrade • Novi Sad • Cluj • Manchester

Editors

Jia-Xi Zhou
School of Earth Sciences
Yunnan University
Kunming
China

Changqing Zhang
Institute of Mineral Resources
Chinese Academy of
Geological Sciences
Beijing
China

Tao Ren
Faculty of Land and
Resources Engineering
Kunming University of
Science and Technology
Kunming
China

Yue Wu
College of Resources and
Environment
Yangtze University
Wuhan
China

Editorial Office
MDPI
St. Alban-Anlage 66
4052 Basel, Switzerland

This is a reprint of articles from the Special Issue published online in the open access journal *Minerals* (ISSN 2075-163X) (available at: www.mdpi.com/journal/minerals/special_issues/D0QECS6P05).

For citation purposes, cite each article independently as indicated on the article page online and as indicated below:

Lastname, A.A.; Lastname, B.B. Article Title. *Journal Name* **Year**, *Volume Number*, Page Range.

ISBN 978-3-0365-8519-2 (Hbk)
ISBN 978-3-0365-8518-5 (PDF)
doi.org/10.3390/books978-3-0365-8518-5

Cover image courtesy of Jia-Xi Zhou

Contents

About the Editors

Jia-Xi Zhou

Dr. Jia-Xi Zhou has been a professor and director at the Key Laboratory of Critical Minerals Metallogeny in Universities of Yunnan Province, School of Earth Sciences, Yunnan University since 2018. Until then, he is an important member of the Stake Key Laboratory of Ore Deposit Geochemistry, Institute of Geochemistry, Chinese Academy of Sciences. As a visiting researcher, he has conducted more than 1 year of substantial academic communications and collaborations with the School of Earth and Environmental Sciences, The University of Queensland, Australia. He has published more than 100 papers in journals indexed in the Web of Science, such as *Economic Geology*, *Mineralium Deposita*, *American Mineralogist*, *Precambrian Research*, *Gondwana Research*, *China Science: Earth Science*, and *Ore Geology Reviews*. His research is mainly focused on the mineralization and prospecting prediction of hydrothermal deposits, the in situ U-Pb chronology of carbonate minerals, and enrichment mechanisms of critical metals (such as Ge, Ga, Cd, In, Se, Tl, Nb, and REE). He is the Associate Editor of *Ore Geology Reviews* and *Journal of Asian Earth Sciences: X*, the Editorial Board of *Minerals*, and is on the Youth Editorial Board of *China Geology* and *Geotectonica et Metallogenia*.

Changqing Zhang

Dr. Zhang Changqing is a researcher at the Ministry of Natural Resources (MNR), Key Laboratory of Metallogeny and Mineral Assessment, Institute of Mineral Resources, Chinese Academy of Geological Sciences since 2008. He has published more than 40 papers in journals indexed in the Web of Science, such as *Journal of Asian Earth Sciences*, *Earth and Planetary Science Letters*, *Journal of Geochemical Exploration*, *Acta geologica sinica*, *Acta Petrologica Sinica*, *Mineral Deposits*, and *International Geology Review*. His research mainly focuses on the metallogenic regularity and prospecting prediction analysis of gold, lead, zinc, and other polymetallic deposits.

Tao Ren

Dr. Tao Ren is a researcher at the Faculty of Land and Resources Engineering, Kunming University of Science and Technology since 2011. He was elected as one of the "Top Young Talents in Yunnan Province's Ten Thousand Talents Plan". His research interests include geology, ore deposit geochemistry, and mineral prospecting and exploration. He has focused on the genesis and metallogenic regularity of copper polymetallic deposits in Zhongdian island arc, tin, and tungsten deposits in SW Yunnan, as well as lead and zinc deposits in NE Yunnan. He has published more than 40 papers in journals indexed in the Web of Science, including *Ore Geology Reviews*, *Precambrian Research*, and *Journal of Asian Earth Sciences*.

Yuc Wu

Dr. Wu Yue completed his PhD degree in ore deposit at the China University of Geosciences (Beijing) in 2013. He then joined the School of Resources and Environment, Yangtze University, where he is currently an associate professor. His research focusses on the low-temperature hydrothermal deposits, especially on carbonated-hosted Zn-Pb deposits as well as non-sulfide Zn-Pb deposits. He has published more than 20 papers in journals indexed in the Web of Science.

Preface

Pb-Zn deposits, and the critical metals within them, are critical for the functioning of human society. Although extensive studies have been carried out on Pb-Zn deposits, several important mechanisms remain unelucidated, including those underlying the detailed ore-forming process, the enrichment of Pb-Zn and associated critical metals and their genetic relationship, and the occurrence of critical metals. Recently, there have been many new advances that are essential for elucidating those important mechanisms. It is timely to gather some thematic papers to update recent developments of geological studies.

This Special Issue comprises 12 articles, concerning ore deposit geology, mineralogy, geochemistry, ore genesis, ore prospecting, computational simulation, big data, and the deep learning of Pb-Zn deposits and associated critical metals.

This preface outlines the papers collected in the Special Issue "Pb-Zn Deposits and Associated Critical Metals".

Sun and Zhou (2022) reported the application of machine learning algorithms to classify Pb-Zn deposit types using LA-ICP-MS data of sphalerite. Cui et al. (2022) presented fractal structure characteristics and prospecting directions of dispersed metals in the Eastern Guizhou Pb-Zn metallogenic belt, SW China. Jia et al. (2022) published a study on the 3D quantitative metallogenic prediction of indium-rich ore bodies in the Dulong Sn-Zn polymetallic deposit, Yunnan Province, SW China. He et al. (2022) analyzed the origin of carbonate components in carbonate-hosted Pb-Zn deposits in the Sichuan-Yunnan-Guizhou Pb-Zn metallogenic province and Southwest China, taking Lekai Pb-Zn deposit as an example. Jiang et al. (2023) discussed the origin of the Caiyuanzi Pb-Zn deposit in the SE Yunnan Province, China (Constraints from in situ S and Pb isotopes). Li et al. (2023) displayed the pore variation characteristics of altered wall rocks in the Huize lead-zinc deposit in Yunnan, China and their geological significance. Chen et al. (2023) studied the genesis of pyrite in the Fule Pb-Zn deposit, Northeast Yunnan Province, China (Evidence from mineral chemistry and in situ sulfur isotopes). An et al. (2023) revealed the trace elements of gangue minerals from the Banbianjie Ge-Zn deposit in Guizhou Province, SW China. Zhang et al. (2023) issued the zircon and garnet U-Pb ages of the Longwan skarn Pb-Zn deposit in Guangxi Province, China and their geological significance. Wu et al. (2023) expressed the LA-ICP-MS trace element geochemistry of sphalerite and metallogenic constraints (A case study from the Nanmushu Zn-Pb deposit in Mayuan district, Shaanxi Province, China). Chen et al. (2023) explained the genesis of the giant Huoshaoyun non-sulfide zinc–lead deposit in Karakoram, Xinjiang (Constraints from mineralogy and trace element geochemistry). Ali et al. (2023) showed the integration of electrical resistivity tomography and induced polarization for the characterization and mapping of (Pb-Zn-Ag) sulfide deposits.

Jia-Xi Zhou, Changqing Zhang, Tao Ren, and Yue Wu

Editors

Article

Application of Machine Learning Algorithms to Classification of Pb–Zn Deposit Types Using LA–ICP–MS Data of Sphalerite

Guo-Tao Sun [1,2,3,*] and Jia-Xi Zhou [4,5]

[1] State Key Laboratory of Public Big Data, Guizhou University, Guiyang 550025, China
[2] College of Resources and Environmental Engineering, Guizhou University, Guiyang 550025, China
[3] Key Laboratory of Karst Georesources and Environment, Ministry of Education, Guiyang 500025, China
[4] School of Earth Sciences, Yunnan University, Kunming 650500, China
[5] Key Laboratory of Critical Minerals Metallogeny in Universities of Yunnan Province, Kunming 650500, China
* Correspondence: gtsun@gzu.edu.cn

Abstract: Pb–Zn deposits supply a significant proportion of critical metals, such as In, Ga, Ge, and Co. Due to the growing demand for critical metals, it is urgent to clarify the different types of Pb–Zn deposits to improve exploration. The trace element concentrations of sphalerite can be used to classify the types of Pb–Zn deposits. However, it is difficult to assess the multivariable system through simple data analysis directly. Here, we collected more than 2200 analyses with 14 elements (Mn, Fe, Co, Ni, Cu, Ga, Ge, Ag, Cd, In, Sn, Sb, Pb, and Bi) from 65 deposits, including 48 analyses from carbonate replacement (CR), 684 analyses from distal magmatic-hydrothermal (DMH), 197 analyses from epithermal, 456 analyses from Mississippi Valley-type (MVT), 199 analyses from sedimentary exhalative (SEDEX), 377 analyses from skarn, and 322 analyses from volcanogenic massive sulfide (VMS) types of Pb–Zn deposits. The critical metals in different types of deposits are summarized. Machine learning algorithms, namely, decision tree (DT), K-nearest neighbors (KNN), naive Bayes (NB), random forest (RF), and support vector machine (SVM), are applied to process and explore the classification. Learning curves show that the DT and RF classifiers are the most suitable for classification. Testing of the DT and RF classifier yielded accuracies of 91.2% and 95.4%, respectively. In the DT classifier, the feature importances of trace elements suggest that Ni (0.22), Mn (0.17), Cd (0.13), Co (0.11), and Fe (0.09) are significant for classification. Furthermore, the visual DT graph shows that the Mn contents of sphalerite allow the division of the seven classes into three groups: (1) depleted in Mn, including MVT and CR types; (2) enriched in Mn, including epithermal, skarn, SEDEX, and VMS deposits; and (3) DMH deposits, which have variable Mn contents. Data mining also reveals that VMS and skarn deposits have distinct Co and Ni contents and that SEDEX and DMH deposits have different Ni and Ge contents. The optimal DT and RF classifiers are deployed at Streamlit cloud workspace. Researchers can select DT or RF classifier and input trace element data of sphalerite to classify the Pb–Zn deposit type.

Keywords: machine learning; sphalerite; LA–ICP–MS; Pb–Zn deposits; web app

Citation: Sun, G.-T.; Zhou, J.-X. Application of Machine Learning Algorithms to Classification of Pb–Zn Deposit Types Using LA–ICP–MS Data of Sphalerite. *Minerals* **2022**, *12*, 1293. https://doi.org/10.3390/min12101293

Academic Editor: George M. Gibson

Received: 18 September 2022
Accepted: 10 October 2022
Published: 14 October 2022

Publisher's Note: MDPI stays neutral with regard to jurisdictional claims in published maps and institutional affiliations.

1. Introduction

Base metal (Pb–Zn) deposits are important sources of critical metals, such as indium (In), germanium (Ge), gallium (Ga), cobalt (Co), and cadmium (Cd). Distinct types of Pb–Zn deposits are enriched in different critical metals due to different sources and ore-forming processes. Distinguishing the Pb–Zn types is essential for identifying new sources of critical metals and enhancing exploration efficiency. Previous studies identified the differences in the trace elements of sphalerite from different types of Pb–Zn deposits [1,2]. However, the studies lack statistical analysis to distinguish the different deposit types reliably. Frenzel et al. [3] applied principal component analysis (PCA) to identify the differences among different types of Pb–Zn deposits. However, PCA is a dimensionality reduction method and is less effective in classification. Therefore, the study did not show classification for different types of Pb–Zn deposits.

Machine learning (ML) is an effective empirical approach for classifying nonlinear systems. Such systems are massively multivariate, involving a few or literally thousands of variables. The types of ML algorithms for classification mainly include K-nearest neighbor (KNN), decision tree (DT), support vector machine (SVM), artificial neural network (ANN), random forest (RF), and naive Bayes (NB). ML has been widely applied to science and engineering problems, such as data mining, artificial intelligence, DNA sequencing, and pattern recognition. The application of ML in geoscience, especially economic geology, is new and limited [4–7].

Here, we collected over 2200 laser ablation–inductively coupled plasma–mass spectrometry (LA–ICP–MS) analyses of sphalerite from 65 deposits (Figure 1) and applied DT, KNN, NB, RF, and SVM algorithms on the Scikit-learn package in Python to classify the types of Pb–Zn deposits using trace element concentrations of sphalerite. Our contribution is twofold: We provide a statistical classification for different types of Pb–Zn deposits and deploy the classifications app online to be accessed by economic geologists.

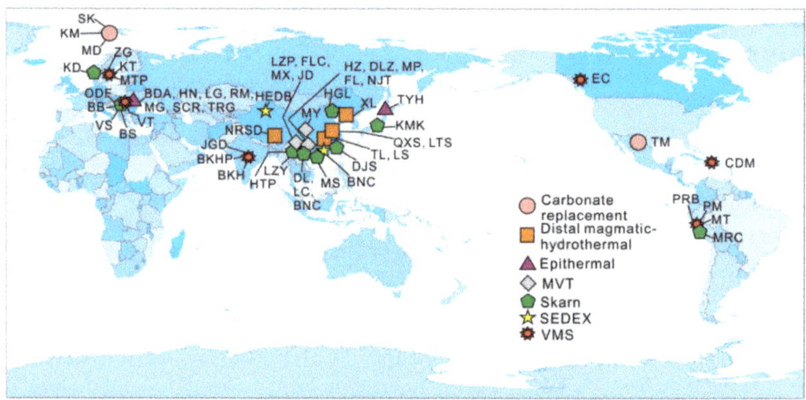

Figure 1. Principal Pb–Zn deposits that reported LA–ICP–MS data of sphalerite. CR type: TM, Tres Marias; SK, Sinkholmen; KM, Kapp Mineral; MD, Melandsgruve. DMH type: TL, Taolin; XL, Xinling; LTS, Luotuoshan; NRSD, Narusongduo; QXS, Qixiashan; LS, Lishan. Epithermal type: BDA, Baia de Aries; HN, Hanes; LG, Larga; RM, Rosia Montana; MG, Magura; SCR, Sacaramb; TRG, Toroiaga; TYH, Toyoha. MVT: DLZ, Daliangzi; HZ, Huize; MX, Mengxing; LZP, Liziping; FLC, Fulongchang; NJT, Niujiaotang; JD, Jinding; MP, Maoping; FL, Fule; MY, Mayuan. Skarn type: HTP, Hetaoping; LZY, Luziyuan; ODF, Ocna de Fier; BB, Baita Bihor; VS, Valea Seaca; BS, Baisoara; KD, Konnerudkollen; KMK, Kamioka; DL, Dulong; LC, Laochang; MS, Miaoshan; MRC, Morococha; BNC, Bainiuchang. SEDEX type: DBS, Dabaoshan; HEDB, Haerdaban. VMS type: VT, Vorta; EC, Eskay Creek; ZG, Zinkgruvan; KT, Kaveltorp; MTP, Marketorp; BKHP, Banskhapa; JGD, Jangaldehri; BKH, Biskhan; MT, María Teresa; PRB, Perubar; PM, Palma; CDM, Cerro de Maimón. Note: Xinling both has DMH and epithermal types of mineralization; Morococha has skarn, epithermal and DMH types of mineralization.

2. Data Preparation and Packages

Data preparation includes data collection and data preprocessing for statistical analyses. Data collection and preprocessing were primarily conducted in Microsoft Excel.

2.1. Data Sources

For data consistency, the collected trace element concentrations of sphalerite were mainly determined by LA–ICP–MS analysis. The LA–ICP–MS data have been collected from published articles [1,2,8–31], leading to a database of 2283 sphalerites from carbonate replacement (CR), distal magmatic-hydrothermal (DMH), epithermal, Mississippi Valley-type (MVT), sedimentary exhalative (SEDEX), skarn, and volcanogenic massive sulfide (VMS) types of Pb–Zn deposits (Table 1).

Table 1. Summary of the collected sphalerite LA–ICP–MS dataset.

Deposit	Country	Type	Number	References	Deposit	Country	Type	Number	References
Tres Marias	Mexico	CR	22	[1]	Mayuan	China	MVT	50	[8]
Sinkholmen	Norway	CR	8	[1]	Hetaoping	China	Skarn	24	[7]
Kapp Mineral	Norway	CR	10	[1]	Luziyuan	China	Skarn	24	[7]
Melandsgruve	Norway	CR	8	[1]	Majdanpek	Serbia	Skarn	8	[1]
Taolin	China	DMH	64	[21]	Ocna de Fier	Romania	Skarn	37	[1]
Xinling	China	DMH	25	[20]	Baita Bihor	Romania	Skarn	30	[1]
Luotuoshan	China	DMH	35	[12]	Valea Seaca	Romania	Skarn	6	[1]
Narusongduo	China	DMH	66	[16]	Baisoara	Romania	Skarn	20	[1]
Qixiashan	China	DMH	122	[9,19]	Lefevre	Canada	Skarn	8	[1]
Morococha	Peru	DMH	323	[28]	Konnerudkollen	Norway	Skarn	5	[1]
Weilasituo	China	DMH	22	[11]	Kamioka	Japan	Skarn	8	[1]
Lishan	China	DMH	27	[21]	Dulong	China	Skarn	57	[23]
Baia de Aries	Romania	Epithermal	6	[1]	Laochang	China	Skarn	16	[26]
Hanes	Romania	Epithermal	8	[1]	Miaoshan	China	Skarn	10	[10]
Larga	Romania	Epithermal	8	[1]	Huanggangliang	China	Skarn	2	[13]
Rosia Montana	Romania	Epithermal	20	[1]	Dingjiashan	China	Skarn	52	[27]
Magura	Romania	Epithermal	8	[1]	Morococha	Peru	Skarn	52	[28]
Sacaramb	Romania	Epithermal	11	[1]	Bainiuchang	China	Skarn	18	[7]
Toroiaga	Romania	Epithermal	6	[1]	Dabaoshan	China	SEDEX	26	[7]
Toyoha	Japan	Epithermal	22	[1]	Haerdaban	China	SEDEX	173	[29]
Wunuer	China	Epithermal	82	[18]	Vorta	Romania	VMS	8	[1]
Xinling	China	Epithermal	19	[20]	Eskay Creek	Canada	VMS	12	[1]
Morococha	Peru	Epithermal	7	[28]	Zinkgruvan	Sweden	VMS	5	[1]
Daliangzi	China	MVT	85	[14]	Kaveltorp	Sweden	VMS	8	[1]
Huize	China	MVT	24	[7]	Marketorp	Sweden	VMS	8	[1]
Mengxing	China	MVT	18	[7]	Sauda Sa	Norway	VMS	10	[1]
Liziping	China	MVT	67	[30]	Banskhapa	Indian	VMS	5	[25]
Fulongchang	China	MVT	48	[30]	Jangaldehri	Indian	VMS	10	[25]
Angouran	Iran	MVT	43	[17]	Biskhan	Indian	VMS	11	[25]
Niujiaotang	China	MVT	26	[7]	María Teresa	Peru	VMS	141	[31]
Jinding	China	MVT	24	[7]	Perubar	Peru	VMS	50	[31]
Maoping	China	MVT	49	[24]	Palma	Peru	VMS	37	[31]
Fule	China	MVT	22	[15]	Cerro de Maimón	Dominican Republic	VMS	17	[31]

2.2. Data Preprocessing

Data preprocessing is a process that fills in missing values, such as some analyses lacking several trace element concentrations and some values below the detection limits.

Most samples included the concentrations of Mn, Fe, Co, Ni, Cu, Ga, Ge, Ag, Cd, In, Sn, Sb, Pb, and Bi. The lack of As, Mo, Hg, Se, Bi, and Tl is significant. Therefore, these elements are excluded from the data. Other unanalyzed data were filled based on the mean value for the elements from others of the same type in the dataset. These data were assumed to be reasonable estimates, as these elements are commonly below detection. This method has been used by Gregory et al., (2019) for data preprocessing. When analyses were below the detection limits, either the detection limit was used, or a value based on nearby values was inserted.

2.3. Library and Package Preparation

Numpy and Pandas are fundamental Python libraries for scientific computing. Streamlit is an open-access library that can easily create custom web apps for machine learning. We use Scikit-learn, a simple and efficient tool for machine learning in Python for classification. The package splits the raw data and trains and tests the DT and RF classifiers. Graphviz, a graph visualization package, is used to represent the structural information of decision trees. The Streamlit Cloud is a workspace for deploying and managing Streamlit apps.

3. Description of ML Methods and Pb–Zn Deposits

3.1. Description of ML Methods

In this study, we applied ML methods, including DT, KNN, NB, RF, and SVM, to establish the classifiers. A simple description of these methods is presented below.

The DT is an effective classification method in data mining classification [32]. It is defined as a process that partitions a dataset into smaller classes. The decision rules are based on the tests defined at each branch [33]. The DT comprises three types of nodes: a root node that has no parent node, some internal nodes (splits) that have both parent and descendant nodes, and a set of terminal nodes (leaves) with no descendant nodes [34].

The KNN method is based on the spatial similarity between a test sample and its k neighbors [35]. The distance is computed in the feature space from the test sample to each sample for which the label is known. The estimate of the test sample is based on the label of k-nearest samples [36]. Therefore, the parameter k is the most important for KNN. As a local method, the KNN is known to be strong in the case of large data and low dimensions [37].

NB is a special form of Bayesian network that is one of the most effective theoretical models in the field of uncertain knowledge expression and reasoning. NB assumes that all variables are mutually independent [38]. Given an unclassified sample x with features $(a_1, a_2, \ldots, a_m)$ and a labeled class C with members $(y_1, y_2, \ldots, y_n)$, if $P(y_k \mid x) = \max\{P(y_1 \mid x), P(y_2 \mid x), \ldots, P(y_n \mid x)\}$, x belongs to y_k. According to the Bayesian principle, the following derivation is made: $P(y_k \mid x) = P(x \mid y_k)\,P(y_k)/P(x)$. NB can further simplify the calculation process to $P(x \mid y_k)^*P(y_k) = P(a_1 \mid y_k)\,P(a_2 \mid y_k), \ldots, P(a_m \mid y_k)\,P(y_k)$ [38].

RF is an ensemble ML algorithm that combines a set of decision trees for classification and prediction [39]. A number of features are randomly chosen for a single DT. The bootstrapping method randomly chooses training data for a single DT. The examples are classified by taking a majority vote cast from all the DT predictors [40].

The SVM method is based on statistical learning theory and is used to determine the location of decision boundaries (optimal hyperplane) that maximizes the distance between the classes [41]. The support vector machine can be linear or nonlinear. In a binary classification problem where classes are linearly separable, the hyperplane corresponds to a linear boundary, and the SVM selects the boundary that produces the maximum margin between the two classes [42]. If the binary classification problem is not linearly separable, the SVM is designed to identify a hyperplane (e.g., plane and sphere) that maximizes the margin. For nonlinear SVM problems, kernel functions, such as polynomial and sigmoid functions, are used to reduce the computational cost of dealing with high-dimensional space by adding an additional dimension to the data [40,41].

3.2. Description of Deposits and Samples

The LA–ICP–MS data were collected from 65 deposits worldwide (Table 1). The origin types and references of each deposit are shown in Supplementary Table S1. The dataset comprises 48 analyses from four CD deposits, 684 analyses from eight DMH deposits, 197 analyses from eleven epithermal deposits, 456 analyses from eleven MVT deposits, 377 analyses from seventeen skarn deposits, 199 analyses from two SEDEX deposits, and 322 analyses from thirteen VMS deposits.

4. Results

4.1. Learning Curves

The learning curves of DT, KNN, NB, RF, and SVM are shown in Figure 2. The learning curve of the DT classifier shows that the accuracy scores of the training data are high, whereas the accuracy scores of cross-validation increase with the increase in the amount of training data (Figure 2a,b). Because the training scores are higher than the cross-validation scores, the classifier is overfitting. The learning curves of KNN increase with increasing training data size. When the training data size is above 1500, the cross-validation scores are up to 0.8 (Figure 2c). The learning curves of the NB classifier converge to 0.38 (Figure 2d), indicating that this classifier is underfitting. The learning curves of the

RF classifier show similar characteristics to those of the DT classifier. The performance of cross-validation is better than that of DT. The cross-validation scores reach 0.95 when the scale of training data is up to 1500 (Figure 2e,f). The learning curves of SVM show that the classifier performs better when the training data are above 1200. The cross-validation scores of this classifier are nearly 0.65 (Figure 2g). The learning curves show that the RF classifier has the best performance for the dataset, whereas the NB has the worst performance.

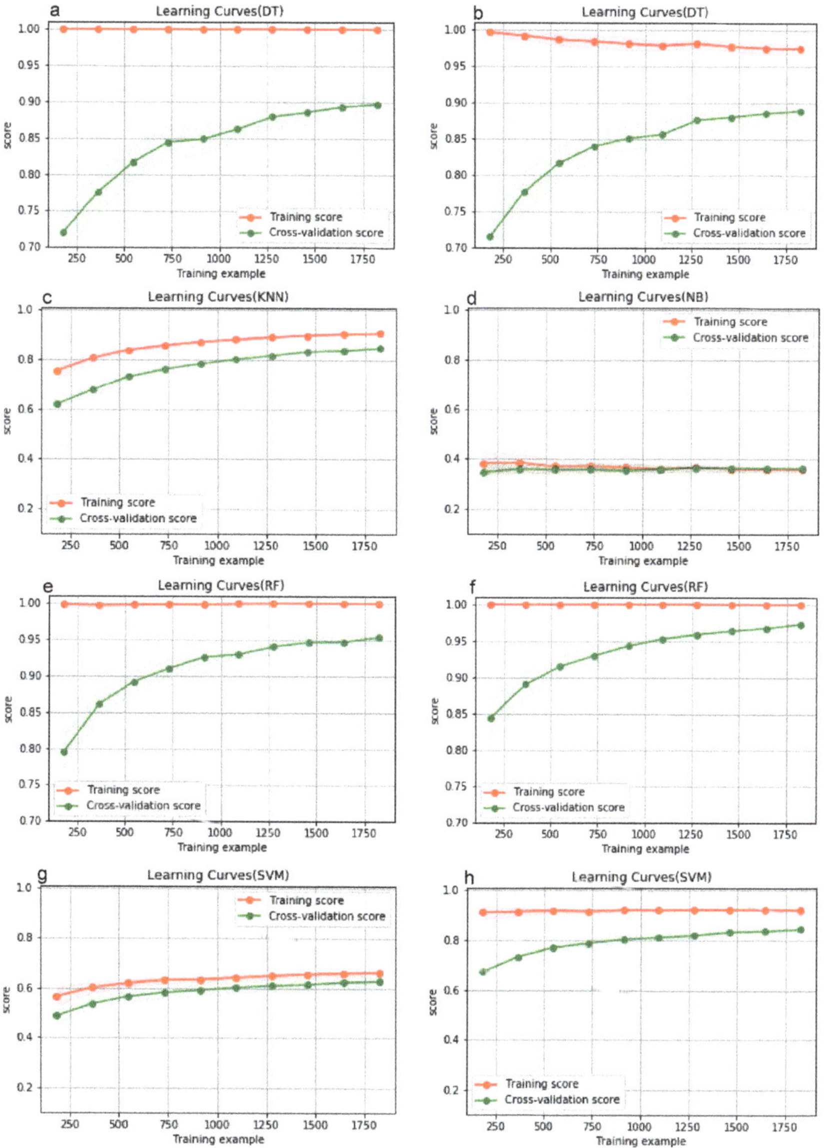

Figure 2. The learning curves of ML-classifiers. (**a**) decision tree classifier, (**b**) decision tree classifier after tuning hyperparameters, (**c**) K-nearest neighbors classifier, (**d**) naive Bayes classifier, (**e**) random forest classifiers, (**f**) random forest classifier after tuning hyperparameters, (**g**) support vector machine classifier, (**h**) support vector machine classifier after tuning hyperparameters.

The DT and RF classifiers have different degrees of overfitting, whereas the KNN, NB, and SVM classifiers have different degrees of underfitting. The scores of the underfitting classifiers cannot be improved by increasing the size of the training dataset. The tuning hyperparameters can somewhat improve the accuracies of the underfitting classifiers. For example, the grid search techniques found that the optimal SVM classifier had an accuracy of up to 0.85 (Figure 2h). The overfitting classifiers can be optimized by tuning the hyperparameters and increasing the size of the dataset. Therefore, the DT and RF classifiers are further tuning the hyperparameters by grid search techniques. Parameter optimization decreases the overfitting or improves the overall accuracies of the classifiers (Figure 2b,f).

4.2. Feature Importances

Feature importances are defined as the total decrease in node impurity. If the value is low, then the feature is not important, and vice versa. The feature importances of elements used for ML methods are shown in Figure 3. Among them, the feature importances of Ni (0.22), Mn (0.17), Cd (0.13), Co (0.11), and Fe (0.09) are higher than those of other elements, suggesting that Ni, Mn, Cd, Co, and Fe are effective in classifying the types of deposits.

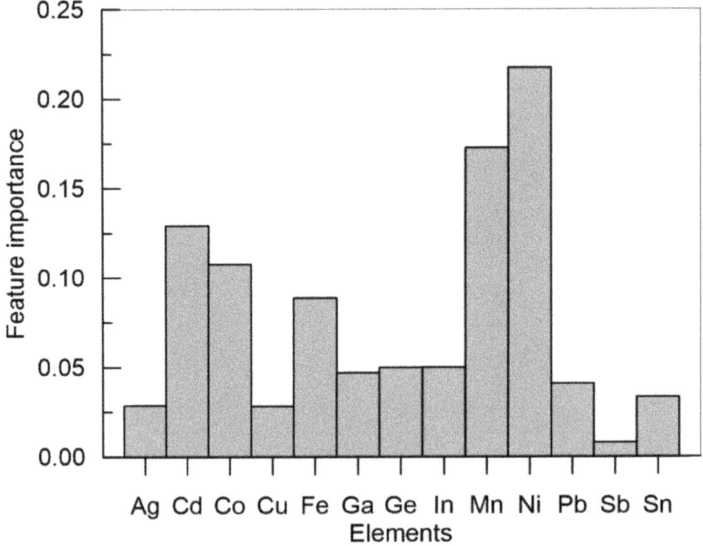

Figure 3. Feature importances of each element that was used in classifiers.

4.3. Accuracies of the DT and RF Classifiers

As presented above, the DT and RF algorithms show good performance for the classification problem. Therefore, the two algorithms are further assessed based on other parameters, such as accuracy score, precision, and recall. The accuracy scores of the DT and RF classifiers are listed in Table 2. In this work, 70% of the data are used as training data to produce the models, whereas the remaining data are used to test the performance of the models. These classifiers were run 10 times with random selections of training data and testing data to assess the effectiveness of the classifiers. The accuracy scores of the DT classifier are between 0.871 and 0.907, with a mean of 0.890 and a standard deviation (SD) of 0.011. The accuracy scores of the RF classifier range from 0.953 to 0.976, with a mean of 0.969 and an SD of 0.007. The accuracy scores show that the RF classifier is effective in distinguishing the different types of deposits.

Table 2. Precision, recall, and F1 score of classifiers.

Classifiers		DT Classifier								RF Classifier						
Types		CR	DMH	Epithermal	MVT	SEDEX	Skarn	VMS		CR	DMH	Epithermal	MVT	SEDEX	Skarn	VMS
Percision	1	0.353	0.835	0.830	0.872	0.917	0.795	0.969	1	1.000	0.976	0.966	0.935	1.000	0.975	1.000
	2	0.692	0.921	0.825	0.917	0.963	0.870	0.960	2	1.000	0.975	0.966	0.963	1.000	0.968	0.990
	3	0.571	0.932	0.764	0.919	0.923	0.826	0.854	3	1.000	0.986	0.965	0.956	1.000	0.928	0.989
	4	0.750	0.911	0.800	0.886	0.849	0.860	0.906	4	1.000	0.952	1.000	0.914	0.980	0.972	0.989
	5	0.667	0.868	0.818	0.954	0.963	0.898	0.920	5	1.000	0.941	0.948	0.970	1.000	0.939	0.989
	Mean	0.607	0.894	0.807	0.910	0.923	0.850	0.922	Mean	1.000	0.966	0.969	0.948	0.996	0.956	0.992
	SD	0.139	0.037	0.024	0.028	0.042	0.036	0.041	SD	0.000	0.017	0.017	0.021	0.008	0.019	0.004
Recall	1	0.375	0.921	0.650	0.848	0.902	0.789	0.939	1	0.563	0.990	0.950	1.000	1.000	0.953	0.970
	2	0.600	0.907	0.825	0.905	0.963	0.934	0.941	2	0.667	0.990	0.889	1.000	1.000	0.984	0.980
	3	0.571	0.894	0.689	0.895	0.923	0.905	0.936	3	0.571	0.986	0.902	1.000	0.969	0.981	0.979
	4	0.429	0.939	0.727	0.918	0.918	0.860	0.897	4	0.381	1.000	0.818	0.994	0.980	0.991	0.969
	5	0.750	0.952	0.652	0.901	1.000	0.882	0.939	5	0.625	0.990	0.797	0.988	1.000	0.973	0.949
	Mean	0.545	0.923	0.709	0.893	0.941	0.874	0.930	Mean	0.561	0.991	0.871	0.996	0.990	0.976	0.969
	SD	0.133	0.021	0.065	0.024	0.036	0.049	0.017	SD	0.098	0.005	0.056	0.005	0.013	0.013	0.011
F1-score	1	0.364	0.876	0.729	0.860	0.909	0.792	0.954	1	0.720	0.983	0.958	0.967	1.000	0.967	0.985
	2	0.643	0.914	0.825	0.911	0.963	0.901	0.950	2	0.800	0.982	0.926	0.981	1.000	0.976	0.985
	3	0.571	0.913	0.724	0.907	0.923	0.864	0.893	3	0.727	0.986	0.932	0.977	0.984	0.954	0.984
	4	0.545	0.925	0.762	0.902	0.882	0.860	0.902	4	0.552	0.975	0.900	0.952	0.980	0.981	0.979
	5	0.706	0.908	0.726	0.927	0.981	0.890	0.929	5	0.769	0.965	0.866	0.979	1.000	0.955	0.969
	Mean	0.566	0.907	0.753	0.901	0.932	0.861	0.926	Mean	0.714	0.978	0.916	0.971	0.993	0.967	0.980
	SD	0.116	0.017	0.039	0.022	0.036	0.038	0.025	SD	0.086	0.008	0.031	0.011	0.009	0.011	0.006

The confusion matrix shows the prediction results and actual types of deposits (Figure 4). Precision, recall, and F1 scores can be calculated from the confusion matrix, and they can evaluate the classification for individual ore deposit types. Precision is defined as the ratio of correctly predicted samples to predicted samples. The recall is defined as the ratio of correctly predicted samples to actual samples. The F1 score is a measurement of precision and recall. The calculation of the F1 score is shown in Formula (1). The precision, recall, and F1 score are calculated five times with random selections of test data. The precision values for individual types of the DT range from 0.607 ± 0.139 to 0.923 ± 0.042. Carbonate replacement, DMH, MVT, SEDEX, skarn, and VMS test data were predicted with precisions of 0.607 ± 0.139, 0.894 ± 0.037, 0.807 ± 0.024, 0.910 ± 0.028, 0.923 ± 0.042, 0.850 ± 0.036, and 0.922 ± 0.041 on average, respectively. The recall mean values of the predicted DMH, epithermal, MVT, SEDEX, skarn, and VMS test data are 0.545 ± 0.133, 0.923 ± 0.021, 0.709 ± 0.065, 0.893 ± 0.024, 0.941 ± 0.036, 0.874 ± 0.049, and 0.930 ± 0.017, respectively. The corresponding F1 scores are 0.566 ± 0.116, 0.907 ± 0.017, 0.753 ± 0.039, 0.901 ± 0.022, 0.932 ± 0.036, 0.861 ± 0.038, and 0.926 ± 0.025 on average, respectively. The results show that the DT classifier has low precision, recall, and F1 score for carbonate replacement deposits.

$$\text{F1 score} = 2 \times \text{Precision} \times \text{Recall}/(\text{Precision} + \text{Recall}) \tag{1}$$

The precision for individual types of the RF ranges from 0.948 ± 0.021 to 0.92 ± 0.04. Carbonate replacement, DMH, epithermal, MVT, SEDEX, skarn, and VMS test data were predicted with precisions of 1.000, 0.966 ± 0.017, 0.969 ± 0.017, 0.948 ± 0.021, 0.996 ± 0.008, 0.956 ± 0.019, and 0.992 ± 0.004 on average, respectively. The mean values of recall are 0.561 ± 0.098, 0.991 ± 0.005, 0.871 ± 0.056, 0.996 ± 0.005, 0.990 ± 0.013, 0.976 ± 0.013, and 0.969 ± 0.011, respectively. The F1 mean scores are 0.714 ± 0.086, 0.978 ± 0.008, 0.916 ± 0.031, 0.971 ± 0.011, 0.993 ± 0.009, 0.967 ± 0.011, and 0.980 ± 0.006, respectively. The results show that the RF classifier has higher prediction accuracies for individual types than the DT classifier.

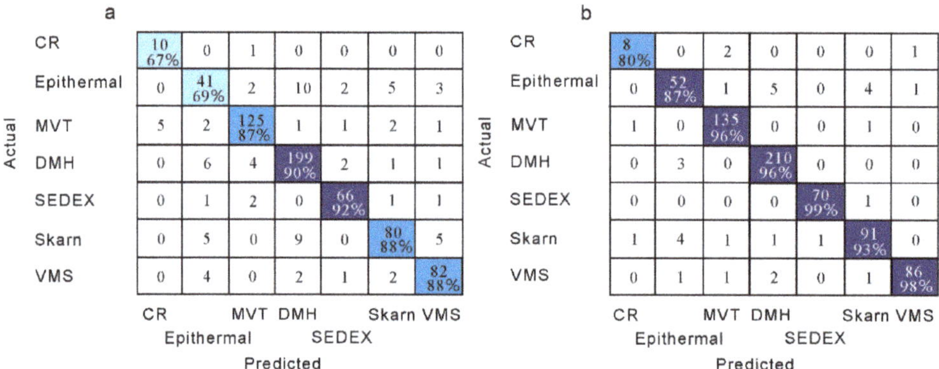

Figure 4. Confusion matrix for decision tree (**a**) and random forest (**b**) classifiers of test data. The numbers in the squares are the numbers of analyses and the percentages are precision values.

5. Discussion

5.1. Critical Metals in Sphalerite

It is well documented that sphalerite is a significant host mineral for critical metals such as Ga, Ge, Cd, In, Co, and Sn [1–3,43–47]. Here, the LA–ICP–MS dataset is firstly used to summarize the critical metal concentrations of sphalerite from different deposit types to investigate the special enrichment of critical metals.

Gallium is a by-product of some MVT deposits. Previous studies have indicated that some MVT deposits produce some Ga metal [46,48]. In this study, sphalerite from some CR,

DMH, epithermal, and VMS deposits also shows enrichment in Ga (Figure 5a), with mean values of 23.1 ppm, 70.5 ppm, 130 ppm, and 27.6 ppm, respectively. Among these types of deposits, high Ga contents are mainly reported from epithermal deposits (Rosia Montana, Romania, 1137 ppm; Sacaramb, Romania, 1126 ppm; Toyoha, Japan, 601 ppm; Xinling, China, 426 ppm; and Morococha, Peru, 1739 ppm) and DMH (Taolin, China, 649 ppm; Morococha, Peru, 2118 ppm; and Lishan, China, 381 ppm) (Supplementary Table S1).

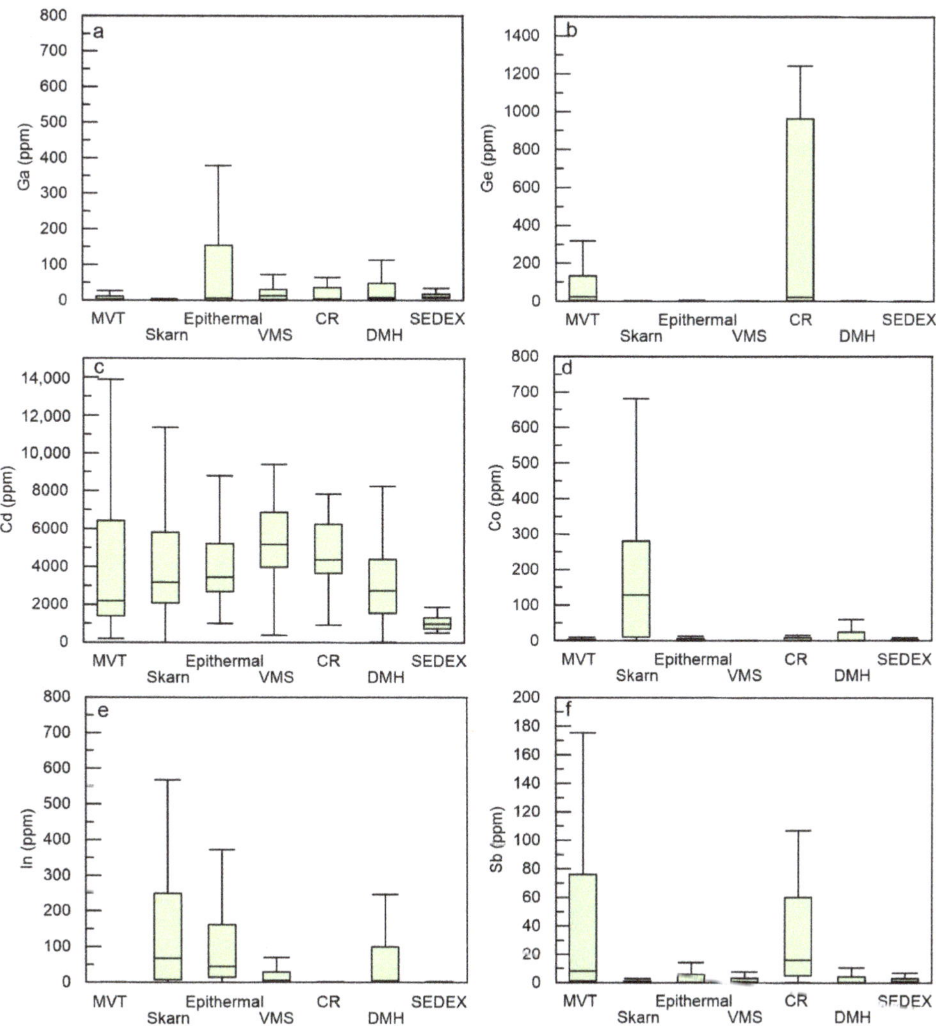

Figure 5. Critical metals in sphalerite from different types of deposits.

Germanium is considered to be mainly hosted in MVT sphalerite (Figure 5b). The Mayuan MVT deposit in China has the highest reported Ge contents (1305 ppm), with a mean value of 609 ppm [8]. Other MVT deposits, such as Dalingzi (China), Huize (China), Angouran (Iran), Niujiaotang (China), Maoping (China), and Fule (China), also have high Ge contents of 328, 354, 339, 288, 652, and 941 ppm, respectively (Supplementary Table S1). Notably, the Tres Marias CR deposits in Mexico show Ge contents ranging from 174 to 1242 ppm with a mean value of 704 ppm [1].

Cadmium is enriched in each type of Pb–Zn deposit (Figure 5c). The mean Cd values of sphalerite from the MVT, skarn, epithermal, VMS, CR, DMH, and SEDEX deposits are 5059 ppm, 5960 ppm, 4436 ppm, 5191 ppm, 4361 ppm, 3115 ppm, and 1572 ppm, respectively. The highest Cd content is reported from the Niujiaotang MVT deposit in China (23400 ppm) [7].

Cook et al. [1] reported that cobalt prefers to be enriched in sphalerite from skarn deposits. In the dataset, the Co contents of sphalerite from skarn deposits have a mean value of 237 ppm, which is higher than those of other types (Figure 5d). The highest Co contents of sphalerite are reported from the Ocna de Fier, Romania (2828 ppm) and Konnerudkollen, Norway (1585 ppm) deposits [1]. Some DMH deposits also reported some sphalerite analyses with high Co contents, such as Taolin, China (830 ppm) and Narusongduo, China (370 ppm) (Supplementary Table S1).

Indium has been reported to be mainly enriched in skarn and VMS deposits [1,48–53]. In the dataset, skarn, epithermal, VMS, and DMH deposits have higher In contents than other types (Figure 5e), with mean values of 225, 2732, 31.7, and 120 ppm. The highest In contents (64818 ppm) are reported from the Toyoha epithermal deposit in Japan [1], resulting in an extremely high mean value. The skarn deposits, such as Baita Bihor, Romania (867 ppm), Dulong, China (4572 ppm), Laochang, China (723 ppm), and Miaoshan, China (562 ppm), report In contents higher than 500 ppm [1,10,23,26]. The DMH deposits, including the Qixiashan, China (794 ppm) and Morococha, Peru (1804 ppm) deposits, have In contents higher than 500 ppm [19,28].

Antimony is mainly enriched in sphalerite from MVT and CR deposits (Figure 5f) with mean values of 103 and 43.3 ppm. The highest Sb contents are reported from the Eskay Creek VMS deposit in Canada (117467 ppm) [1]. For MVT deposits, the contents in the Furongchang, China (max 1131 ppm) [30] and Fule, China (max 1403 ppm) [15] deposits are higher than those of other deposits.

Overall, the critical metals in the MVT, skarn, epithermal, VMS, CR, DMH, and SEDEX deposits are Ge–Cd–Sb, Cd–Co–In, Ga–Cd–In, Ga–Cd–In, Ga–Ge–Cd–Sb, Ga–Cd–In, and Cd, respectively (Figure 5). Due to a lack of data, the dataset does not include data from some important deposits, such as Red Dog, leading to the summary being incomplete. Although the dataset has shortcomings, the suggestion that critical metals can correspond with further ML classifiers could somewhat facilitate exploration for critical metals.

5.2. Assessment of Different ML Methods for Sphalerite LA–ICP–MS Data

As shown in the learning curves, the NB classifier is the worst for distinguishing the different types of deposits (Figure 2d). The mechanism of NB may result in poor performance. The NB method is based on the hypothesis that the features are unrelated. However, some trace elements in sphalerite are related. For example, Fe and Mn are reported to be negatively related [54]. Correlations between Cu and Ge, Cu and In, and Ag and Sn are common in some deposits due to coupled substitution [1,2,23,55–57]. Therefore, the NB method may not be suitable for the sphalerite trace element data.

The SVM classifier has accuracies of approximately 0.6, as shown in Figure 2g. The problem may be due to the shortcomings of the SVM method. The classical SVM algorithm was originally designed for binary classification [42]. The multiclass classification in this study needs to be solved by combining several binary SVM classifiers, such as one-against-one and one-against-rest [58]. The one-against-rest method was applied in this study. The hyperparameters C and *gamma* are significant for the accuracy of the SVM classifier. The hyperparameter C is a penalty coefficient for misclassified samples. Higher C values will lead to fewer misclassified samples, narrower margins, and higher accuracies. The hyperparameter *gamma* represents the influence distance of the samples. Higher *gamma* values will result in smaller influence distances and higher accuracies. Therefore, we increase the values of C and *gamma* values. The learning curves of the optimal SVM ($C = 10$, *gamma* = 0.5) show that the accuracies are up to 0.85 and are better than those of the original SVM classifier ($C = 1$, *gamma* = auto).

The KNN algorithm has a poor effect in the case of unbalanced data, which easily results in misclassification. The numbers of analyses from carbonate replacement, DMH, epithermal, MVT, SEDEX, skarn, and VMS deposits are 48, 684, 197, 527, 199, 377, and 322, respectively, suggesting that the classes in this study are unbalanced. This may explain the accuracies being lower than 0.90 (Figure 2c). The accuracies may be improved by adding the analyses from carbonate replacement, epithermal, and SEDEX deposits.

The DT and RF classifiers produce better accuracies than other classifiers. The investigation for individual types finds that the predictions of the DT classifier for carbonate replacement deposits reveal that the precision and recall are lower than 0.65. The predictions of the RF classifier for carbonate replacement deposits reveal that the recall is lower than 0.60. The small size of the carbonate replacement data may lead to worse precisions and recalls. Further work can increase the size of the carbonate replacement data to improve accuracy.

According to the learning curves, the NB algorithm has the lowest accuracy (<0.4) and is not suitable for the classification of deposits based on trace element data. The SVM, KNN, and DT classifiers have accuracies between 0.8 and 0.9. They can be improved by modifying the data structure. The RF algorithm has the highest accuracies (>0.95) and may be the most suitable for the case in this study.

5.3. Statistical Element Characteristics of Different Types of Pb–Zn Deposits

Previous researchers have noticed that trace element concentrations of sphalerite are variable in different types of Pb–Zn deposits. For example, sphalerite from MVT deposits is enriched in Ga, Ge, and Cd, whereas sphalerite from skarn deposits is enriched in In. However, statistical analyses are rarely conducted on the trace elements of sphalerite. Here, the DT graph shows the statistical characteristics of trace elements (Mn, Fe, Co, Ni, Cu, Ga, Ge, Ag, Cd, In, Sn, Sb, Pb, and Bi) for classification (Supplementary Figure S1). For example, Mn is significant for distinguishing the different types of deposits. A total of 92% of analyses from MVT deposits and 94% of analyses from carbonate replacement deposits have Mn contents lower than 62.8 ppm, whereas 96% of skarn, 91% of SEDEX, 100% of VMS, and 99% of epithermal deposits have Mn contents higher than 62.8 ppm. Seventy percent of DMH deposits have values higher than 62.8 ppm. Several explanations can be invoked for the difference in Mn contents, such as formation temperatures and metal sources. Although the MVT and SEDEX deposits may form under similar temperatures [59], the SEDEX deposits have Mn contents higher than 62.8 ppm. Temperature is unlikely to be the reason for the difference. The epithermal, skarn, and VMS deposits mainly form from magmatic-hydrothermal fluids. The MVT deposits are considered to be unrelated to magmatic-hydrothermal fluids [60,61]. Therefore, metal sources may cause different Mn contents. The high Mn contents of sphalerite from SEDEX deposits may also result from different sources.

For the DMH deposits, the Mn contents of sphalerite are variable at the deposit, generation, and sample scales. At the deposit scale, sphalerite with low Mn contents (< 62.8 ppm) forms as the second or third generation in the deposit. For example, the first generation of sphalerite from the Taolin deposit in China has Mn contents between 63.8 and 549 ppm [21], whereas the second generation of sphalerite has Mn contents mainly between 12.9 and 58.4 ppm (Supplementary Table S1). The variance may result from fluid evolution. On a generation scale, the different samples from the second generation show variable Mn contents. For example, some samples of second-generation sphalerite from the Morococha district in Peru show high Mn contents (520–1949 ppm), whereas some samples of the second sphalerite show low Mn contents (<1.4–22 ppm) [28]. Sphalerite can be zoned, showing variable Mn concentrations in the same crystal. On a sample scale, the Mn contents of the second sphalerite can range from <1.4 to 1830 ppm [28]. The Mn contents of sphalerite at the deposit scale may reflect that the Mn concentrations of first-stage ore-forming fluids are high, whereas the Mn concentrations can be low due to the evolution of fluids or mixing with meteoric water. The variance in the generation scale may be caused by heterogeneous fluids. The heterogeneous Mn contents at the sample scale may result from self-organization processes [54,62].

The data mining reveals that sphalerite from the skarn and VMS deposits has distinct Co and Ni contents. Compared with that from skarn deposits, sphalerite from the VMS deposits has low Co (<33 ppm) and high Ni contents (>0.85 ppm). Although they both occur in magmatic-hydrothermal systems, the metal sources that control the composition of the orebodies may be different. The metal sources of most VMS deposits are from two sources: i) high-temperature reaction zones and ii) magmatic fluids [63]. The high-temperature reaction zones can release abundant Ni from ferromagnesian minerals [64]. The input of magmatic fluids may lead to partial enrichment in Co [65]. The metals of skarn deposits are mainly from felsic magmatic fluids [66,67], which result in significant enrichment in Co and depletion in Ni.

Data mining also finds that the sphalerite from SEDEX has higher Ni (>0.16 ppm) and Ge (>0.86 ppm) contents than that from DMH deposits. It is well documented that pyrite from SEDEX deposits has high Ni contents [5,42,68], and sphalerite from some SEDEX deposits (Red Dog) has Ge contents of approximately 100 ppm [48]. As discussed above, the data-driven ML method can discover the intrinsic structures of data, which are proven to be reasonable by geochemical features. Therefore, ML methods are suitable for exploring the statistical characteristics of geochemical data.

5.4. Sphalerite Prediction Application

Utilizing the Streamlit cloud workspace, we deploy the prediction app at https://share.streamlit.io/sun199908/sphalerite--prediction/main/sp-pr-app.py (assecced on 1 September 2022). Two ML algorithms (DT and RF) are provided for testing. Users can select the algorithms and input sphalerite trace element data in the sidebar. The "predict" button is used to start the prediction and display the results. The web app can be used to suggest the origin of some deposits that are debated or newly discovered by drilling. Because different origins of deposits have distinct mineralization regularity, a quick judgment of the origin of deposits is essential to guide further exploration. For example, the origin of Laochang Pb–Zn–Ag–Cu deposit in SW China is debated between VMS [69,70] and magmatic-hydrothermal mineralization [71,72]. The trace element data of sphalerite from the Laochang deposit [70] are inputted into the app. The web app automatically predicts that the deposit may be skarn in origin, which is consistent with the geochronologic evidence (the Re–Os age of pyrite and U–Pb age of hydrothermal titanite are consistent with the zircon U–Pb age) [71]. Then the exploration industry can explore the deposit as a skarn type rather than a VMS type. Furthermore, classification of origin can also timely indicate which critical metals in the Pb–Zn deposits may be recovered by the exploration industry.

Although the application can provide online services, the current version has two shortcomings. First, some important deposits were not included in the dataset, such as the Red Dog deposit in the USA or Mt. Isa in Australia, due to few reports or lack of access. The lack of some data may lead to a partially subjective classification model. The application cannot perform well for all Pb–Zn deposits worldwide. Second, the application only considers the trace elements of sphalerite, which is one aspect of ore genesis. Ore genesis can also be reflected by other geochemical data, such as formation temperature, salinity, and sulfur isotopic composition. These data will be involved in the prediction application in future versions. Furthermore, the host rocks, structure, and other geological characteristics can be transformed into available data and included in future models, as these characteristics could be useful for distinguishing ore deposit types.

6. Conclusions

Based on the trace element (Mn, Fe, Co, Ni, Cu, Ga, Ge, Ag, Cd, In, Sn, Sb, and Pb) contents of sphalerite, the DT, KNN, NB, RF, and SVM algorithms were applied to train classifiers that distinguish ore deposit type. The RF algorithm is most suitable for the classification case, with an overall accuracy of 0.969 ± 0.007. The significant critical metals hosted in different types of deposits are summarized based on our dataset. The data mining reveals three statistical characteristics of the trace element data of sphalerite: (1) carbonate replacement and MVT deposits mainly have Mn contents lower than 62.8 ppm, whereas epithermal, SEDEX, skarn, and VMS deposits have Mn contents higher than 62.8 ppm; (2) compared to skarn deposits, VMS deposits have lower Co and higher Ni contents; and (3) compared to DMH deposits, SEDEX deposits have higher Ni and Ge contents. To enable economic geologists to access predictions online, a web app has been created and deployed at https://share.streamlit.io/sun199908/sphalerite-prediction/main/sp-pr-app.py, accessed on 1 September 2022.

Supplementary Materials: The following supporting information can be downloaded at: https://www.mdpi.com/article/10.3390/min12101293/s1. Figure S1: Decision tree graph showing the decision processes. Gini values and the color of nodes represent the degree of confusion. The smaller Gini values and deeper color mean a lower degree of confusion. Values present the numbers of analyses from CR, epithermal, MVT, DMH, SEDEX, skarn, VMS at the nodes. Table S1: Complete LA–ICP–MS data of sphalerite.

Author Contributions: Investigation, G.-T.S.; Methodology, G.-T.S.; Writing—original draft, G.-T.S.; Writing—review & editing, J.-X.Z. All authors have read and agreed to the published version of the manuscript.

Funding: This study was funded by the National Natural Science Foundation of China (42263010, 42202086), the Applied Basic Research Foundation of Yunnan Province (202001BB050020), and the Natural Science Special (special post) scientific research fund project of Guizhou University (No. 2022-24).

Acknowledgments: We would like to thank Kai Luo and Hao Zhang, Min Wang, Ni Peng, Ruifeng Zhu, Ye He, Yunlin An, Zhimou Yang at Yunnan University for their assistance in collecting trace element data of sphalerite. We are grateful to Zhilong Huang and Lin Ye for their suggestions.

References

1. Cook, N.J.; Ciobanu, C.L.; Pring, A.; Skinner, W.; Shimizu, M.; Danyushevsky, L.; Saini-Eidukat, B.; Melcher, F. Trace and minor elements in sphalerite: A LA-ICPMS study. *Geochim. Cosmochim. Acta* **2009**, *73*, 4761–4791. [CrossRef]
2. Ye, L.; Cook, N.J.; Ciobanu, C.L.; Yuping, L.; Qian, Z.; Tiegeng, L.; Wei, G.; Yulong, Y.; Danyushevskiy, L. Trace and minor elements in sphalerite from base metal deposits in South China: A LA-ICPMS study. *Ore Geol. Rev.* **2011**, *39*, 188–217. [CrossRef]
3. Frenzel, M.; Hirsch, T.; Gutzmer, J. Gallium, germanium, indium, and other trace and minor elements in sphalerite as a function of deposit type—A meta-analysis. *Ore Geol. Rev.* **2016**, *76*, 52–78. [CrossRef]
4. Lary, D.J.; Alavi, A.H.; Gandomi, A.H.; Walker, A.L. Machine learning in geosciences and remote sensing. *Geosci. Front.* **2016**, *7*, 3–10. [CrossRef]
5. Gregory, D.D.; Cracknell, M.J.; Large, R.R.; McGoldrick, P.; Kuhn, S.; Maslennikov, V.V.; Baker, M.J.; Fox, N.; Belousov, I.; Figueroa, M.C.; et al. Distinguishing Ore Deposit Type and Barren Sedimentary Pyrite Using Laser Ablation-Inductively Coupled Plasma-Mass Spectrometry Trace Element Data and Statistical Analysis of Large Data Sets. *Econ. Geol.* **2019**, *114*, 771–786. [CrossRef]
6. Wang, Y.; Qiu, K.-F.; Müller, A.; Hou, Z.-L.; Zhu, Z.-H.; Yu, H.-C. Machine Learning Prediction of Quartz Forming-Environments. *J. Geophys. Res. Solid Earth* **2021**, *126*, e2021JB021925. [CrossRef]
7. Zhong, R.; Deng, Y.; Li, W.; Danyushevsky, L.V.; Cracknell, M.J.; Belousov, I.; Chen, Y.; Li, L. Revealing the multi-stage ore-forming history of a mineral deposit using pyrite geochemistry and machine learning-based data interpretation. *Ore Geol. Rev.* **2021**, *133*, 104079. [CrossRef]
8. Hu, P.; Wu, Y.; Zhang, C.Q.; Hu, M.Y. Trace and Minor Elements in Sphalerite from the Mayuan Lead-Zinc Deposit, Northern Margin of the Yangtze Plate: Implications from LA-ICP-MS Analysis. *Acta Mineral. Sin.* **2014**, *34*, 461–468. (In Chinese with English Abstract)
9. Yu, H.H. Study on Mineralization of Qixiashan Pb-Zn Deposit, Nanjing, China. Master's Thesis, Hefei University of Technology, Hefei, China, 2016. (In Chinese with English Abstract)
10. Xing, B.; Zheng, W.; Ouyang, Z.X.; Wu, X.D.; Lin, W.P.; Tian, Y. Sulfide microanalysis and S isotope of the Miaoshan Cu polymetallic deposit in western Guandong Province, and its constraints on the ore genesis. *Acta Geol. Sin.* **2016**, *90*, 971–986. (In Chinese with English Abstract)

11. Tao, L.C. In situ LA-ICP-MS trace element analysis of sulfides from Weilasituo polymetallic deposit and its significance. Master's Thesis, China University of Geosciences (Beijing), Beijing, China, 2017. (In Chinese with English Abstract)

12. Xing, B.; Xiang, J.F.; Ye, H.S.; Chen, X.D.; Zhang, G.S.; Yang, C.Y.; Jin, X.; Hu, Z.Z. Genesis of Luotuoshan sulfur polymetallic deposit in western Henan Province: Evidence from trace elements of sulfide revealed by using LA-ICP-MS in lamellar ores. *Miner. Depos.* **2017**, *36*, 83–106. (In Chinese with English Abstract)

13. Xu, Z.B.; Shao, Y.J.; Yang, Z.A.; Liu, Z.F.; Wang, W.X.; Ren, X.M. LA-ICP-MS analysis of trace elements in sphalerite from the Huanggangliang Fe-Sn deposit, Inner Mongolia, and its implications. *Acta Petrol. Miner.* **2017**, *36*, 360–370. (In Chinese with English Abstract)

14. Yuan, B.; Zhang, C.; Yu, H.; Yang, Y.; Zhao, Y.; Zhu, C.; Ding, Q.; Zhou, Y.; Yang, J.; Xu, Y. Element enrichment characteristics: Insights from element geochemistry of sphalerite in Daliangzi Pb–Zn deposit, Sichuan, Southwest China. *J. Geochem. Explor.* **2018**, *186*, 187–201. [CrossRef]

15. Ren, T.; Zhou, J.X.; Wang, D.; Yang, G.S.; Lv, C.L. Trace elemental and S-Pb isotopic geochemistry of the Fule Pb-Zn deposit, NE Yunnan Province. *Acta Petrol. Sin.* **2019**, *35*, 3493–3505. (In Chinese with English Abstract)

16. Gong, X.J.; Yang, Z.S.; Zhuang, L.L.; Ma, W. Genesis of Narusongduo Pb-Zn deposit, Tibet: Constraint from in-situ LA-ICPMS analyses of minor and trace elements in sphalerite. *Miner. Depos.* **2019**, *38*, 1365–1378. (In Chinese with English Abstract)

17. Zhuang, L.; Song, Y.; Liu, Y.; Fard, M.; Hou, Z. Major and trace elements and sulfur isotopes in two stages of sphalerite from the world-class Angouran Zn–Pb deposit, Iran: Implications for mineralization conditions and type. *Ore Geol. Rev.* **2019**, *109*, 184–200. [CrossRef]

18. Fan, X.; Lü, X.; Wang, X. Textural, Chemical, Isotopic and Microthermometric Features of Sphalerite from the Wunuer Deposit, Inner Mongolia: Implications for Two Stages of Mineralization from Hydrothermal to Epithermal. *Geol. J.* **2020**, *55*, 6936–6958. [CrossRef]

19. Sun, X.; Ni, P.; Yang, Y.; Chi, Z.; Jing, S. Constraints on the Genesis of the Qixiashan Pb-Zn Deposit, Nanjing: Evidence from Sulfide Trace Element Geochemistry. *J. Earth Sci.* **2020**, *31*, 287–297. [CrossRef]

20. Sun, G.; Zeng, Q.; Zhou, J.-X.; Zhou, L.; Chen, P. Genesis of the Xinling vein-type Ag-Pb-Zn deposit, Liaodong Peninsula, China: Evidence from texture, composition and in situ S-Pb isotopes. *Ore Geol. Rev.* **2021**, *133*, 104120. [CrossRef]

21. Yu, D.; Xu, D.; Zhao, Z.; Huang, Q.; Wang, Z.; Deng, T.; Zou, S. Genesis of the Taolin Pb-Zn deposit in northeastern Hunan Province, South China: Constraints from trace elements and oxygen-sulfur-lead isotopes of the hydrothermal minerals. *Min. Depos.* **2020**, *55*, 1467–1488. [CrossRef]

22. Yu, D.; Xu, D.; Wang, Z.; Xu, K.; Huang, Q.; Zou, S.; Zhao, Z.; Deng, T. Trace element geochemistry and O-S-Pb-He-Ar isotopic systematics of the Lishan Pb-Zn-Cu hydrothermal deposit, NE Hunan, South China. *Ore Geol. Rev.* **2021**, *133*, 104091. [CrossRef]

23. Xu, J.; Cook, N.J.; Ciobanu, C.L.; Li, X.; Kontonikas-Charos, A.; Gilbert, S.; Lv, Y. Indium distribution in sphalerite from sulfide-oxide–silicate skarn assemblages: A case study of the Dulong Zn–Sn–In deposit, Southwest China. *Min. Depos.* **2021**, *56*, 307–324. [CrossRef]

24. Wei, C.; Ye, L.; Hu, Y.; Huang, Z.; Danyushevsky, L.; Wang, H. LA-ICP-MS analyses of trace elements in base metal sulfides from carbonate-hosted Zn-Pb deposits, South China: A case study of the Maoping deposit. *Ore Geol. Rev.* **2021**, *130*, 103945. [CrossRef]

25. Mishra, B.P.; Pati, P.; Dora, M.L.; Baswani, S.R.; Meshram, T.; Shareef, M.; Pattanayak, R.S.; Suryavanshi, H.; Mishra, M.; Raza, M.A. Trace-element systematics and isotopic characteristics of sphalerite-pyrite from volcanogenic massive sulfide deposits of Betul belt, central Indian Tectonic Zone: Insight of ore genesis to exploration. *Ore Geol. Rev.* **2021**, *134*, 104149.

26. Zhao, Y.; Chen, S.; Tian, H.; Zhao, J.; Tong, X.; Chen, X. Trace element and S isotope characterization of sulfides from skarn Cu ore in the Laochang Sn-Cu deposit, Gejiu district, Yunnan, China: Implications for the ore-forming process. *Ore Geol. Rev.* **2021**, *134*, 104155. [CrossRef]

27. Xing, B.; Mao, J.; Xiao, X.; Liu, H.; Jia, F.; Wang, S.; Huang, W.; Li, H. Genetic discrimination of the Dingjiashan Pb-Zn deposit, SE China, based on sphalerite chemistry. *Ore Geol. Rev.* **2021**, *135*, 104212. [CrossRef]

28. Benites, D.; Torró, L.; Vallance, J.; Kouzmanov, K.; Chelle-Michou, C.; Fontboté, L. Distribution of indium, germanium, gallium and other minor and trace elements in polymetallic ores from a porphyry system: The Morococha district, Peru. *Ore Geol. Rev.* **2021**, *136*, 104236. [CrossRef]

29. Jiang, Z.; Zhang, Z.; Duan, S.; Lv, C.; Dai, Z. Genesis of the sediment-hosted Haerdaban Zn-Pb deposit, Western Tianshan, NW China: Constraints from textural, compositional and sulfur isotope variations of sulfides. *Ore Geol. Rev.* **2021**, *139*, 104527. [CrossRef]

30. Liu, S.; Zhang, Y.; Ai, G.; Xue, X.; Li, H.; Shah, S.A.; Wang, N.; Chen, X. LA-ICP-MS trace element geochemistry of sphalerite: Metallogenic constraints on the Qingshuitang Pb–Zn deposit in the Qinhang Ore Belt, South China. *Ore Geol. Rev.* **2021**, *141*, 104659. [CrossRef]

31. Torró, L.; Benites, D.; Vallance, J.; Laurent, O.; Ortiz-Benavente, B.A.; Chelle-Michou, C.; Proenza, J.A.; Fontboté, L. Trace element geochemistry of sphalerite and chalcopyrite in arc-hosted VMS deposits. *J. Geochem. Explor.* **2022**, *232*, 106882. [CrossRef]

32. Brijain, M.; Patel, R.; Kushik, M.; Rana, K. A Survey on Decision Tree Algorithm for Classification. *Int. J. Eng. Dev. Res.* **2014**, *2*, 1–5.

33. Friedl, M.A.; Brodley, C.E. Decision tree classification of land cover from remotely sensed data. *Remote Sens. Environ.* **1997**, *61*, 399–409. [CrossRef]

34. Safavian, S.R.; Landgrebe, D. A survey of decision tree classifier methodology. *IEEE Trans. Syst. Man Cybern.* **1991**, *21*, 660–674. [CrossRef]
35. McRoberts, R.E.; Tomppo, E.O.; Finley, A.O.; Heikkinen, J. Estimating areal means and variances of forest attributes using the k-Nearest Neighbors technique and satellite imagery. *Remote Sens. Environ.* **2007**, *111*, 466–480. [CrossRef]
36. Franco-Lopez, H.; Ek, A.R.; Bauer, M.E. Estimation and mapping of forest stand density, volume, and cover type using the k-nearest neighbors method. *Remote Sens. Environ.* **2001**, *77*, 251–274. [CrossRef]
37. Kramer, O. (Ed.) K-Nearest Neighbors. In *Dimensionality Reduction with Unsupervised Nearest Neighbors*; Springer: Berlin/Heidelberg, Germany, 2013; pp. 13–23.
38. Lowd, D.; Domingos, P. Naive Bayes models for probability estimation. In Proceedings of the 22nd International Conference on Machine Learning, Bonn, Germany, 7–11 August 2005; Association for Computing Machinery: New York, NY, USA, 2005; pp. 529–536.
39. Breiman, L. Random Forests. *Mach. Learn.* **2001**, *45*, 5–32. [CrossRef]
40. Pal, M. Random forest classifier for remote sensing classification. *Int. J. Remote Sens.* **2005**, *26*, 217–222. [CrossRef]
41. Noble, W.S. What is a support vector machine? *Nat. Biotechnol.* **2006**, *24*, 1565–1567. [CrossRef] [PubMed]
42. Sun, G.; Zeng, Q.; Zhou, J.-X. Machine learning coupled with mineral geochemistry reveals the origin of ore deposits. *Ore Geol. Rev.* **2022**, *142*, 104753. [CrossRef]
43. Shaw, D.M. The geochemistry of gallium, indium, thallium—A review. *Phys. Chem. Earth* **1957**, *2*, 164–211. [CrossRef]
44. Johan, Z. Indium and germanium in the structure of sphalerite: An example of coupled substitution with copper. *Mineral. Petrol.* **1988**, *39*, 211–229. [CrossRef]
45. Cave, B.; Lilly, R.; Hong, W. The Effect of Co-Crystallising Sulphides and Precipitation Mechanisms on Sphalerite Geochemistry: A Case Study from the Hilton Zn-Pb (Ag) Deposit, Australia. *Minerals* **2020**, *10*, 797. [CrossRef]
46. Liu, Y.C.; Hou, Z.Q.; Yue, L.L.; Ma, W.; Tang, B.L. Critical metals in sediment-hosted Pb-Zn deposits in China. *Chin. Sci. Bull.* **2022**, *67*, 406–424, (In Chinese with English Abstract). [CrossRef]
47. Mondillo, N.; Herrington, R.; Boyce, A.J.; Wilkinson CSantoro, L.; Rumsey, M. Critical elements in nonsulphide Zn deposits: A re-analysis of the Kabwe Zn-Pb ores. *Mineral. Mag.* **2018**, *82* (Suppl. S1), 89–114. [CrossRef]
48. Paradis, S. *Indium, Germanium and Gallium in Volcanic-and Sediment-Hosted Base-Metal Sulphide Deposits*; British Columbia Ministry of Energy and Mines: Victoria, BC, Canada, 2015; pp. 23–29.
49. Murakami, H.; Ishihara, S. Trace elements of Indium-bearing sphalerite from tin-polymetallic deposits in Bolivia, China and Japan: A femto-second LA-ICPMS study. *Ore Geol. Rev.* **2013**, *53*, 223–243. [CrossRef]
50. Liu, J. Indium Mineralization in a Sn-Poor Skarn Deposit: A Case Study of the Qibaoshan Deposit, South China. *Minerals* **2017**, *7*, 76. [CrossRef]
51. Xu, J.; Li, X.F. Spatial and temporal distributions, metallogenic backgrounds and processes of indium deposits. *Acta Petrol. Sin.* **2018**, *34*, 3611–3626, (In Chinese with English Abstract).
52. Li, X.F.; Xu, J.; Zhu, Y.T.; Lv, Y.H. Critical minerals of indium: Major ore types and scientific issues. *Acta Petrol. Sin.* **2019**, *35*, 3292–3302, (In Chinese with English Abstract).
53. Bauer, M.E.; Seifert, T.; Burisch, M.; Krause, J.; Richter, N.; Gutzmer, J. Indium-bearing sulfides from the Hämmerlein skarn deposit, Erzgebirge, Germany: Evidence for late-stage diffusion of indium into sphalerite. *Min. Depos.* **2019**, *54*, 175–192. [CrossRef]
54. Benedetto, F.D.; Bernardini, G.P.; Costagliola, P.; Plant, D.; Vaughan, D.J. Compositional zoning in sphalerite crystals. *Am. Mineral.* **2005**, *90*, 1384–1392. [CrossRef]
55. Bauer, M.E.; Burisch, M.; Ostendorf, J.; Krause, J.; Frenzel, M.; Seifert, T.; Gutzmer, J. Trace element geochemistry of sphalerite in contrasting hydrothermal fluid systems of the Freiberg district, Germany: Insights from LA-ICP-MS analysis, near-infrared light microthermometry of sphalerite-hosted fluid inclusions, and sulfur isotope geochemistry. *Min. Depos.* **2019**, *54*, 237–262.
56. Belissont, R.; Boiron, M.-C.; Luais, B.; Cathelineau, M. LA-ICP-MS analyses of minor and trace elements and bulk Ge isotopes in zoned Ge-rich sphalerites from the Noailhac—Saint-Salvy deposit (France): Insights into incorporation mechanisms and ore deposition processes. *Geochim. Cosmochim. Acta* **2014**, *126*, 518–540. [CrossRef]
57. Cook, N.J.; Ciobanu, C.L.; Brugger, J.; Etschmann, B.; Howard, D.L.; de Jonge, M.D.; Ryan, C.; Paterson, D. Determination of the oxidation state of Cu in substituted Cu In Fe bearing sphalerite via µ-XANES spectroscopy. *Am. Mineral.* **2012**, *97*, 476–479. [CrossRef]
58. Hsu, C.-W.; Lin, C.-J. A comparison of methods for multiclass support vector machines. *IEEE Trans. Neural Netw.* **2002**, *13*, 415–425. [PubMed]
59. Leach, D.L.; Sangster, D.F.; Kelley, K.D.; Large, R.R.; Garven, G.; Allen, C.R.; Gutzmer, J.; Walters, S. Sediment-Hosted Lead-Zinc Deposits: A Global Perspective. In *Economic Geology: One Hundredth Anniversary Volume*; Society of Economic Geologists: Littleton, CO, USA, 2005.
60. Brannon, J.C.; Podosek, F.A.; McLimans, R.K. Alleghenian age of the Upper Mississippi Valley zinc–lead deposit determined by Rb–Sr dating of sphalerite. *Nature* **1992**, *356*, 509–511. [CrossRef]
61. Leach, D.L.; Sangster, D.F. Mississippi Valley-type lead-zinc deposits. *Geol. Assoc. Can. Spec. Pap.* **1993**, *40*, 289–314.
62. Bernardini, G.P.; Borgheresi, M.; Cipriani, C.; Di Benedetto, F.; Romanelli, M. Mn distribution in sphalerite: An EPR study. *Phys. Chem. Miner.* **2004**, *31*, 80–84. [CrossRef]

63. Franklin, J.M.; Gibson, H.L.; Jonasson, I.R.; Galley, A.G. Volcanogenic Massive Sulfide Deposits. In *Economic Geology: One Hundredth Anniversary Volume*; Society of Economic Geologists: Littleton, CO, USA, 2005. [CrossRef]
64. Hannington, M.D.; Ronde, C.E.J.D.; Petersen, S. Sea-Floor Tectonics and Submarine Hydrothermal Systems. In *Economic Geology: One Hundredth Anniversary Volume*; Society of Economic Geologists: Littleton, CO, USA, 2005. [CrossRef]
65. Hannington, M.D.; Poulsen, K.H.; Thompson, J.F.H.; Sillitoe, R.H. Volcanogenic Gold in the Massive Sulfide Environment. In *Reviews in Economic Geology: Volcanic Associated Massive Sulfide Deposits: Processes and Examples in Modern and Ancient Settings*; Society of Economic Geologists: Littleton, CO, USA, 1997. [CrossRef]
66. Meinert, L.D. Skarns and Skarn Deposits. *Geosci. Can.* **1992**, *19*, 145–162.
67. Meinert, L.D.; Dipple, G.M.; Nicolescu, S. World Skarn Deposits. In *Economic Geology: One Hundredth Anniversary Volume*; Society of Economic Geologists: Littleton, CO, USA, 2005. [CrossRef]
68. Bralia, A.; Sabatini, G.; Troja, F. A revaluation of the Co/Ni ratio in pyrite as geochemical tool in ore genesis problems. *Miner. Depos.* **1979**, *14*, 353–374. [CrossRef]
69. Li, G.; Deng, J.; Wang, Q.; Liang, K. Metallogenic model for the Laochang Pb–Zn–Ag–Cu volcanogenic massive sulfide deposit related to a Paleo-Tethys OIB-like volcanic center, SW China. *Ore Geol. Rev.* **2015**, *70*, 578–594. [CrossRef]
70. Wei, C.; Ye, L.; Huang, Z.; Gao, W.; Hu, Y.; Li, Z.; Zhang, J. Ore Genesis and Geodynamic Setting of Laochang Ag-Pb-Zn-Cu Deposit, Southern Sanjiang Tethys Metallogenic Belt, China: Constraints from Whole Rock Geochemistry, Trace Elements in Sphalerite, Zircon U-Pb Dating and Pb Isotopes. *Minerals* **2018**, *8*, 516. [CrossRef]
71. Deng, X.-D.; Li, J.-W.; Zhao, X.-F.; Wang, H.-Q.; Qi, L. Re–Os and U–Pb geochronology of the Laochang Pb–Zn–Ag and concealed porphyry Mo mineralization along the Changning–Menglian suture, SW China: Implications for ore genesis and porphyry Cu–Mo exploration. *Min. Depos.* **2016**, *51*, 237–248. [CrossRef]
72. Sun, G.; Zhou, J.-X.; Long, H.-S.; Zhou, L.; Luo, K. Vertical evolution of Ag-Pb-Zn-(Cu)-Mo in porphyry system: A case study from the Laochang deposit, SW China. *Ore Geol. Rev.* **2021**, *139*, 104419. [CrossRef]

Article

Fractal Structure Characteristics and Prospecting Direction of Dispersed Metals in the Eastern Guizhou Pb–Zn Metallogenic Belt, SW China

Zhongliang Cui [1], Jiaxi Zhou [2,3,*], Kai Luo [2,3] and Maoda Lu [4]

1   Jiangxi Institute of Applied Science and Technology, Nanchang 330100, China
2   School of Earth Sciences, Yunnan University, Kunming 650500, China
3   Key Laboratory of Critical Minerals Metallogeny in Universities of Yunnan Province, Kunming 650500, China
4   104 Geological Team, Guizhou Bureau of Geology and Mineral Exploration and Development, Duyu 558000, China
*   Correspondence: zhoujiaxi@ynu.edu.cn

Abstract: The eastern Guizhou Pb–Zn metallogenic belt (EGMB) is an important source of Pb–Zn resources and other critical minerals (including dispersed metals, such as Ge, Cd and Ga) in China. In order to ensure the continuous resource supply of Pb–Zn and associated dispersed metals, it is urgent to explore the direction of further prospecting for them. Fractal theory can realize the fractal structure characterization of fault structures and the spatial distribution of mineral deposits, which is helpful for mineral exploration. However, the fault fractal research and prospecting application are still seldom covered in the EGMB. We used fractal theory to determine fine-scale fractal structure characteristics of fault structures and ore deposits in the EGMB, and Fry analysis to delineate favorable metallogenic areas. The results show that within a scale range of 3.670–58.716 km, the integrated faults capacity dimension (CPD) is 1.5095, the information dimension (IND) is 1.5391, and the correlation dimension (CRD) is 1.5436, indicating fault structures with high maturity, which are conducive to the migration and accumulation of ore-forming fluids. The multi-fractal spectrum width and height are 0.3203 and 1.5355, respectively, implying a significant metallogenic potential. The spatial distribution fractal dimensions (SDD) of Pb–Zn specifically and metal deposits in general are 1.0193 and 1.0709, respectively; the quantity distribution fractal dimensions (QDD) are 1.4225 and 1.4716, respectively, and the density distribution fractal dimensions (DDD) are 1.422 and 1.472, respectively, indicating strong clustering. Hence, the favorable metallogenic regions can be divided into four grades, among which grade I region is continuously distributed in space and has the greatest prospecting potential.

Keywords: fault structure; Pb–Zn deposits; dispersed metals; fractal structure characteristic; Fry analysis; prospecting direction; eastern Guizhou metallogenic belt; SW China

Citation: Cui, Z.; Zhou, J.; Luo, K.; Lu, M. Fractal Structure Characteristics and Prospecting Direction of Dispersed Metals in the Eastern Guizhou Pb–Zn Metallogenic Belt, SW China. *Minerals* **2022**, *12*, 1567. https://doi.org/10.3390/min12121567

Academic Editor: Behnam Sadeghi

Received: 10 November 2022
Accepted: 2 December 2022
Published: 5 December 2022

Publisher's Note: MDPI stays neutral with regard to jurisdictional claims in published maps and institutional affiliations.

1. Introduction

The western Hubei–western Hunan–eastern Guizhou metallogenic belt is an important source of Pb–Zn metals in China [1,2]. Within this belt, the eastern Guizhou Pb–Zn metallogenic belt (EGMB) hosts a large number of Pb–Zn and other metal deposits/ore fields, including the Niujiaotang Cd-rich Pb–Zn ore field [2,3]. In recent years, many researchers have systematically studied the geological characteristics of the Pb–Zn deposits [4–8], a source of metallogenic materials [9–13] and ore-forming fluids [12,14–17], the ore genesis of deposits [2–4,6,12,17], and the metallogenic model [3,18–21]. These Pb–Zn deposits are obviously controlled by faults [2,12,22,23], and they belong to the Mississippi Valley-type (MVT) Pb–Zn deposits [2–4,12,14,24].

A previously developed fracture–lithology–fluid coupling metallogenic model has guided Pb–Zn exploration in this area and identified significant supernormal enrichment of Ge (more than 1000 times enrichment compared with the crustal abundance of Ge), including large (e.g., Zhulingou) to super-large (e.g., Banbianjie) Ge-Zn deposits [21,25–30]. Dispersed elements are those that have very low abundance in the crust (mostly grades 10^{-9}) and are dispersed in rocks [31]. They are critical minerals that have great practical significance to national security and the development of emerging industries [32], especially in the development of "high-tech" technology and future energy [21,33]. Statistics show that: Ge, Cd, Ga, Tl, and other dispersed elements are enriched in Pb–Zn deposits and Pb–Zn poly-metallic deposits [34], and mainly exist in the form of symbiotic associations [34–39]; and MVT Pb–Zn deposits are enriched with one or more dispersed elements compared with other types of Pb–Zn deposits or Pb–Zn poly-metallic deposits [2,31,38,40–46]. In summary, the huge metallogenic potential of the EGMB offers a potential production base for scarce resource minerals (e.g., Pb–Zn) and critical minerals (dispersed metals) in China. To ensure the continuous resource supply, it is of great theoretical and practical significance to strengthen research on the metallogenic law of dispersed metals, and to explore potential directions for future prospecting.

Although there have been many research achievements in the EGMB, so far, the fault fractal research and prospecting application are seldom covered. Fractal theory, which was proposed by the famous mathematician, Mandelbrot [47], can reveal the inner connection between the part and the whole of things; it can describe complex structures in detail and quantitatively reveal the hidden laws [48]. Fractal theory has been applied in the quantitative characterization of faults [49–54], the spatial distribution of deposits [48,55–59], metallogenic laws, and prospecting prediction [51,60–70]. Currently, three basic conclusions are generally recognized in the study of fault fractal [55]: (1) the fault system has fractal characteristics; (2) the fractal dimension of the fault structure is related to the connectivity of the fault (geological body); and (3) the fractal dimension of the fault structure is closely related to geological mineralization, which can be used as an indicator of metallogenic prediction. However, in the coupling study of fault fractal and deposit distribution, there are few reports on the research results of deposit spatial location prediction, which needs further exploration.

In this study, we applied fractal theory to quantitatively describe the coupling relationship between the fractal structure of faults of the EGMB and the spatial distribution of ore deposits. In addition, we performed Fry analysis of the ore deposits. Based on the results, we identified favorable prospecting directions for the exploration of Pb–Zn and associated critical minerals (dispersed metals).

2. Geological Background

The EGMB is located on the southeastern margin of the Yangtze block, and extends from the Bamianshan intra-continental deformation belt to the northwest to the Xuefengshan structural belt to the southeast (Figure 1a) [3]. Magmatic activity in the EGMB is not obvious [12,23], and magmatic rocks (e.g., potassium–magnesium lamprophyre) are sporadic (Figure 1b). The EGMB may have experienced varies periods of orogeny, including those during the Caledonian, Hercynian, Indosinian–Yanshanian, and other periods [23,71–73], and folds and fault structures are widely developed. Fault structures are mainly NE-trending (including NNE-trending), but are NW-trending and near-NS-trending in part. The basement is Neoproterozoic shallow metamorphic rocks. In the sedimentary cover, except for the Upper Paleozoic Carboniferous and the Mesozoic-Cenozoic Jurassic, Paleogene, and Neogene missing, the others are exposed. Among them, Cambrian carbonate strata are widely exposed and are the most important ore-bearing horizon [3,12,23]. Ore deposits in the EGMB are obviously controlled by faults (especially NE-trending fault structures) [3,12,22,23]. Of the 61 metal deposits, all are medium-low temperature hydrothermal deposits, including 53 Pb–Zn deposits, 5 Sb deposits, and 3 Hg deposits. From the perspectives of ore-bearing horizon, main ore-controlling factors, and genesis types of deposits, the deposits in the area are highly similar. As such, the study area has the prerequisites for quantitatively exploring the coupling relationship between fault structures and deposit distribution.

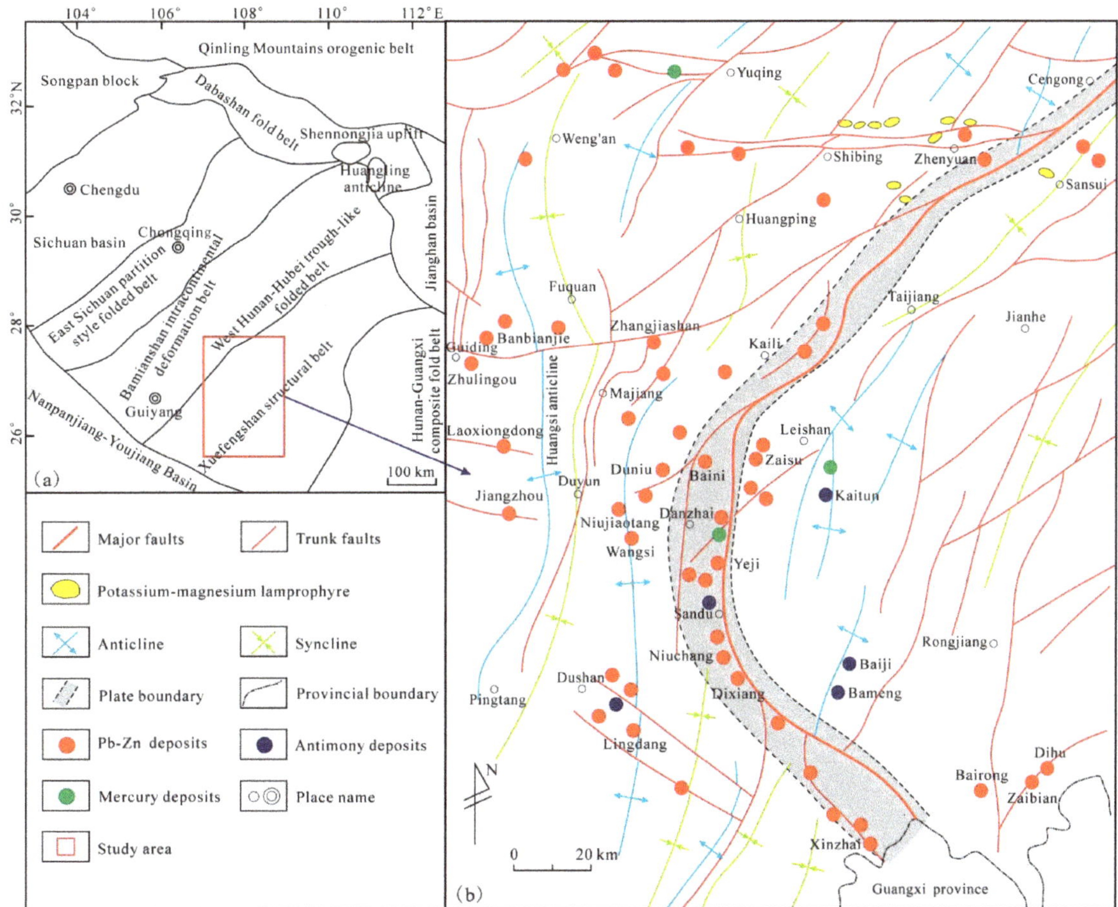

Figure 1. Geotectonic position of the study area: (**a**) according to Ref. [3] and fault structures, deposits distribution map (**b**): according to Ref. [26].

3. Single Fractal Characteristics of Fault Structures

3.1. Calculation Method

3.1.1. Capacity Dimension, D_0 Calculation

At present, there are many calculation methods for the capacity dimension of linear structures, including the box-counting dimension method, the circle covering method, and the length-frequency statistics method [74–76]. Among them, the box-counting dimension method is intuitive and easy to understand, and offers accurate statistics and strong operability. Therefore, we adopted the box-counting dimension method for the calculation of the capacity dimension (CPD) based on the fault structures and ore-deposit distribution map (Figure 1b). The algorithm was as follows: square grids with different side lengths r (r = L, L/2, L/4, and L/8 . . . , which are proportional sequences with a common ratio equal to 0.5) were used to cover the study area and the number of grids $N(r)$ covering the faults was calculated. If $N(r)$ and r satisfied the following power-law relationship (Equation (1)), the research object was fractal:

$$N(r) = Cr^{-D_0} \tag{1}$$

where C is a constant and D_0 is the CPD value that attempted to acquire. Taking the logarithm of Equation (1) yields Equation (2), from which the CPD value D_0 was obtained by taking the absolute value of the slope of the straight line:

$$\ln N(r) = -D_0 \ln r + \ln c \tag{2}$$

The specific steps of the calculation process were as follows: (1) two-dimensional orthogonal grids with side lengths of 58.716, 29.358, 14.679, 7.340, and 3.670 km were used to cover the study area. Then, the numbers of grids $N(r)$ covered by integrated faults, NE-trending faults, NW-trending faults, near-SN-trending faults, near-EW-trending faults, major faults, and the plate contact transition zone were counted. In Excel, we took $\ln r$ as the horizontal axis and $\ln N(r)$ as the vertical axis, and drew a straight regression line to obtain CPD values of different types of faults; (2) numbering the two-dimensional orthogonal grid with a side length of 58.716 km, the study area was divided into 12 divisions (Figure 2). For each division, the number $N(r)$ of different two-dimensional orthogonal grids covering faults with side lengths of 29.358, 14.679, 7.340, and 3.670 km were counted. Use Excel, we drew a straight regression line to obtain the CPD value of the overall faults of the division; and (3) numbering the two-dimensional orthogonal grid with a side length of 29.358 km, the study area was divided into 48 subdivisions (Figure 2). For each subdivision, the numbers $N(r)$ of different two-dimensional orthogonal grids covering faults with side lengths of 14.679, 7.340, and 3.670 km were counted. Using Excel, we drew a regression fitting line to obtain the overall CPD value of faults within the subdivision.

3.1.2. Information Dimension, D_1 Calculation

The fault information dimension (IND) not only considers whether a two-dimensional grid is crossed by faults, but also considers the number (or probability) of crossing faults. The study area was covered by a two-dimensional orthogonal grid with side length r, and it was assumed that faults were divided into $N(r)$ parts. If faults appeared in the i-th orthogonal grid, the probability was $P_i(r)$ (Equation (3)) and the total amount of information at this time was $I(r)$ (Equation (4)).

$$P_i(r) = \frac{n_i}{\sum\limits_{i=1}^{N(r)} n_i} \tag{3}$$

$$I(r) = -\sum_{i=1}^{N(r)} P_i(r) \ln P_i(r) \tag{4}$$

After transforming the side length r of the two-dimensional orthogonal grid, if there is the following linear relationship between $I(r)$ and $\ln r$ (Equation (5)), the IND value, D_1, can be obtained from the slope of the straight line:

$$I(r) = -D_1 \ln r + I_0 \tag{5}$$

The specific steps of the calculation process were as follows: (1) two-dimensional orthogonal grids with side lengths of 58.716, 29.358, 14.679, 7.340, and 3.670 km were used to cover the study area. Then, the information contents $I(r)$ of integrated faults, NE-trending faults, NW-trending faults, near-SN-trending faults, near-EW-trending faults, and major faults were calculated. In Excel, we took $\ln r$ as the horizontal axis and $I(r)$ as the vertical axis, and drew a straight regression line to obtain the IND values of different types of faults; (2) for each division (Figure 2), the overall information content $I(r)$ of the faults was calculated when the two-dimensional orthogonal grids with side lengths of 29.358, 14.679, 7.340, and 3.670 km were covered. We used Excel to draw a straight regression line to obtain the overall IND value of the faults in the division; and (3) for each subdivision (Figure 2), the overall information content $I(r)$ of the faults was calculated when covered by two-dimensional orthogonal grids with side lengths of 14.679, 7.340, and 3.670 km. We used Excel to draw a regression fitting line to obtain the overall IND value of faults in the subdivision.

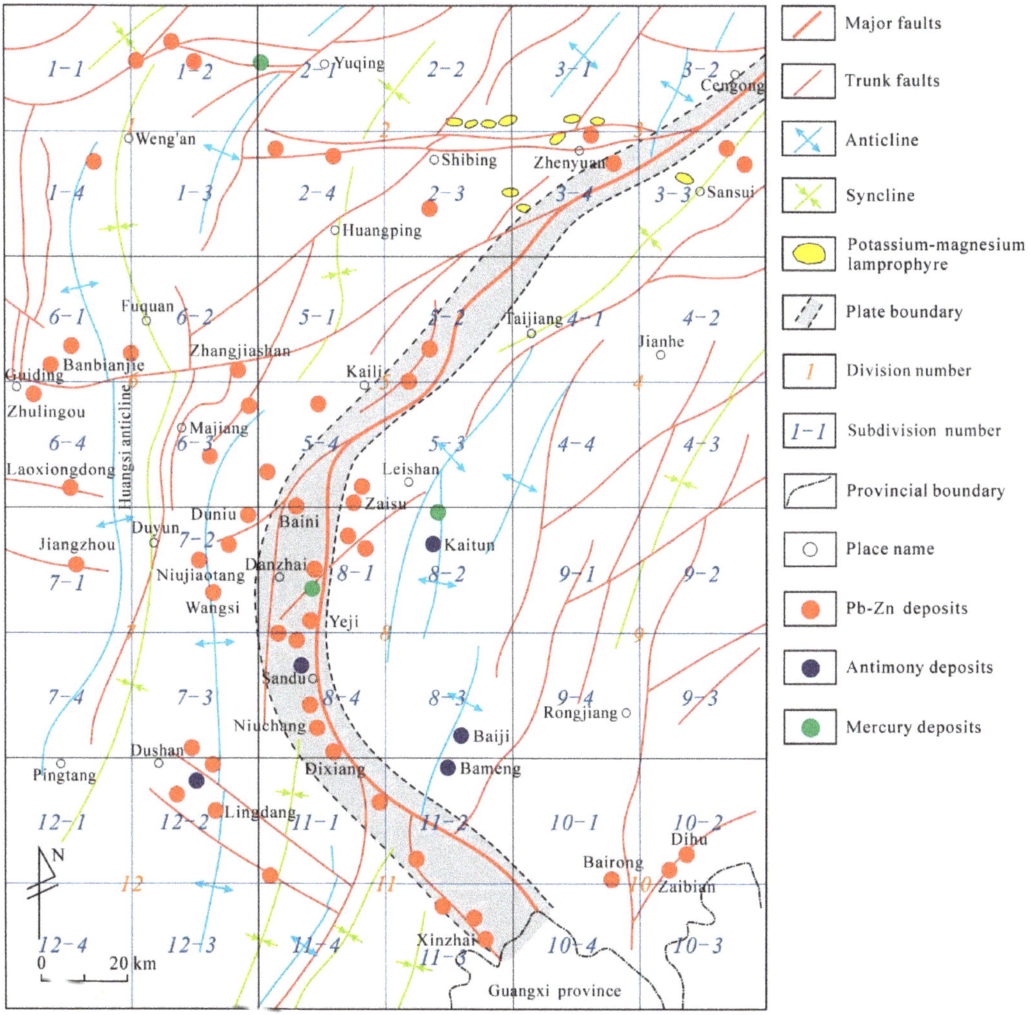

Figure 2. Computation partition map of fractal dimension (modified from Ref. [26]).

3.1.3. Correlation Dimension, D_2 Calculation

The calculation process for the correlation dimension (CRD) values was similar to that of the IND, and followed Equation (6):

$$I(r) = -\ln \sum_{i=1}^{N(r)} P_i^2(r) \tag{6}$$

3.2. Single Fractal Characteristics of Faults across the Whole Area

Statistical parameters of fault fractal dimension values are listed in Table 1; ln$r$ versus ln$N(r)$ plots for integrated faults, NW-trending faults, NE-trending faults, near-SN-trending faults, near-EW-trending faults, major faults, and the plate contact transition zone with their linear regression parameters are shown in Figures 3–5.

Table 1. Statistical table of calculation parameters of fractal dimensions for fault structures in the eastern Guizhou Pb–Zn metallogenic belt (EGMB).

Category	CPD, D_0				IND, D_1			CRD, D_2		
	r (km)	$N(r)$	ln$r$	ln$N(r)$	r (km)	ln$r$	$I(r)$	r (km)	ln$r$	$I(r)$
Integrated faults	58.716	12	4.073	2.485	58.716	4.073	2.366	58.716	4.073	2.291
	29.358	45	3.380	3.807	29.358	3.380	3.692	29.358	3.380	3.606
	14.679	146	2.686	4.984	14.679	2.686	4.874	14.679	2.686	4.767
	7.340	368	1.993	5.908	7.340	1.993	5.836	7.340	1.993	5.743
	3.670	785	1.300	6.666	3.670	1.300	6.628	3.670	1.300	6.572
NE-trending faults	58.716	11	4.073	2.398	58.716	4.073	2.257	58.716	4.073	2.165
	29.358	37	3.380	3.611	29.358	3.380	3.475	29.358	3.380	3.369
	14.679	114	2.686	4.736	14.679	2.686	4.628	14.679	2.686	4.518
	7.340	281	1.993	5.638	7.340	1.993	5.582	7.340	1.993	5.509
	3.670	570	1.300	6.346	3.670	1.300	6.316	3.670	1.300	6.275
NW-trending faults	58.716	6	4.073	1.792	58.716	4.073	1.676	58.716	4.073	1.569
	29.358	11	3.380	2.398	29.358	3.380	2.307	29.358	3.380	2.213
	14.679	25	2.686	3.219	14.679	2.686	3.170	14.679	2.686	3.114
	7.340	53	1.993	3.970	7.340	1.993	3.957	7.340	1.993	3.937
	3.670	95	1.300	4.554	3.670	1.300	4.550	3.670	1.300	4.544
Near-SN-trending faults	58.716	4	4.073	1.386	58.716	4.073	1.273	58.716	4.073	1.176
	29.358	6	3.380	1.792	29.358	3.380	1.676	29.358	3.380	1.569
	14.679	13	2.686	2.565	14.679	2.686	2.479	14.679	2.686	2.392
	7.340	30	1.993	3.401	7.340	1.993	3.370	7.340	1.993	3.329
	3.670	64	1.300	4.159	3.670	1.300	4.122	3.670	1.300	4.040
Near-EW-trending faults	58.716	4	4.073	1.386	58.716	4.073	1.311	58.716	4.073	1.259
	29.358	10	3.380	2.303	29.358	3.380	2.211	29.358	3.380	2.120
	14.679	18	2.686	2.890	14.679	2.686	2.834	14.679	2.686	2.774
	7.340	34	1.993	3.526	7.340	1.993	3.469	7.340	1.993	3.409
	3.670	71	1.300	4.263	3.670	1.300	4.248	3.670	1.300	4.226
Major faults	58.716	6	4.073	1.792	58.716	4.073	1.792	58.716	4.073	1.792
	29.358	13	3.380	2.565	29.358	3.380	2.565	29.358	3.380	2.565
	14.679	25	2.686	3.219	14.679	2.686	3.219	14.679	2.686	3.219
	7.340	51	1.993	3.932	7.340	1.993	3.932	7.340	1.993	3.932
	3.670	100	1.300	4.605	3.670	1.300	4.605	3.670	1.300	4.605
Plate contact transition zone	58.716	8	4.073	2.079					-	
	29.358	17	3.380	2.833						
	14.679	43	2.686	3.761						
	7.340	112	1.993	4.718						
	3.670	328	1.300	5.793						

From Figures 3–5: (1) the coefficient of determination (degree of fitting) R^2 of the 19 regression lines range from 0.9854 to 0.9994, with most >0.99. The overall fitting degree of the straight lines is relatively high, showing that the fault structures have good statistical self-similarity on scales 3.670–58.716 km; (2) the integrated faults, NE-trending faults, NW-trending faults, near-SN-trending faults, near-EW-trending faults, and major fault CPD values are 1.5095, 1.4316, 1.0239, 1.0322, 1.0065, and 1.0090, respectively; the IND values are 1.5391, 1.4752, 1.0673, 1.0665, 1.0290, and 1.0090, respectively; and the CRD values are 1.5436, 1.4947, 1.1072, 1.0803, 1.0421, and 1.0090, respectively. The CPD value of the plate contact transition zone is 1.3435; (3) CPD values decrease as follows: integrated faults > NE-trending faults > plate contact transition zone > near-SN-trending faults > NW-trending faults > major faults > near-EW-trending faults. IND and CRD values decrease as follows: integrated faults > NE-trending faults > NW-trending faults > near-SN-trending faults > near-EW-trending faults > major faults; (4) based on the CPD, IND, and CRD, the fractal dimension value of integrated faults is the largest, closely followed by that of NE-trending faults; this reflects the dominance of NE-trending faults in the EGMB and is consistent with regional tectonic characteristics; and (5) only one major fault was involved in the calculation of the fractal dimension, and so the values of the CPD, IND, and CRD were equal; however, we believe that major faults are still of great significance to the mineralization of the study area.

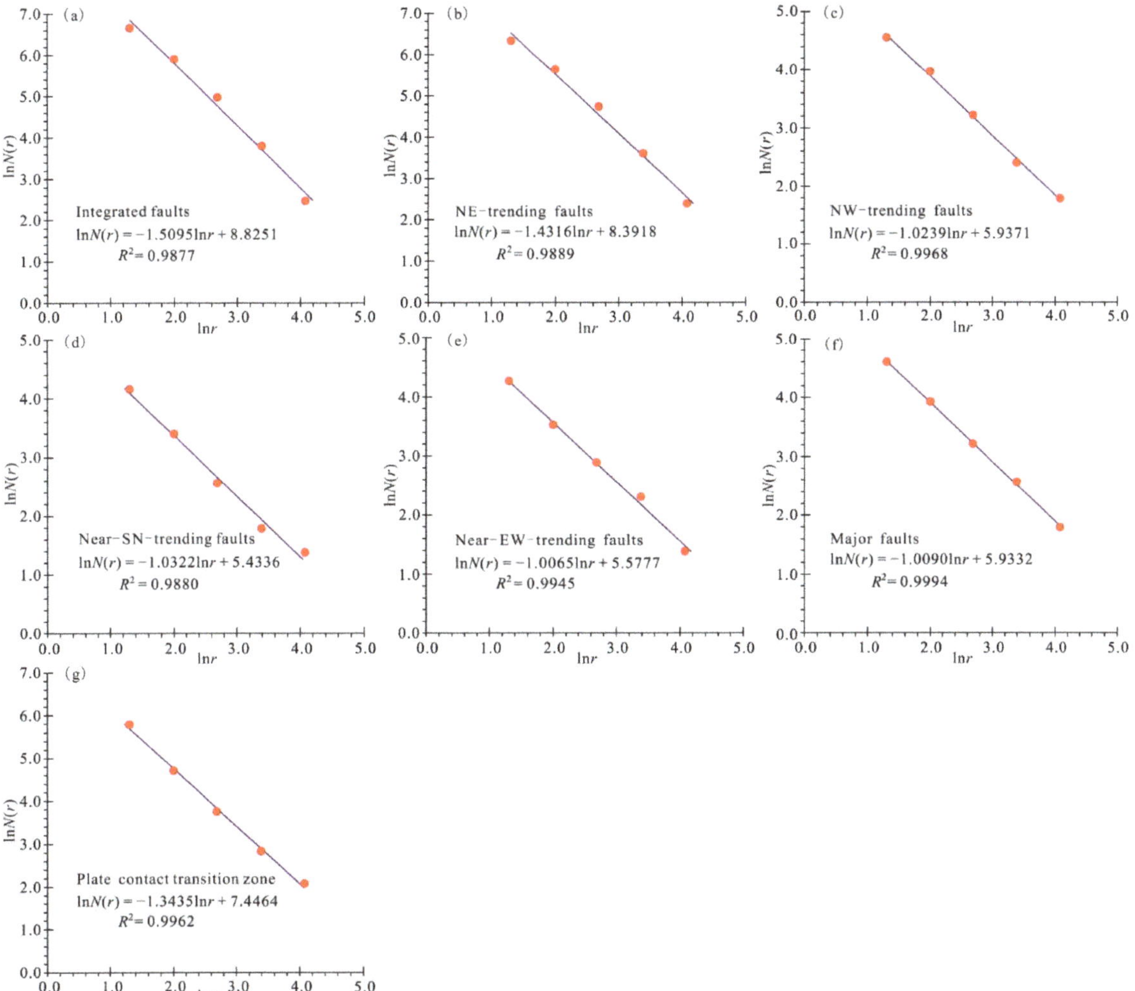

Figure 3. Linear fitting diagrams of the capacity dimension (CPD) calculation for faults in the eastern Guizhou Pb–Zn metallogenic belt (EGMB). The ln$r$ versus ln$N(r)$ plots of CPD data for (**a**) Integrated faults; (**b**) NE-trending faults; (**c**) NW-trending faults; (**d**) Near-SN-trending faults; (**e**) Near-EW-trending faults; (**f**) Major faults; and (**g**) Plate contact transition zone, showing their linear regression parameters.

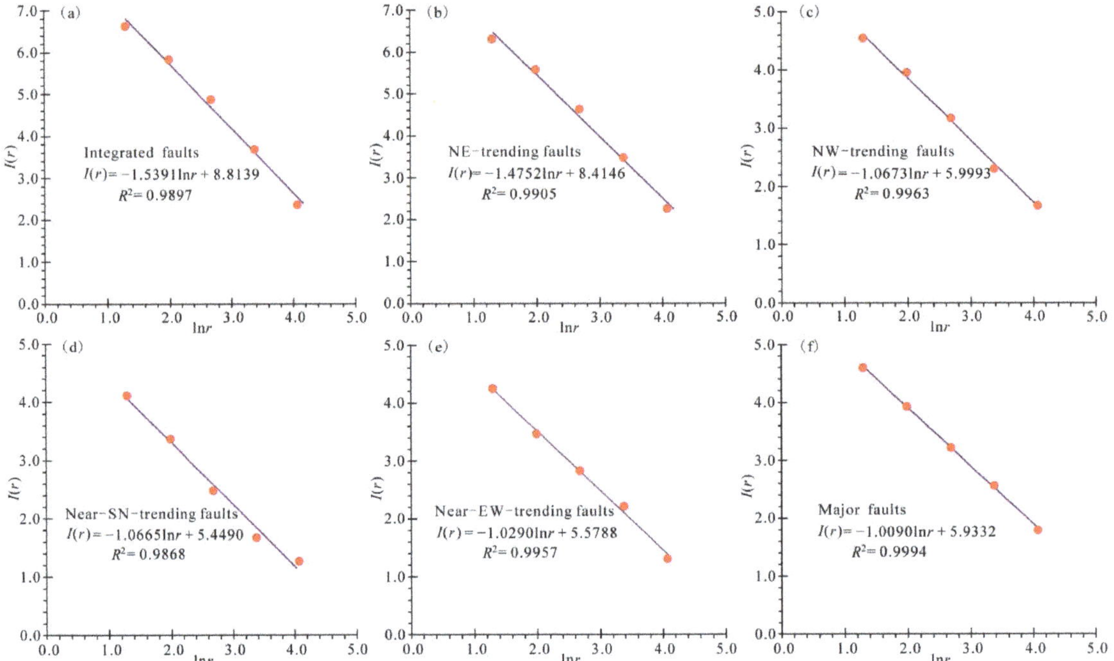

Figure 4. Linear fitting diagrams of the information dimension (IND) calculation for faults in the eastern Guizhou Pb–Zn metallogenic belt (EGMB). The ln$r$ versus $I(r)$ plots of IND data for (**a**) Integrated faults; (**b**) NE-trending faults; (**c**) NW-trending faults; (**d**) Near-SN-trending faults; (**e**) Near-EW-trending faults; and (**f**) Major faults, showing their linear regression parameters.

The fractal dimension value of a fault structure is related to the connectivity of the fault (geological body). That is to say, with an increasing fractal dimension value of the fault structure, the spatial distribution of the fault structure becomes increasingly complex, the permeability of the fault (geological body) becomes stronger, and the connectivity improves. Therefore, an increasing fractal dimension value is more conducive to the activation of ore-forming elements and the migration and accumulation of ore-forming fluids. Based on the critical fractal dimension of faults (1.22–1.38) [77], the CPD values of integrated faults, NE-trending faults, and the plate contact transition zone are 1.5095, 1.4316, and 1.3435, respectively; the values for NW-trending faults, near-SN-trending faults, and near-EW-trending faults are all <1.22. Based on these results, we concluded that the overall metallogenic geological conditions of the EGMB are good. In particular, the plate contact transition zone is conducive to ore formation, which is consistent with the belt-like distribution of ore deposits along this zone.

The fault CPD of the EGMB is larger than those of most areas in China (Table 2), including ore fields, metallogenic belts, and ore concentration areas, and is close to the upper limit of the active area (Diwa area) CPD in mainland China. Among regions with smaller fractal scales (upper limit) than the EGMB, the fault CPD of the EGMB is larger than that of the Zhaoyuan gold ore concentration area, but smaller than those of the southeastern Guangxi gold and silver mineralization area, the Qitianling ore concentration area of southern Hunan, and the Yadu–Mangdong metallogenic belt of northwest Guizhou Province. Among regions with the same fractal scales (upper limit) as the EGMB, the fault CPD of the EGMB is larger than that of the Kangguertage gold belt in east Tianshan, but smaller than those of Xikuangshan–Longshan, Dashenshan, and Simingshan Sb belts in central Hunan Province. Among regions with larger fractal scales (upper limit) than the

EGMB, the fault CPD of the EGMB is larger than those of Southern China and Sichuan Province, but is close to those of Pb–Zn ore concentration regions bordering the three provinces of Sichuan, Yunnan, and Guizhou.

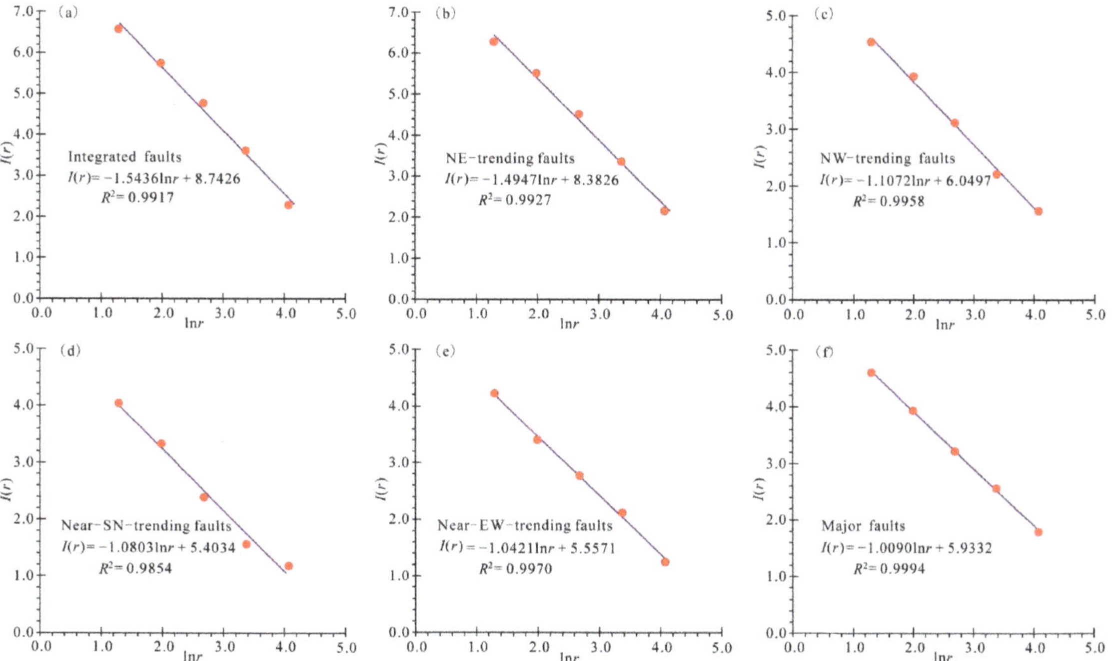

Figure 5. Linear fitting diagrams for the correlation dimension (CRD) calculation of faults in the eastern Guizhou Pb–Zn metallogenic belt (EGMB). The $\ln r$ versus $I(r)$ plots of CRD data for (**a**) Integrated faults; (**b**) NE-trending faults; (**c**) NW-trending faults; (**d**) Near-SN-trending faults; (**e**) Near-EW-trending faults; and (**f**) Major faults, showing their linear regression parameters.

Table 2. Statistical table of fractal dimension for fault structures in some areas of China.

Region	Scale Interval (km)	CPD, D_0	IND, D_1	CRD, D_2	Reference
Activity Area of Continent in China (Diwa Area)	8–256	1.236–1.624	-	-	
Stable Area of Continent in China (platform area)	8–256	0.827–1.074	-	-	[49]
Yungui Activity Area	8–256	1.332	-	-	
China Continent	8–256	1.493	-	-	
Shell Binding Site	8–256	>1.5	-	-	
Sichuan-Yunnan-Guizhou Pb–Zn Metallogenic Province	9.336–149.373	1.5395	-	-	[54]
Zhaxikang Ore Concentration Area	0.073–4.7	1.249	-	-	[75]
Gudui–Longzi Region, Tibet	1.875–30	1.678	-	-	[78]
Tongling Ore Concentration Area	0.1–3	1.29	-	-	[52]
Jiaojia District, Jiaodong	0.50–16.00	1.3507	-	-	
Sanshandao-Cangshang Gold Mine Field in Jiaojia District	0.25–4.00	1.0103	-	-	[79]

Table 2. *Cont.*

Region	Scale Interval (km)	CPD, D_0	IND, D_1	CRD, D_2	Reference
Jiaojia Gold Mine Field in Jiaojia District	0.25–4.00	1.3198	-	-	
Canzhuang-Lingshangou Gold Mine Field in Jiaojia District	0.25–4.00	1.3656	-	-	
Xiyou–Zhuqiao Area in Jiaojia District (Mineral-free Area)	0.25–4.00	1.1315	-	-	
Kangguertage Gold Belt in East Tianshan	1.69412–54.2118	0.716	-	-	[48]
South China	25–400	1.4142	-	-	[80]
Jiangnan Diwa Region in South China	10–160	1.5939	-	-	[81]
Southeast Diwa Region in South China	10–160	1.6800	-	-	
Xikuangshan–Longshan Sb Ore Belt in Central Hunan	5–60	1.8183	1.8102	-	
Simingshan Sb Ore Belt in Central Hunan	5–60	1.7346	1.7067	-	[82]
Damshenshan Sb Ore Belt in Central Hunan	5–60	1.5975	1.5933	-	
Zhaoyuan Gold Ore Concentration Area	1–5	1.4806	-	-	[51]
Gold and Silver Metallogenic Area in Southeast Guangxi	1.25–40	1.61	-	-	[83]
Qitianling Ore Concentration Area in Southern Hunan	0.625–10	1.656	-	-	[84]
Hutouya Polymetallic Ore Collection Area, Qinghai Province	0.15–0.7	1.085	-	-	[85]
Gejiu Mining Area in Southeast Yunnan	0.5–5	1.432	-	-	
Malage Ore Field	0.5–5	1.093	-	-	[86]
Laochang Ore Field	0.5–5	1.263	-	-	
Kafang Ore Field	0.5–5	1.121	-	-	
Southern Jiangxi Province	0.5–10	1.2797	-	-	[87]
Faults of Maokou Formation in Southeast Sichuan	2.5–40	1.423	1.467	1.468	[50]
Xiciwa Area in Bozhong Sag	0.5–8	1.2137	1.2903	1.3582	[88]
Sichuan Area	3.75–120	1.4524	1.5136	1.5455	[89]
Shuiyanba Ore Field, Hezhou, Guangxi Province	0.171875–5.5	1.3475	-	-	[90]
Yadu-Mangdong Metallogenic Belt in NW Guizhou Province	3.371–26.965	1.6052	1.6051	-	[91]
EGMB	3.670–58.716	1.5095	1.5391	1.5436	this article

The IND and CRD of faults in China are less well studied than the CPD. From the limited data available, IND and CRD of fault structures in the EGMB are larger than those in areas with smaller fractal scales (upper limit), including the Maokou Formation in southeastern Sichuan and Xiciwa in the Bozhong sag. Among regions with larger fractal scales (upper limit) than the EGMB, the fractal dimension values are similar (e.g., the Sichuan area). Finally, among regions with similar fractal scales (upper limit) compared to the EGMB, the IND of fault structures in the EGMB is smaller than those of the Xikuangshan–Longshan, Dashenshan, and Simingshan Sb belts of Central Hunan.

3.3. Single Fractal Characteristics of Fault Divisions

The coefficient of determination of the straight line fitted by the division's fractal dimension values is 0.9648–0.9983 (Tables 3–5), and the straight line has a high degree of fit, indicating that fault structures within the divisions have good statistical self-similarity. The CPD values of the divisions range from 0.9230 to 1.5095 (Table 3), and the median is 1.4003. The IND values of the divisions range from 0.9746 to 1.5262 (Table 4), and the median is 1.4164. The CRD values of the divisions range from 1.0222 to 1.5410 (Table 5), and the median is 1.4195. Pb–Zn deposits are developed in 10 of the 12 divisions, and the fractal dimension value (CPD, IND, CRD) interval of the developed Pb–Zn deposit area covers the fractal dimension values (CPD, IND, CRD) of all divisions. Therefore, in order to explore the coupling relationship between fractal dimension values and ore deposit distribution, subdivisions were divided on the basis of divisions (Figure 2), and the fractal dimension values of subdivisions were calculated.

Table 3. Statistical table of division for the capacity dimension (CPD) calculation parameters.

Division Number		Fractal Scale, r (km)				CPD, D_0	Coefficient of Determination (R^2)
		29.358	14.679	7.340	3.670		
	1	4	12	34	68	1.3765	0.9901
	2	4	14	36	92	1.4934	0.9946
	3	4	15	42	91	1.5009	0.9863
	4	4	15	35	76	1.3967	0.9827
	5	4	14	43	90	1.5095	0.9875
$N(r)$	6	4	14	33	77	1.4038	0.9899
	7	4	9	18	30	0.9721	0.9901
	8	4	11	25	50	1.2116	0.9928
	9	4	16	41	93	1.4976	0.9843
	10	4	8	14	28	0.9230	0.9983
	11	4	14	36	71	1.3812	0.9823
	12	1	4	11	19	1.4204	0.9648

Table 4. Statistical table of division for the information dimension (IND) calculation parameters.

Division Number		Fractal Scale, r (km)				IND Values, D_1	Coefficient of Determination (R^2)
		29.358	14.679	7.340	3.670		
	1	1.309	2.415	3.481	4.188	1.4001	0.9905
	2	1.339	2.563	3.496	4.472	1.4907	0.9961
	3	1.362	2.622	3.685	4.480	1.5028	0.9901
	4	1.334	2.636	3.498	4.299	1.4077	0.9855
	5	1.382	2.575	3.694	4.461	1.4943	0.9909
$I(r)$	6	1.305	2.487	3.403	4.279	1.4192	0.9947
	7	1.277	2.164	2.871	3.379	1.0119	0.9856
	8	1.288	2.322	3.166	3.892	1.2486	0.9936
	9	1.310	2.651	3.650	4.503	1.5262	0.9891
	10	1.242	2.043	2.599	3.309	0.9746	0.9956
	11	1.295	2.582	3.545	4.240	1.4135	0.9820
	12	0.000	1.332	2.398	2.944	1.4282	0.9689

Table 5. Statistical table of division for the correlation dimension (CRD) calculation parameters.

Division Number		Fractal Scale, r (km)				CRD Values, D_2	Coefficient of Determination (R^2)
		29.358	14.679	7.340	3.670		
	1	1.232	2.354	3.427	4.146	1.4161	0.9907
	2	1.294	2.498	3.406	4.408	1.4791	0.9966
	3	1.339	2.546	3.626	4.440	1.4980	0.9927
	4	1.289	2.572	3.420	4.249	1.4035	0.9875
	5	1.378	2.526	3.620	4.406	1.4686	0.9931
$I(r)$	6	1.241	2.358	3.301	4.192	1.4131	0.9972
	7	1.184	2.120	2.844	3.348	1.0412	0.9823
	8	1.185	2.254	3.107	3.863	1.2822	0.9936
	9	1.242	2.537	3.573	4.457	1.5410	0.9925
	10	1.099	1.997	2.549	3.276	1.0222	0.9918
	11	1.240	2.534	3.495	4.207	1.4228	0.9828
	12	0.000	1.273	2.398	2.944	1.4367	0.9723

3.4. Single Fractal Characteristics of Fault Subdivisions

The subdivisions were squares with sides of 29.358 km, and so a reasonable upper limit of the study scale was 29.358 km. To ensure that fault structures can be regarded as an ideal straight line or curve shape to the greatest extent, theoretically speaking, the lower limit of the research scale should be as large as possible. Taking into account the control scale of the fault structure on the deposit space, the research scale interval used in the calculation of the fractal dimension value of the subdivisions was 3.670–29.358 km. However, considering the calculation characteristics of the CPD, IND, and CRD, a scale interval of 3.670–29.358 km was used for the calculation of the subdivision CPD, and a scale interval of 3.670–14.679 km was used for the calculation of IND and CPD. The statistics of the calculation parameters of the fractal dimension value of the subdivisions are shown in Tables 6 and 7. The subdivision CPD values are 0 to 1.6834, with a median of 1.3712. The subdivision IND values are 0 to 1.6091, with a median of 1.1797. The subdivision CRD values are 0 to 1.6179, with a median of 1.2010.

Table 6. Statistical table of subdivision for the capacity dimension (CPD) calculation parameters.

Sub-Division Number/Serial Number	Fractal Scale, r (km)				CPD Values, D_0	Coefficient of Determination (R^2)
	29.358	14.679	7.340	3.670		
1-1/1	1	4	10	19	1.4066	0.9713
1-2/2	1	4	11	23	1.5031	0.9809
1-3/11	1	2	5	10	1.1288	0.9968
1-4/12	1	2	8	16	1.4001	0.9800
2-1/3	1	3	9	22	1.4964	0.9977
2-2/4	1	4	9	20	1.4136	0.9792
2-3/9	1	3	9	27	1.5850	1.0000
2-4/10	1	4	9	23	1.4741	0.9859
3-1/5	1	3	11	20	1.4841	0.9808
3-2/6	1	4	9	20	1.4136	0.9792
3-3/7	1	4	9	18	1.3680	0.9718
3-4/8	1	4	13	33	1.6834	0.9925
4-1/17	1	4	9	18	1.3680	0.9718
4-2/18	1	4	8	15	1.2721	0.9597
4-3/19	1	4	11	26	1.5561	0.9878
4-4/20	1	3	7	17	1.3485	0.9965
5-1/15	1	4	12	27	1.5850	0.9865
5-2/16	1	4	15	31	1.6770	0.9821
5-3/21	1	3	8	14	1.2838	0.9809
5-4/22	1	3	8	18	1.3925	0.9955
6-1/13	1	3	8	21	1.4593	0.9990
6-2/14	1	4	10	23	1.4893	0.9845
6-3/23	1	3	10	21	1.4914	0.9911
6-4/24	1	4	5	12	1.1077	0.9276
7-1/25	1	3	5	8	0.9737	0.9524
7-2/26	1	3	7	11	1.1601	0.9684
7-3/35	1	1	1	2	0.3000	0.6000
7-4/36	1	2	5	9	1.0832	0.9937
8-1/27	1	4	12	29	1.6160	0.9899
8-2/28	1	2	2	2	0.3000	0.6000
8-3/33	1	2	3	4	0.6585	0.9608
8-4/34	1	3	8	15	1.3136	0.9862

The row-label column group is headed $N(r)$.

Table 6. *Cont.*

Sub-Division Number/Serial Number	Fractal Scale, r (km)				CPD Values, D_0	Coefficient of Determination (R^2)
	29.358	14.679	7.340	3.670		
9-1/29	1	4	13	33	1.6834	0.9925
9-2/30	1	4	8	19	1.3744	0.9773
9-3/31	1	4	12	27	1.5850	0.9865
9-4/32	1	4	8	14	1.2423	0.9521
10-1/41	1	2	4	9	1.0510	0.9984
10-2/42	1	3	6	9	1.0510	0.9565
10-3/43	1	1	1	3	0.4755	0.6000
10-4/44	1	2	3	7	0.9007	0.9836
11-1/39	1	4	11	23	1.5031	0.9809
11-2/40	1	4	12	22	1.4964	0.9721
11-3/45	1	3	5	10	1.0703	0.9749
11-4/46	1	3	8	16	1.3416	0.9903
12-1/37	0	0	0	0	0.0000	-
12-2/38	1	4	11	19	1.4204	0.9648
12-3/47	0	0	0	0	0.0000	-
12-4/48	0	0	0	0	0.0000	-

Table 7. Statistical table of calculation parameters for the information dimension (IND) and correlation dimension (CRD) of subdivision.

Subdivision Number/Serial Number	r (km)	$\ln r$	$I(r)$ for IND	IND, D_1	R^2	$I(r)$ for CRD	CRD, D_2	R^2
1-1/1	14.679	2.686	1.332			1.273		
	7.340	1.993	2.272	1.1501	0.9894	2.231	1.1740	0.9896
	3.670	1.300	2.926			2.900		
1-2/2	14.679	2.686	1.321			1.269		
	7.340	1.993	2.338	1.2723	0.9922	2.281	1.2684	0.9924
	3.670	1.300	3.085			3.027		
1-3/11	14.679	2.686	0.637			0.588		
	7.340	1.993	1.561	1.1797	0.9944	1.504	1.1853	0.9956
	3.670	1.300	2.272			2.231		
1-4/12	14.679	2.686	0.693			0.693		
	7.340	1.993	2.079	1.5001	0.9643	2.079	1.5001	0.9643
	3.670	1.300	2.773			2.773		
2-1/3	14.679	2.686	1.079			1.059		
	7.340	1.993	2.133	1.4111	0.9980	2.079	1.3828	0.9986
	3.670	1.300	3.035			2.976		
2-2/4	14.679	2.686	1.277			1.184		
	7.340	1.993	2.254	1.2274	0.9927	2.197	1.2767	0.9930
	3.670	1.300	2.979			2.954		
2-3/9	14.679	2.686	0.995			0.898		
	7.340	1.993	2.079	1.6091	0.9997	1.962	1.6179	0.9991
	3.670	1.300	3.226			3.141		
2-4/10	14.679	2.686	1.352			1.327		
	7.340	1.993	2.023	1.2669	0.9818	1.974	1.2578	0.9784
	3.670	1.300	3.108			3.070		
3-1/5	14.679	2.686	1.055			1.022		
	7.340	1.993	2.272	1.3876	0.9771	2.231	1.3936	0.9793
	3.670	1.300	2.979			2.954		

Table 7. *Cont.*

Subdivision Number/Serial Number	r (km)	ln$r$	$I(r)$ for IND	IND, D_1	R^2	$I(r)$ for CRD	CRD, D_2	R^2
3-2/6	14.679 7.340 3.670	2.686 1.993 1.300	1.330 2.272 2.965	1.1797	0.9923	1.281 2.231 2.924	1.1853	0.9919
3-3/7	14.679 7.340 3.670	2.686 1.993 1.300	1.332 2.146 2.857	1.1001	0.9985	1.273 2.088 2.813	1.1112	0.9989
3-4/8	14.679 7.340 3.670	2.686 1.993 1.300	1.311 2.505 3.461	1.5512	0.9959	1.259 2.449 3.415	1.5555	0.9964
4-1/17	14.679 7.340 3.670	2.686 1.993 1.300	1.330 2.164 2.890	1.1259	0.9984	1.281 2.120 2.890	1.1610	0.9994
4-2/18	14.679 7.340 3.670	2.686 1.993 1.300	1.330 2.043 2.686	0.9784	0.9991	1.281 1.997 2.655	0.9911	0.9994
4-3/19	14.679 7.340 3.670	2.686 1.993 1.300	1.369 2.303 3.194	1.3162	0.9998	1.350 2.197 3.107	1.2674	0.9996
4-4/20	14.679 7.340 3.670	2.686 1.993 1.300	1.099 1.946 2.833	1.2513	0.9998	1.099 1.946 2.833	1.2513	0.9998
5-1/15	14.679 7.340 3.670	2.686 1.993 1.300	1.321 2.393 3.245	1.3881	0.9956	1.269 2.287 3.165	1.3680	0.9982
5-2/16	14.679 7.340 3.670	2.686 1.993 1.300	1.373 2.649 3.404	1.4651	0.9785	1.362 2.590 3.364	1.4442	0.9831
5-3/21	14.679 7.340 3.670	2.686 1.993 1.300	1.040 2.043 2.639	1.1537	0.9788	0.981 1.997 2.639	1.1962	0.9834
5-4/22	14.679 7.340 3.670	2.686 1.993 1.300	1.079 1.850 2.839	1.2696	0.9950	1.059 2.297 2.781	1.2424	0.9400
6-1/13	14.679 7.340 3.670	2.686 1.993 1.300	0.974 1.951 2.968	1.4385	0.9999	0.901 1.834 2.878	1.4264	0.9990
6-2/14	14.679 7.340 3.670	2.686 1.993 1.300	1.215 2.098 3.066	1.3353	0.9993	1.099 1.994 2.976	1.3540	0.9993
6-3/23	14.679 7.340 3.670	2.686 1.993 1.300	1.099 2.243 2.997	1.3695	0.9861	1.099 2.187 2.941	1.3292	0.9891
6-4/24	14.679 7.340 3.670	2.686 1.993 1.300	1.386 1.792 2.485	0.7925	0.9777	1.386 1.792 2.485	0.7925	0.9777
7-1/25	14.679 7.340 3.670	2.686 1.993 1.300	1.099 1.609 2.079	0.7076	0.9994	1.099 1.609 2.079	0.7076	0.9994

Table 7. *Cont.*

Subdivision Number/Serial Number	r (km)	$\ln r$	$I(r)$ for IND	IND, D_1	R^2	$I(r)$ for CRD	CRD, D_2	R^2
7-2/26	14.679	2.686	1.040	0.9592	0.9703	0.981	0.9738	0.9714
	7.340	1.993	1.906			1.856		
	3.670	1.300	2.369			2.331		
7-3/35	14.679	2.686	0.000	0.5000	0.7500	0.000	0.5000	0.7500
	7.340	1.993	0.000			0.000		
	3.670	1.300	0.693			0.693		
7-4/36	14.679	2.686	0.693	1.0850	0.9843	0.693	1.0850	0.9843
	7.340	1.993	1.609			1.609		
	3.670	1.300	2.197			2.197		
8-1/27	14.679	2.686	1.369	1.4188	0.9980	1.350	1.4023	0.9988
	7.340	1.993	2.428			2.380		
	3.670	1.300	3.336			3.294		
8-2/28	14.679	2.686	0.693	<0.5	-	0.693	<0.5	-
	7.340	1.993	0.693			0.693		
	3.670	1.300	0.693			0.693		
8-3/33	14.679	2.686	0.693	0.5000	0.9905	0.693	0.5000	0.9905
	7.340	1.993	1.099			1.099		
	3.670	1.300	1.386			1.386		
8-4/34	14.679	2.686	1.055	1.1925	0.9812	1.022	1.2165	0.9789
	7.340	1.993	2.079			2.079		
	3.670	1.300	2.708			2.708		
9-1/29	14.679	2.686	1.358	1.4154	0.9932	1.332	1.5472	1.0000
	7.340	1.993	2.479			2.392		
	3.670	1.300	3.320			3.477		
9-2/30	14.679	2.686	1.330	1.1418	0.9951	1.281	1.1464	0.9937
	7.340	1.993	2.025			1.966		
	3.670	1.300	2.912			2.870		
9-3/31	14.679	2.686	1.352	1.3929	0.9945	1.327	1.3968	0.9970
	7.340	1.993	2.441			2.388		
	3.670	1.300	3.283			3.263		
9-4/32	14.679	2.686	1.386	0.9037	0.9962	1.386	0.9037	0.9962
	7.340	1.993	2.079			2.079		
	3.670	1.300	2.639			2.639		
10-1/41	14.679	2.686	0.693	1.0850	0.9980	0.693	1.0850	0.9980
	7.340	1.993	1.386			1.386		
	3.670	1.300	2.197			2.197		
10-2/42	14.679	2.686	1.099	0.7925	0.9776	1.099	0.7925	0.9776
	7.340	1.993	1.792			1.792		
	3.670	1.300	2.197			2.197		
10-3/43	14.679	2.686	0.000	0.7925	0.7500	0.000	0.7925	0.7500
	7.340	1.993	0.000			0.000		
	3.670	1.300	1.099			1.099		
10-4/44	14.679	2.686	0.637	0.9036	0.9646	0.588	0.8958	0.9707
	7.340	1.993	1.055			1.022		
	3.670	1.300	1.889			1.829		
11-1/39	14.679	2.686	1.352	1.2669	0.9936	1.327	1.2578	0.9958
	7.340	1.993	2.352			2.297		
	3.670	1.300	3.108			3.070		

Table 7. *Cont.*

Subdivision Number/Serial Number	r (km)	ln$r$	$I(r)$ for IND	IND, D_1	R^2	$I(r)$ for CRD	CRD, D_2	R^2
11-2/40	14.679	2.686	1.352	1.2341	0.9757	1.327	1.2242	0.9795
	7.340	1.993	2.441			2.388		
	3.670	1.300	3.063			3.024		
11-3/45	14.679	2.686	1.099	0.8685	0.9924	1.099	0.8685	0.9924
	7.340	1.993	1.609			1.609		
	3.670	1.300	2.303			2.303		
11-4/46	14.679	2.686	1.040	1.2350	0.9902	0.981	1.2560	0.9908
	7.340	1.993	2.043			1.997		
	3.670	1.300	2.752			2.722		
12-1/37	14.679	2.686	0.000	0.0000	-	0.000	0.0000	-
	7.340	1.993	0.000			0.000		
	3.670	1.300	0.000			0.000		
12-2/38	14.679	2.686	1.332	1.1631	0.9666	1.273	1.2058	0.9616
	7.340	1.993	2.398			2.398		
	3.670	1.300	2.944			2.944		
12-3/47	14.679	2.686	0.000	0.0000	-	0.000	0.0000	-
	7.340	1.993	0.000			0.000		
	3.670	1.300	0.000			0.000		
12-4/48	14.679	2.686	0.000	0.0000	-	0.000	0.0000	-
	7.340	1.993	0.000			0.000		
	3.670	1.300	0.000			0.000		

The determination coefficient R^2 value of the fitting straight line in the calculation of fractal dimension value was small for a small number of subdivisions (e.g., subdivision 10-3). In addition, the slope of the fitted straight line in the calculation of fractal dimension value for a small number of partitions was zero, and only the value range could be judged (e.g., the IND and CRD of subdivision 8-2). However, these phenomena had little effect on the coupling relationship between the fractal dimensions of subdivisions and the spatial distribution of deposit.

The main reasons are as follows: (1) among the 48 capacity-dimensional data in the subdivisions, only three have determination coefficients R^2 of <0.9; (2) compared with other data, the size relationship of such data is still very reliable. For example, the calculated value of the CPD of subdivision 10-3 is 0.4755, which is larger than the calculated values of subdivisions 7-3, 8-2, 12-1, 12-3, and 12-4, but is smaller than the calculated values of other subdivisions. This result is consistent with the original meaning of the capacity-dimensional representation. The CPD, also known as the box dimension, was originally used to characterize the ability of a fractal to occupy a box under the corresponding research scale. The larger the fractal dimension value, the stronger the ability to occupy the box. The relationship between the calculated CPD value of subdivision 10-3 and those of other subdivisions is consistent with the original meaning of the representation of the capacity dimension; and (3) the value of such data is relatively low in the overall data, and has no effect on the judgment of the favorable fractal dimension interval of the coupling relationship between fractal dimension value and the spatial distribution of ore deposit.

4. Multi-Fractal Characteristics of Fault Structure Spatial Distribution

4.1. Calculation Method

Multi-fractal is the mutual entanglement and mosaic of multiple single fractals in space. It is a generalization of single fractals [92,93] that can reflect more complex spatial structures [94]. The calculation of a multi-fractal function spectrum is the core of multi-fractal research, and is usually expressed as the functional relationship between the holder singularity exponent and fractal dimension. It is generally described by the curve between α–$f(\alpha)$. Methods of calculating the multi-fractal function spectrum include the quadratic moment method, moment method, multiplier method, histogram method, and wavelet method, among others [63,95–99]. The most mature and widely used method is the moment method.

The steps for calculating the multi-fractal spectral function $f(\alpha)$ by the moment method are as follows:

(1) Define the fractal measure $P_i(r)$:

$$P_i(r) = \frac{n_i}{N(r) \sum\limits_{i=1} n_i} \tag{7}$$

where r is the side length of the square grid covering the study area, i is the serial number of the grid at the r scale, n_i is the number of faults in the i-th grid, and $N(r)$ is the number of grids at the r scale.

(2) Build the multi-fractal partition function $X_q(r)$:

$$X_q(r) = \sum_{i=1}^{N(r)} P_i^q(r) \tag{8}$$

where q is an arbitrary number defined as the q-order moment of the fractal measure $P_i(r)$.

(3) Calculate the quality index $\tau(q)$:

$$\tau(q) = \lim_{r \to 0} \frac{\ln X_q(r)}{\ln r} = \lim_{r \to 0} \frac{\ln \sum\limits_{i=1}^{N(r)} P_i^q(r)}{\ln r} \tag{9}$$

In actual calculation, for an arbitrarily determined q value, the quality index $\tau(q)$ is obtained by calculating the slope of the best straight line fitted by the projected points $(\ln r, \ln X_q(r))$ at different scales r.

(4) Calculate the singularity index $\alpha(q)$:

$$\alpha(q) = \frac{d\tau(q)}{dq} = \lim_{r \to 0} \frac{\sum\limits_{i=1}^{N(r)} P_i^q(r) \ln P_i(r)}{\ln r \sum\limits_{i=1}^{N(r)} P_i^q(r)} \tag{10}$$

In the actual calculation, for an arbitrarily determined q value, the singularity index $\alpha(q)$ is obtained by calculating the slope of the best straight line fitted by the projected points $\left(\ln r \dfrac{\sum\limits_{i=1}^{N(r)} P_i^q(r) \ln P_i(r)}{\sum\limits_{i=1}^{N(r)} P_i^q(r)}\right)$ at different scales r.

(5) Calculate the multi-fractal spectral function $f(\alpha)$:

$$f(\alpha) = q\alpha(q) - \tau(q) = q\frac{d\tau(q)}{dq} - \tau(q) \tag{11}$$

The singularity index $\alpha(q)$ and the multi-fractal spectral function $f(\alpha)$ reflect the local characteristics of the multi-fractal. The singularity index $\alpha(q)$ represents the fractal dimension of the small area of the fractal body, and its increment $\Delta\alpha$ (multi-fractal spectral width) describes the degree of inhomogeneity of the distribution of the subsets formed by the relevant physical quantities on the multi-fractal set. That is to say, it reflects the unevenness of the probability measure distribution on the entire fractal structure, and is used to describe the fluctuation range of the data set. The multi-fractal spectral function $f(\alpha)$ is a spectrum composed of infinite sequences composed of different singularity exponents $\alpha(q)$, which can describe the changing trend of the number of elements in the subset formed by the multi-fractal and related physical quantities. Its increment $\Delta f(\alpha)$ (multi-fractal spectrum height) describes the magnitude of variation in the number of elements in the subset formed by the relevant physical quantity.

4.2. Multi-Fractal Characteristics of Fault Structures

When carrying out the multi-fractal spectrum calculation for the fault structures in the study area, the fractal scale interval used was 3.670–58.716 km, the q-order moment was -10 to 10, and the step size was 0.5. The calculation results are shown in Table 8. According to the fault multi-fractal spectrum data in the study area (Table 8), we drew the multi-fractal spectrum in the study area (Figure 6).

Table 8. Fault multifractal spectrum data table for the study area.

Serial Number	q	$\alpha\,(q)$	$f\,(\alpha)$	Serial Number	q	$\alpha\,(q)$	$f\,(\alpha)$
1	-10.0	1.0141	2.4020	22	0.5	1.5130	1.5203
2	-9.5	1.0150	2.3935	23	1.0	1.5390	1.5390
3	-9.0	1.0161	2.3831	24	1.5	1.5463	1.5475
4	-8.5	1.0175	2.3713	25	2.0	1.5383	1.5328
5	-8.0	1.0191	2.3572	26	2.5	1.5184	1.4876
6	-7.5	1.0212	2.3410	27	3.0	1.4900	1.4092
7	-7.0	1.0237	2.3233	28	3.5	1.4564	1.2999
8	-6.5	1.0267	2.3031	29	4.0	1.4211	1.1675
9	-6.0	1.0306	2.2787	30	4.5	1.3867	1.0214
10	-5.5	1.0355	2.2506	31	5.0	1.3553	0.8724
11	-5.0	1.0418	2.2176	32	5.5	1.3279	0.7288
12	-4.5	1.0503	2.1773	33	6.0	1.3048	0.5961
13	-4.0	1.0618	2.1286	34	6.5	1.2858	0.4775
14	-3.5	1.0782	2.0673	35	7.0	1.2703	0.3730
15	-3.0	1.1018	1.9910	36	7.5	1.2580	0.2840
16	-2.5	1.1360	1.8974	37	8.0	1.2481	0.2073
17	-2.0	1.1843	1.7894	38	8.5	1.2404	0.1434
18	-1.5	1.2480	1.6786	39	9.0	1.2343	0.0907
19	-1.0	1.3235	1.5845	40	9.5	1.2296	0.0472
20	-0.5	1.4003	1.5268	41	10.0	1.2260	0.0120
21	0.0	1.4662	1.5095				

The graph connecting points $(q, \alpha(q), f(\alpha))$ in the three-dimensional coordinate system is a spiral curve (Figure 6a), and the nonlinear relationship is obvious. When the q-order moment is -10 to 10, the singularity index $\alpha(q)$ ranges from 1.0141 to 1.5463; it first increases and then decreases with the increase of the order moment q (Figure 6e). At the same time, $f(\alpha)$ ranges from 0.0120 to 2.4020, and decreases as a whole and increases locally with the increase of the order moment q (Figure 6d). The curve connected by points $(\alpha(q), f(\alpha))$ is not a common parabolic (or hook) shape with downward opening, but a combination of two semi-parabolic shapes with opposite opening directions (i.e., a bifurcation; Figure 6b). When the q-order moment ranges from 1 to 10, the curve connecting the points $(\alpha(q), f(\alpha))$ is a typical semi-parabolic shape (Figure 6c).

When the q-order moment is between -10 and 10, the shape of the multi-fractal spectrum is quite different from that reported in most previous literatures. Most multi-fractal spectrum parameter calculations in the literature adopted the fitting method, such as the singularity index $\alpha(q)$, etc., and did not strictly use the limit method for calculation (which cannot be realized); this increases the multi-fractal spectrum shape diversity to a certain extent. Various shapes of multi-fractal spectra have been reported. In addition to the typical downward-opening parabola or hook, there can also be zigzag [100] and bifurcated [101–103]. The main reasons for the diverse shapes of multi-fractal spectra are as follows: (1) the characteristics and differences of the calculation method itself; (2) differences in tectonic distribution characteristics (or element enrichment methods) in different regions; (3) differences in the value range of the q-order moment; and (4) buried fault structures were not discovered. When using the same calculation method to calculate the multi-fractal spectrum, in addition to the characteristics of the fractal itself, the value of the q-order moment is also an important factor.

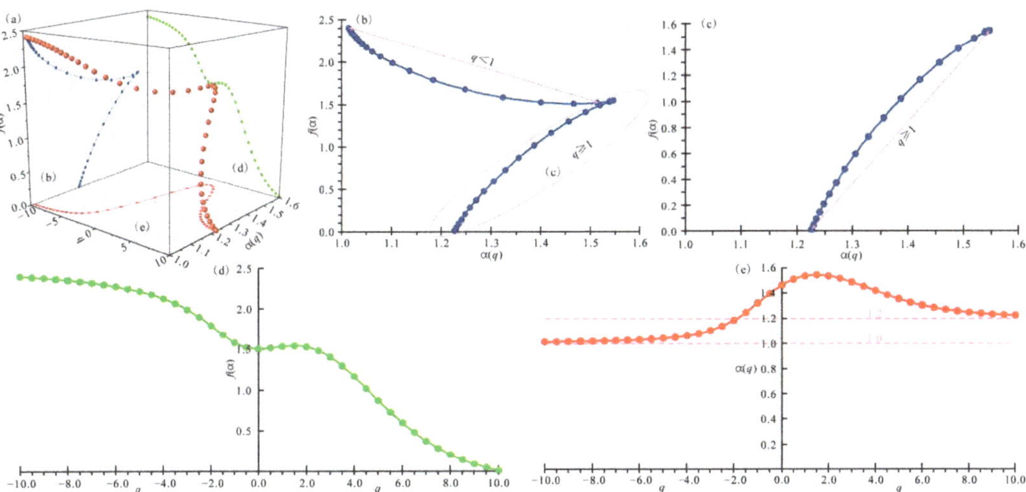

Figure 6. Multi-fractal spectrum of faults in the study area. (**a**) 3-D plot of q, $\alpha(q)$, $f(\alpha)$, showing points $(q, \alpha(q), f(\alpha))$ in the three-dimensional coordinate system is a spiral; (**b**) 2-D plot of $\alpha(q)$, $f(\alpha)$, showing the curve connected by points $(\alpha(q), f(\alpha))$ is not a common parabolic (or hook) shape with downward opening, but a combination of two semi-parabolic shapes with opposite opening directions; (**c**) When the q-order moment ranges from 1 to 10, the curve connecting the points $(\alpha(q), f(\alpha))$ is a typical semi-parabolic shape; (**d**) $f(\alpha)$ ranges from 0.0120 to 2.4020, and decreases as a whole and increases locally with the increase of the order moment q; and (**e**) the singularity index $\alpha(q)$ first increases and then decreases with the increase of the order moment q.

Theoretically, the value of the q-order moment can be any real number, but there is no uniform standard when carrying out multi-fractal spectrum calculations, and the numerical ranges used in different studies vary greatly. The multi-fractal spectrum uses different q-order moment values to describe the characteristics of different levels of the fractal body. When q is greater than 0, the multi-fractal spectrum can describe the basic characteristics of the fractal body. When q is less than 0, the multi-fractal spectrum focuses on the properties of low-probability regions, reflecting small structural changes in the fractal structure. The smaller q is, the more easily affected it is by measurement errors or interference factors. As most previous studies carried out parameter calculation on parabolic or hooked fractal spectrum, in order to ensure the reliability of this calculation, we only calculated $\Delta f(\alpha)$ and $\Delta\alpha$ of the multi-fractal spectrum when the q-order moment was between 1 and 10. Where $\Delta\alpha = \alpha_{max} - \alpha_{min}$, $\Delta f(\alpha) = f(\alpha)_{max} - f(\alpha)_{min}$. When $\alpha_{max} = 1.5463$, $\alpha_{min} = 1.2260$, $f(\alpha)_{max} = 1.5475$, and $f(\alpha)_{min} = 0.0120$, then $\Delta\alpha = \alpha_{max} - \alpha_{min} = 0.3203$, $\Delta f(\alpha) = f(\alpha)_{max} - f(\alpha)_{min} = 1.5355$, implying that the study area has great metallogenic potential.

5. Fractal Clustering Characteristics of Ore Deposits

5.1. Fractal Characteristics of Spatial Distribution of Ore Deposits

The 48 subdivisions of the EGMB contain 61 metal deposits, including 53 Pb–Zn deposits. Metal deposits are distributed in 27 subdivisions, including 1-2, 2-4, 3-3, 3-4, 6-1, 6-2, 6-3, and 6-4, etc., accounting for 56.3% of the total subdivisions. Among them, there are 30 metal ore deposits in seven subdivisions including 1-2, 5-4, 7-2, 8-1, 8-4, 12-2, and 11-3, accounting for ~49.2% of the total number of metal deposits. Pb–Zn deposits are distributed in 24 subdivisions, including 1-2, 2-4, 3-3, 3-4, 6-1, 6-2, 6-3, and 6-4, accounting for 50% of the total number of subdivisions. Among them, seven subdivisions including 1-2, 5-4, 7-2, 8-1, 8-4, 12-2, and 11-3 have developed 27 Pb–Zn deposits, accounting for 50.9% of all Pb–Zn deposits. In summary, deposits are mainly distributed in a small number of subdivisions, and from a qualitative perspective, deposit distribution has significant clustering.

Taking ore deposits as a point set, the spatial distribution fractal dimension (SDD) of ore deposits can be calculated by the counting-box method, similar to that applied to the treatment of CPD values for fault systems described in the previous sections. The statistics of the calculation parameters of the SDD values are shown in Table 9. According to Table 9, Figure 7 shows the ln$r$–ln$N(r)$ regression fitting line graph. From the statistical table of the SDD value of mineral deposits in some areas of China (Table 10) and the linear fitting diagram of the SDD value calculation of mineral deposits (Figure 7), it can be seen that: (1) the fitting degrees of the linear fitting lines of Sb deposits, Pb–Zn deposits, and metal deposits in the study area are all greater than 0.97, indicating that their spatial distributions have fractal cluster structures; (2) the SDD value of metal deposits is greater than that of Pb–Zn deposits, while the SDD value of Pb–Zn deposits is greater than that of Sb deposits; (3) the metal deposits and Pb–Zn deposits in the study area have a smaller SDD than most other regions in China with larger fractal scales (upper limit). Compared with the same fractal scale (upper limit) of Pb–Zn deposits in the Yadu–Mangdong metallogenic belt, the SDD is also smaller, indicating that the metal deposits and Pb–Zn deposits in the study area are more clustered; and (4) the clustering of the ore deposits results in decreasing SDD, while the SDD of the ore deposits is much smaller than those of the integrated faults and NE-trending faults in the study area.

Table 9. Statistical table of calculation parameters for the spatial distribution fractal dimensions (SDD) of deposits.

Deposit Category	Fractal Scale, r (km)	$N(r)$	ln$r$	ln$N(r)$
Metal deposits	58.716	10	4.073	2.303
	29.358	30	3.380	3.401
	14.679	55	2.686	4.007
	7.340	97	1.993	4.575
Pb–Zn deposits	58.716	10	4.073	2.303
	29.358	27	3.380	3.296
	14.679	48	2.686	3.871
	7.340	87	1.993	4.466
Sb deposits	58.716	3	4.073	1.099
	29.358	5	3.380	1.609
	14.679	7	2.686	1.946
	7.340	9	1.993	2.197

5.2. Fractal Characteristics of Deposit Quantity and Density

An important step in the exploration of ore deposits is to investigate the distribution characteristics of known ore deposits within an area delineated by a finite distance [58,104].

To quantitatively determine the distribution character of ore deposits within a circular area of radius r, we normally adopted a probability density function defined as:

$$d(r) = Kr^{D_D - 2} \ (2 > D_D > 0) \tag{12}$$

where $d(r)$ is the probability density function, denoting the number of ore deposits per unit area within radius r, taking a known ore deposit as the center of the circle; K is a constant; and D_D is the density distribution fractal dimension (DDD). In a non-scale section, the higher the D_D value, the greater the number of ore deposits [58,106].

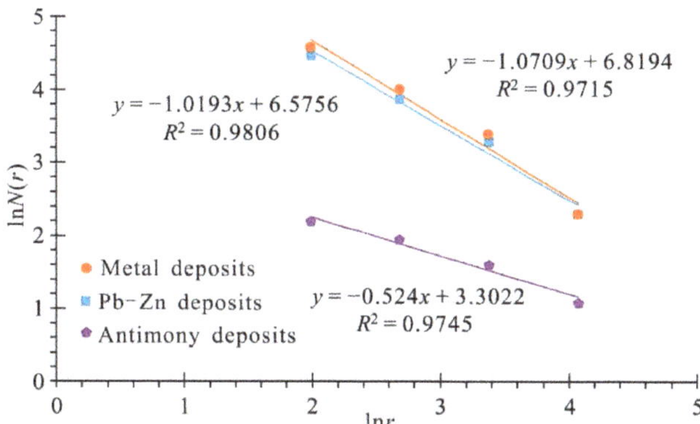

Figure 7. Linear fitting graph for spatial distribution fractal dimensions (SDD) calculation of ore deposits.

Table 10. Statistical table of spatial distribution fractal dimensions (SDD) values of deposits in some areas of China.

Location	Kinds of Minerals	Scale Interval (km)	SDD	References
Anhui Province	Coal, Copper, Iron, etc	17. 8125–285	1.3371	[59]
South China	Uranium	20–400	1.0468	[80]
Western and Northern Yunkai Uplift	Gold	1.25–10	0.3552	[58]
		10–160	1.2418	
China	Gold	20–150	0.2293	[104]
		150–5000	1.3073	
	Gold	1–20	0.1923	
		20–750	0.7168	
Zhejiang Province	Fluorite	1–20	0.3778	[105]
		20–750	1.1851	
	Pb and Zn	1–20	0.1459	
		20–750	1.1723	
Altai Region of Xinjiang	Gold, Copper, Pb, Zn, etc.	1.25–16.32	0.2305	[56]
		16.32–150	1.512	
Yadu–Mangdong metallogenic belt	Pb and Zn	6.741–53.930	1.3262	this article
	Pb and Zn	7.34–58.716	1.0193	this article
EGMB	Sb	7.34–58.716	0.5240	this article
	Pb, Zn, Sb, etc.	7.34–58.716	1.0709	this article

The quantity fractal distribution function is proposed to represent quantitatively the number of possible ore deposits $N(r)$ that is likely to be explored within a definite radius from the center:

$$N(r) = Lr^{D_S} \tag{13}$$

where $N(r)$ is the quantity distribution function, denoting the number of ore deposits within radius r, taking a known ore deposit as the center of the circle; L is a constant; and D_S is the quantity distribution fractal dimension (QDD).

In practical calculation, we took 10 ore deposits with a relatively uniform distribution as the center of the circle. The number and density of the ore deposits covered by areas of various radius, r, were calculated and we took the averaged values of 10 deposit centers (Table 11). Finally, the data were fitted (Figure 8).

Within a research scale of 20 to 80 km, Pb–Zn and all metals deposits versus the average number of deposits show power-law relationships, and the coefficients of determination are 0.9906 and 0.9966, respectively, indicating a high degree of fit. The number distributions of Pb–Zn and all metals deposits have fractal structures, and the QDD values are 1.4225 and 1.4716, respectively (Figure 8a). The Pb–Zn and all metals deposits versus the deposit density also have power-law relationships, and the determination coefficients are 0.9454 and 0.9742, respectively, indicating a high degree of fit. The density distributions of the Pb–Zn and all metals deposits have fractal structures, and the DDD values are 1.422 and 1.472, respectively (Figure 8b). For the fractal distribution of both number of deposits and density of deposits, the fractal dimension values of Pb–Zn and all metals deposits are high (>1.42), indicating high clustering.

Table 11. Statistics of fractal distribution data of deposit number and density.

Fractal Scale, r (km)	Pb–Zn Deposits		Metal Deposits	
	Average Number	Density (No./km^2)	Average Number	Density (No./km^2)
20	3.1	0.00248	3.3	0.00264
30	4.5	0.00160	5.4	0.00192
40	7.1	0.00142	8.5	0.00170
50	10.3	0.00132	12	0.00153
60	14.2	0.00126	16.9	0.00150
70	17	0.00111	20.1	0.00131
80	20.4	0.00102	23.8	0.00119

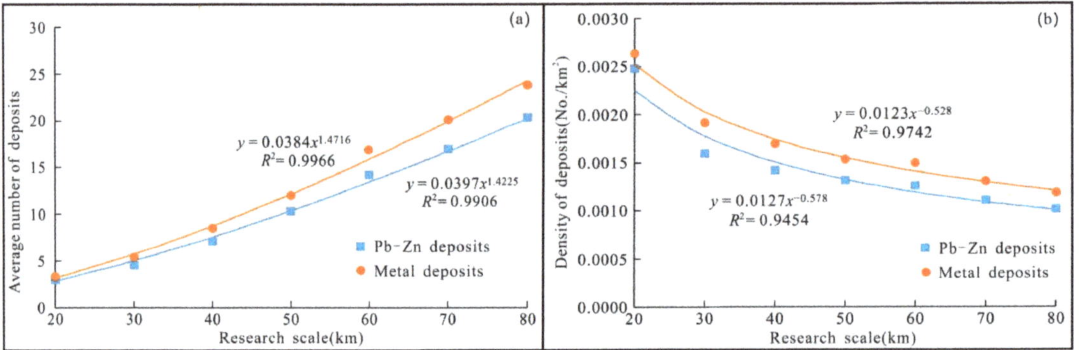

Figure 8. Fitting diagram of fractal distribution for deposit number (**a**) and density (**b**).

6. Coupling Law of Fault Fractal Characteristics and Spatial Distribution of Ore Deposits

According to the theory of self-organized criticality, a fracture system with fractal characteristics is formed by the connection, evolution, and spontaneous organization of small fractures to a point during the dissipation process of the fracture, so that the strain is concentrated on the main fault zone with fractal geometry. The fault fractal dimension value is related to its connectivity. When the fractal dimension value of the fault is lower than the critical value, the deformation and permeability are low, the fault is isolated, and the fault connectivity is poor. When the fractal dimension of the fault reaches or exceeds the critical value, deformation is strong, permeability increases, and the connectivity of the fault is good, which is conducive to the migration and accumulation of ore-forming fluids and the formation of hydrothermal deposits. Numerical simulations of biaxial compression tests of rock blocks show that the critical value of the fractal dimension is 1.22 to 1.38 [77]. The fault fractal dimension value has some locality (relative to the study scale). However, since the scale of the study roughly matches the scale of the study area, and the scale of the study area basically matches the scale of the structure, the critical value of the fractal dimension of the fault still has certain reference significance for this study; that is, areas with the fault CPD values of >1.22 are conducive to mineralization.

The distribution of hydrothermal deposits is not only controlled by fault factors, but also by favorable lithology (or lithologic combination) and others. In this study, the fault CPD of the subdivisions with developed ore deposits is mostly greater than 1.22, but for some subdivisions with ore deposits it is less than and very close to 1.22. This confirms that the distribution of metal deposits is mainly controlled by faults, and confirms the reliability of distribution analysis using fractal dimension value. From a qualitative perspective, the neighborhood fractal dimension values of the subdivisions of various metal deposits are generally relatively low, which may be because neighborhood areas with relatively low fractal dimension values are conducive to blocking and sealing ore-forming fluids. In fact, owing to the clustered distribution of ore deposits, ore deposits are often developed in two or more consecutive subdivisions. Such subdivisions should to be regarded as a whole so as to understand the role of adjacent regions in blocking and sealing ore-forming fluids. Taking the CPD values as an example, we drew a horizontal and vertical fluctuation diagram of subdivision CPD values (Figure 9). The subdivision or subdivision complex of developed deposits are adjacent to at least one subdivision with a relatively low fractal dimension value in a two-dimensional perspective.

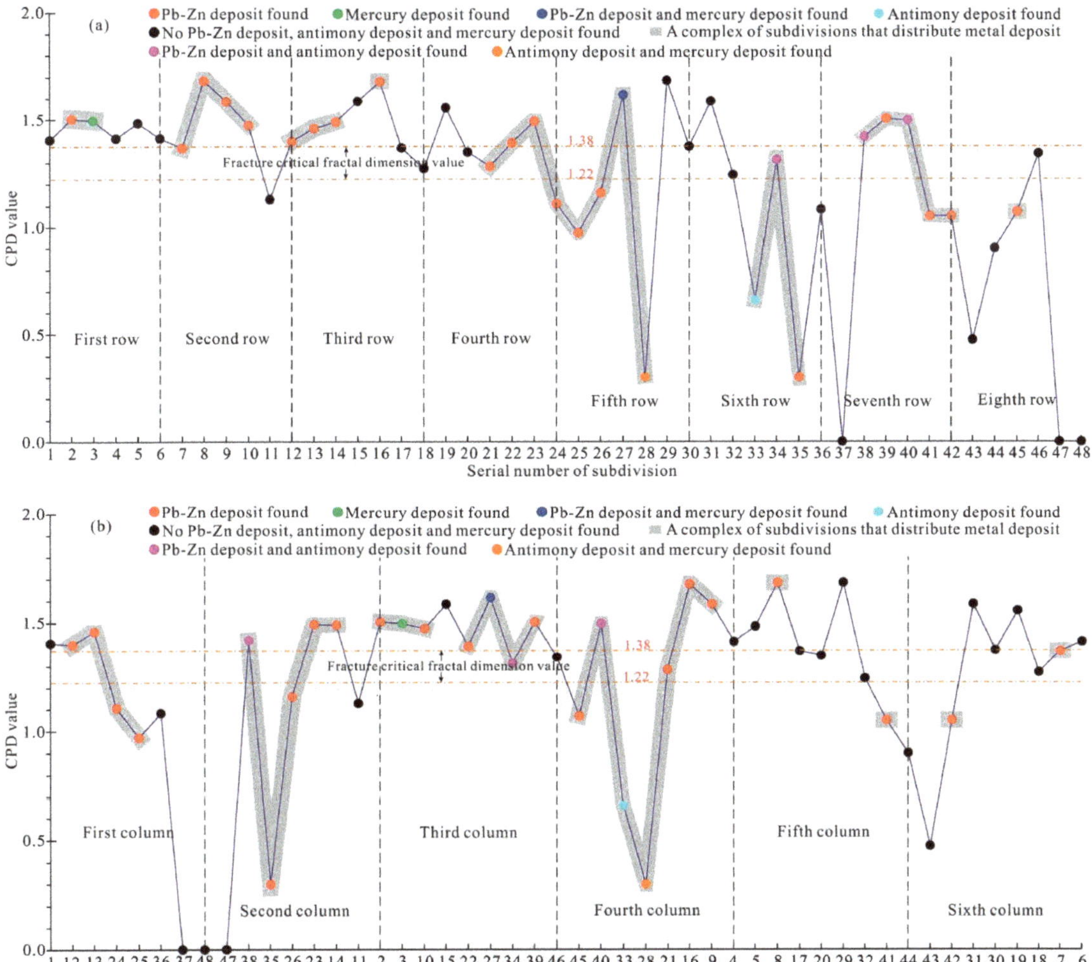

Figure 9. Vertical and horizontal wave graphs of fractal dimensions. (**a**) Horizontal wave graphs of fractal dimensions; and (**b**) Vertical wave graphs of fractal dimensions.

To systematically explore the relationship between the distribution of deposits and the fractal dimension value, a projection map of the fractal dimension value of the subdivision and number of deposits was drawn (Figure 10), along with a projection map of different types of fractal dimension values of the subdivision (Figure 11). The favorable fractal dimension distribution intervals of Pb–Zn and all metals deposits are basically the same; both are mainly distributed in ore-bearing subdivisions that simultaneously satisfy three conditions: CPD > 1.16, IND > 0.95, and CRD > 0.97.

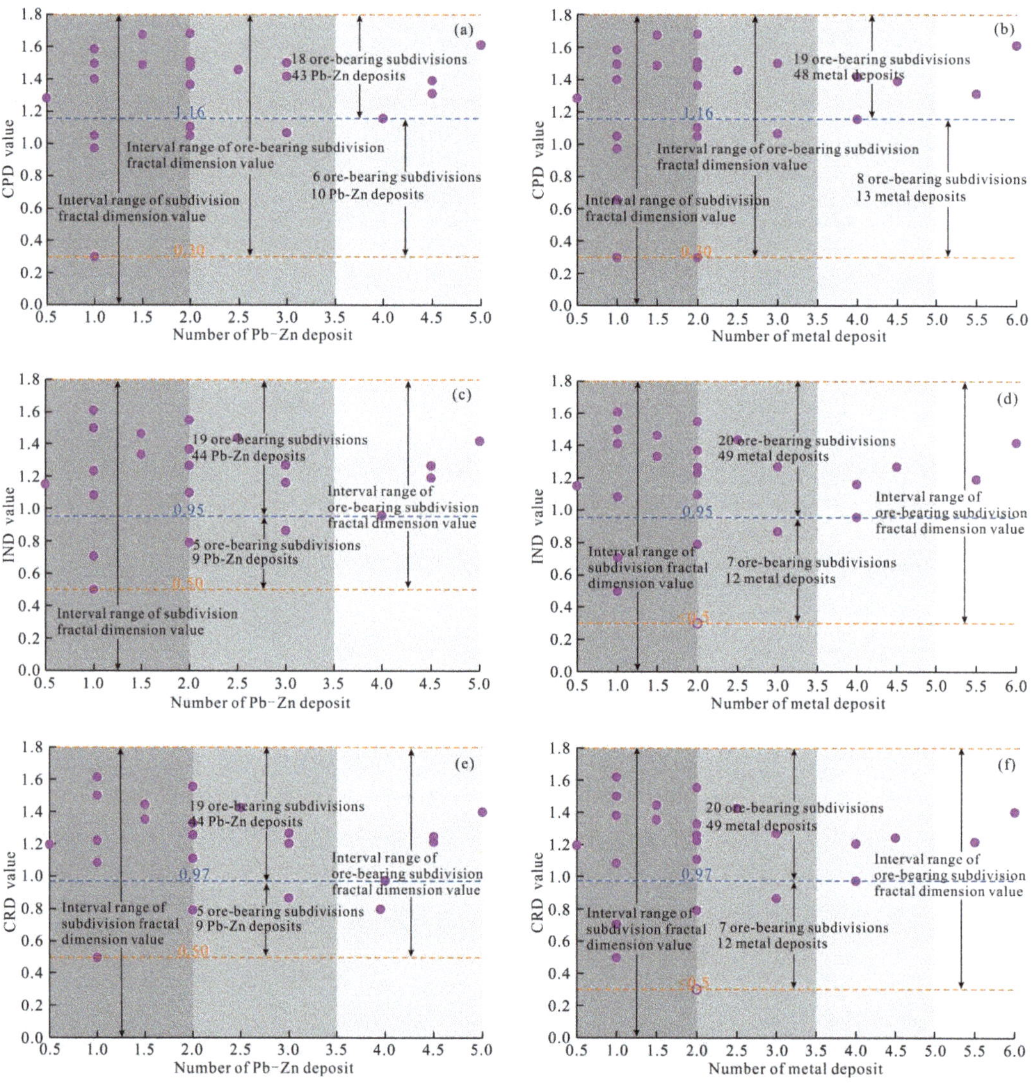

Figure 10. Subdivision projection maps of deposit quantity versus fractal dimension. (**a**) Subdivision projection maps of Pb–Zn deposit quantity versus CPD; (**b**) Subdivision projection maps of all metals deposits quantity versus CPD; (**c**) Subdivision projection maps of Pb–Zn deposit quantity versus IND; (**d**) Subdivision projection maps of all metals deposits quantity versus IND; (**e**) Subdivision projection maps of Pb–Zn deposit quantity versus CRD; and (**f**) Subdivision projection maps of all metals deposits quantity versus CRD.

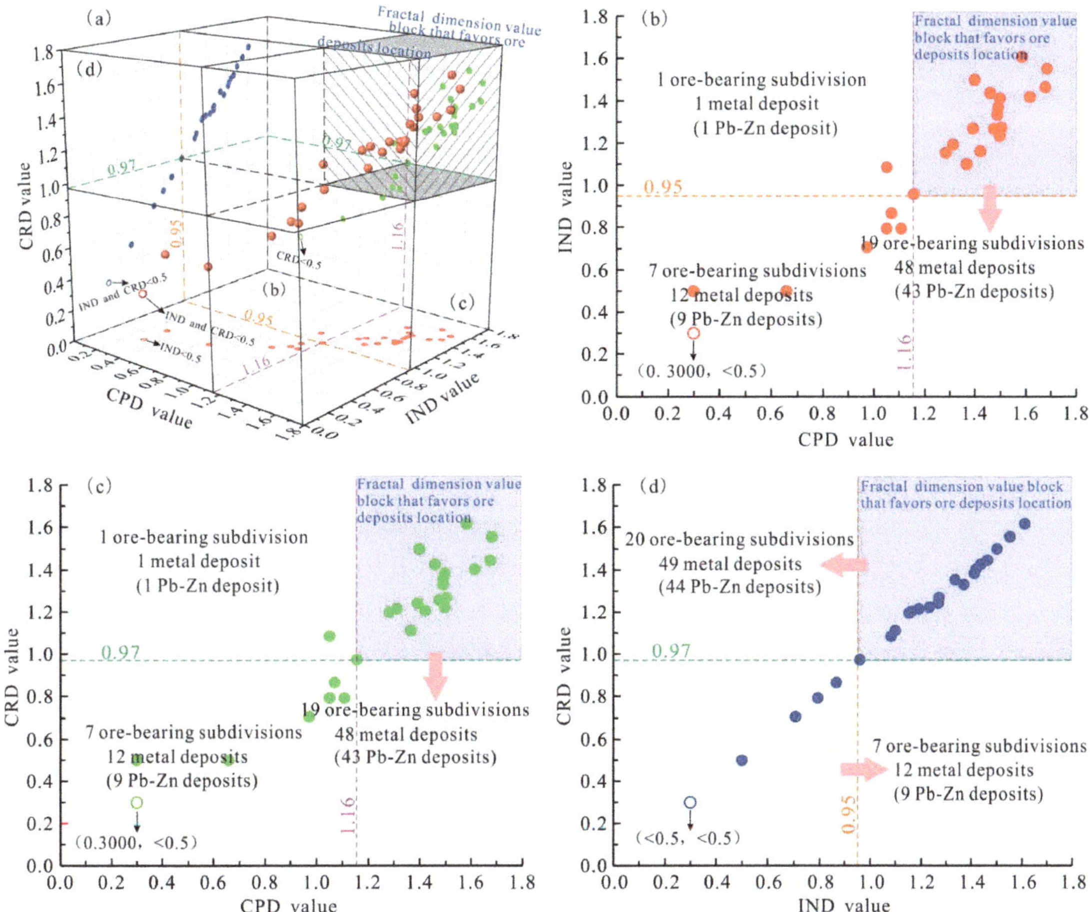

Figure 11. Different types of fractal dimension projection maps of subdivisions. (**a**) 3-D plot of CPD, IND, CRD of ore-bearing subdivision; (**b**) CPD versus IND plot of ore-bearing subdivision; (**c**) CPD versus CRD plot of ore-bearing subdivision; and (**d**) IND versus CRD plot of ore-bearing subdivision.

7. Prediction of Favorable Areas for Prospecting

7.1. Fractal Dimension Value Analysis

From the perspective of fractal dimension value, areas favorable for the distribution of ore deposits should satisfy two conditions: (1) CPD > 1.16, IND > 0.95, and CRD > 0.97; and (2) on a two-dimensional plane, there are adjacent regions that relatively block fluid flow. Favorable prospecting areas based on CPD, IND, and CRD were delineated (Figure 12a–c), and their overlapping area was taken as the comprehensive favorable metallogenic area (Figure 12d).

7.2. Fry Analysis

Fry analysis was first developed for mineral rock stress analysis [107–110], and was subsequently extended to measure the spatial distribution of ore deposits and infer potential ore-controlling structures [111–113]. The basic principles of the Fry analysis method are as follows. Assuming that there are n points in a known plane A, copying the plane n times can obtain n identical planes $A_1, A_2, \ldots A_n$. Select a point in plane A_1 as a reference point to establish a rectangular coordinate system, and arbitrarily select a point other than the reference point in plane A_2, and place it at the coordinate origin of plane A_1. Similarly, a point is arbitrarily selected from the remaining n-2 points in plane A_3 and placed at the coordinate origin of plane A_1. The above process is repeated until each point coincides with the coordinate origin of plane A_1, and finally n(n-1) points are generated in the plane. Fry analysis is a spatial autocorrelation method used to study the distribution trend of spatial points. In practical application, the areas with more deposit distribution tend to have dense Fry points, so favorable metallogenic areas can be divided according to the relative number of Fry points' distribution at the macro scale. In this study, we applied Fry analysis to the 61 known metal deposits, including 53 Pb–Zn deposits, 5 Sb deposits, and 3 Hg deposits. The Fry projection was obtained by 61 shots (Figure 13a,b). According to the number of projected ore deposit points in the subdivision, the favorable metallogenic areas of Pb–Zn ores and Sb–Hg ores were identified (Figure 13c,d). From Figure 14, the metallogenic potential of Pb–Zn deposits decreases in the following order: subdivisions 5-4 and 8-1 > 9 subdivisions, including 6-3, 6-2; and 5-1 > 8 subdivisions, including 1-2, 1-3, and 7-2. For Sb–Hg deposits, nine subdivisions, including 5-2, 5-3, and 8-2, have relatively great metallogenic potential. Most favorable metallogenic subdivisions of Sb–Hg deposits are also favorable metallogenic areas of Pb–Zn deposits, indicating strong spatial consistency.

7.3. Prediction of Comprehensive Favorable Metallogenic Areas

Dispersed metals in the study area are mainly enriched in Pb–Zn deposits or Pb–Zn poly-metallic deposits, and so favorable metallogenic areas of Pb–Zn are also favorable metallogenic area of dispersed metals. According to importance, the comprehensive favorable metallogenic areas of Pb–Zn were divided into four grades (Figure 14a), while those of Sb–Hg were divided into two grades (Figure 14b). Grade I denote common overlap of fractal dimension value comprehensive favorable metallogenic areas and subdivisions with > 70 Pb–Zn deposits after Fry's projection. Most known deposits are distributed in this area, which has the greatest prospecting potential. Grade II areas are those comprehensive favorable metallogenic area with >100 Pb–Zn deposits after Fry's projection; these include seven known ore deposits. Grade III deposits are those subdivisions with 70–99 Pb–Zn deposits after Fry's projection. In space, these are mainly adjacent to grade I and grade II areas. The remaining subdivisions (50–69 Pb–Zn deposits after Fry's projection) are grade IV and have low prospecting potential.

The common overlap of the fractal dimension value analysis comprehensive favorable metallogenic area and subdivisions of the favorable distribution of Sb (or Hg) deposits after Fry's projection form the grade I favorable metallogenic areas of Sb (or Hg). This includes three areas with the greatest prospecting potential. The remaining comprehensive favorable metallogenic areas and subdivisions of the favorable distribution of Sb (or Hg) deposits after Fry's projection are grade II.

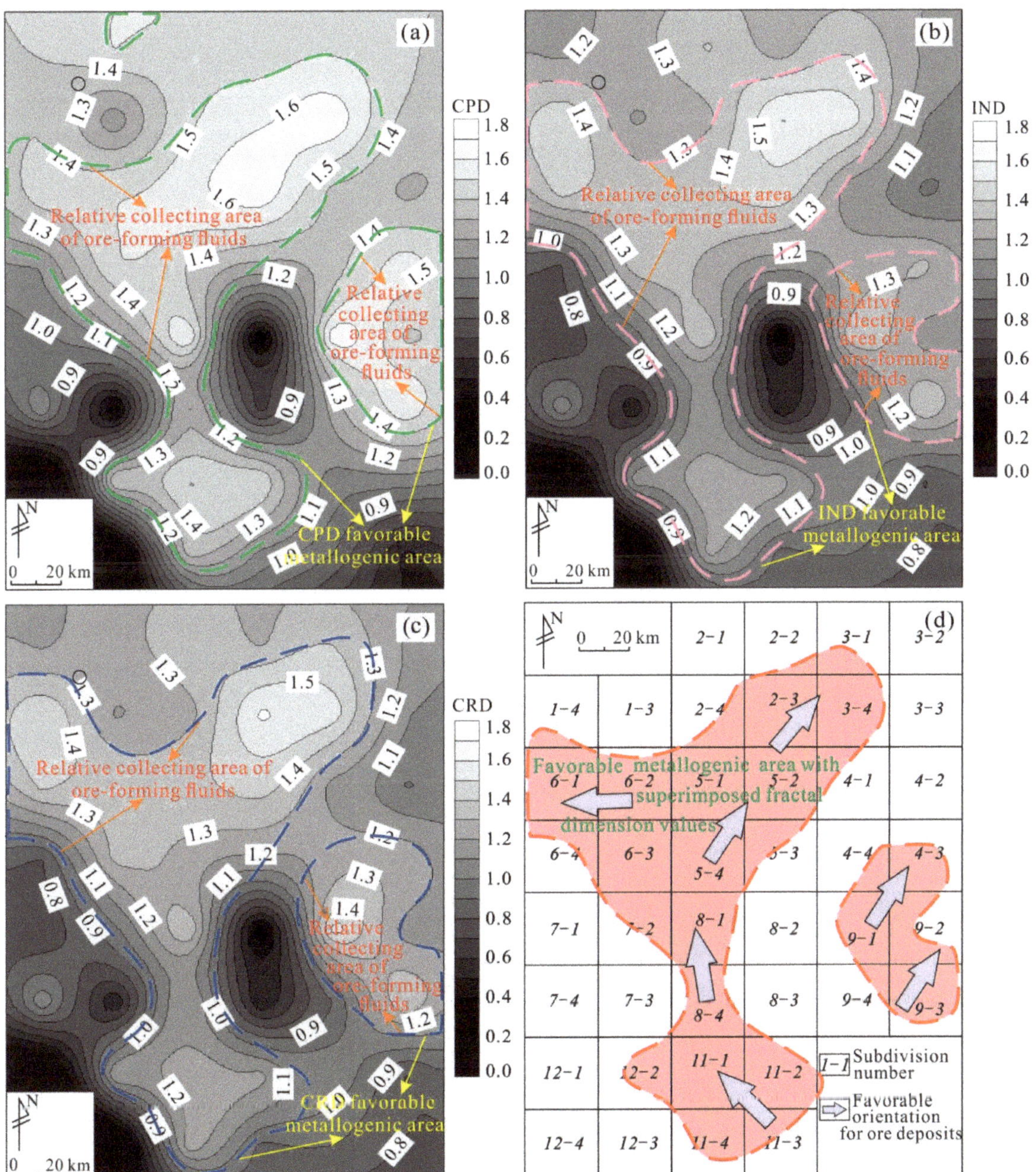

Figure 12. Fractal dimension favorable mining area. (**a**) CPD of fault favorable mining area; (**b**) IND of fault favorable mining area; (**c**) CRD of fault favorable mining area; and (**d**) Comprehensive consideration of fault fractal dimension value for favorable metallogenic area.

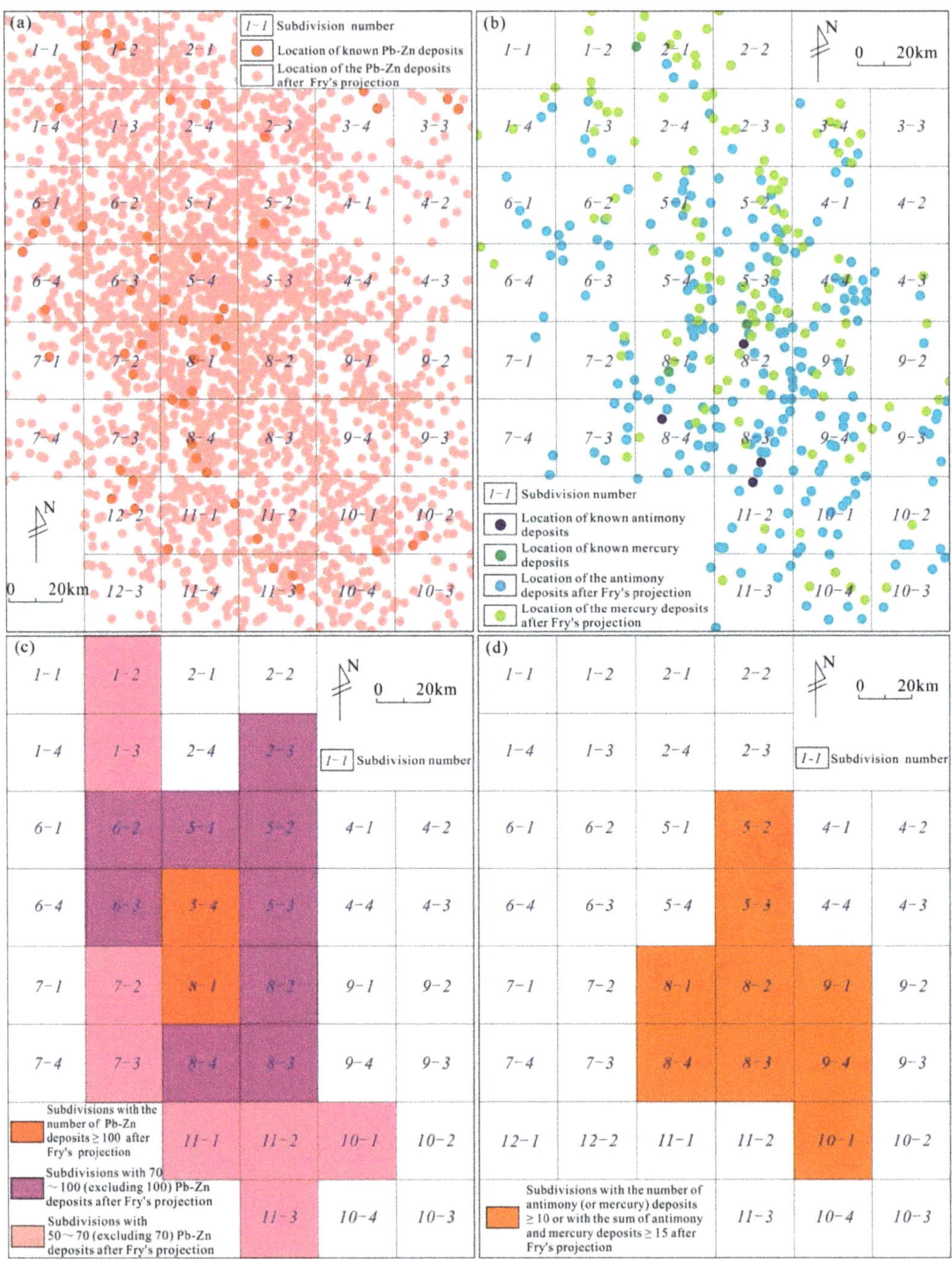

Figure 13. Fry analysis map of the deposit. (**a**) Fry analysis diagram of Pb–Zn deposits; (**b**) Fry analysis diagram of Sb–Hg deposits; (**c**) Fry point distribution map of Pb–Zn deposits; and (**d**) Fry point distribution map of Sb–Hg deposits.

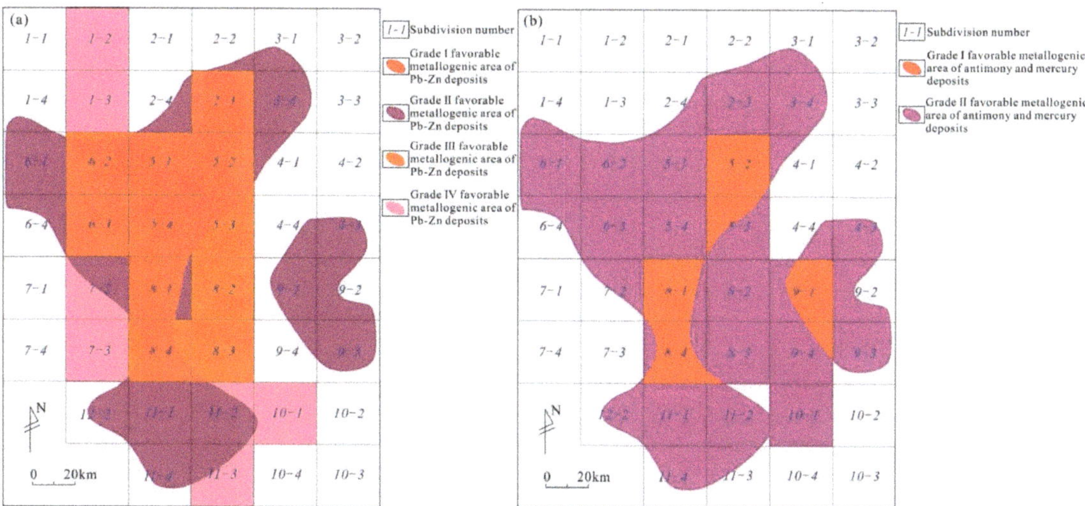

Figure 14. Distribution map of comprehensive favorable metallogenic areas. (**a**) Distribution map of Pb–Zn deposits comprehensive favorable metallogenic areas and (**b**) Distribution map of Sb–Hg deposits comprehensive favorable metallogenic areas.

8. Conclusions

(1) In the scale range of 3.670–58.716 km, fault structures in the EGMB have good statistical self-similarity. The integrated faults CPD is 1.5095, the IND is 1.5391, and the CRD is 1.5436, indicating fault structures with high maturity, which are conducive to the migration and accumulation of ore-forming fluids.

(2) When the q-order moment ranges from 1 to 10, $\Delta\alpha$ is 0.3203 and $\Delta f(\alpha)$ is 1.5355, implying that the study area has great metallogenic potential.

(3) Within the scale range of 7.340–58.716 km, the SDD values of Sb, Pb–Zn, and other metal deposits are 0.5240, 1.0193, and 1.0709, respectively. Within the scale of 20–80 km, the number and density distributions of Pb–Zn and metal deposits are all fractal structures; the QDD values are 1.4225 and 1.4716, respectively, and the DDD values are 1.422 and 1.472, respectively, indicating high clustering of both Pb–Zn and other metal deposits.

(4) From the perspective of fractal dimension value, areas favorable for the distribution of ore deposits should satisfy two conditions: (1) CPD > 1.16, IND > 0.95, and CRD > 0.97; and (2) on the two-dimensional plane, the fractal dimension value of the adjacent area is lower (i.e., adjacent regions relatively block fluid flow).

(5) The comprehensive favorable metallogenic areas of Pb–Zn and associated dispersed metals are divided into four grades. Among them, favorable metallogenic region of grade I is continuously distributed in space. Most known deposits are distributed in this area, and the prospecting potential is the greatest.

Author Contributions: Methodology, data creation, investigation, writing-original draft, Z.C.; methodology, supervision, writing-review and editing, J.Z.; investigation, formal analysis, writing-review and editing, K.L.; formal analysis, investigation, M.L. All authors have read and agreed to the published version of the manuscript.

Funding: This research was funded by the National Natural Science Foundation of China [grant numbers U1812402 and 42172082], Yunnan University Scientific Research Start-up Project [grant number YJRC4201804], and the Science and Technology Project of Department of Education, Jiangxi Province [grant number GJJ213014].

Data Availability Statement: The data presented in this study are available on reasonable request from the corresponding authors.

Conflicts of Interest: The authors declare no conflict of interest.

References

1. Zhang, C.Q.; Rui, Z.Y.; Chen, Y.C.; Wang, D.H.; Chen, Z.H.; Lou, D.B. The main successive strategic bases of resources for Pb–Zn deposits in China. *Geol. China* **2013**, *40*, 248–272. (In Chinese)
2. Ye, L.; Hu, Y.S.; Yang, S.P.; Wei, C.; Yang, X.Y.; Li, Z.L.; An, Q.; Lu, M.D. A discussion on the Pb–Zn mineralization of the Qiandong metallogenic belt. *Acta Mineral. Sin.* **2018**, *38*, 709–715. (In Chinese)
3. Li, K.; Liu, F.; Zhao, W.Q.; Zhao, S.R.; Tang, Z.Y.; Duan, Q.F.; Cao, L. Metallogenic Model of Carbonate-Hosted Pb–Zn Deposits in West Hunan and East Guizhou Provinces, South China. *Earth Sci.* **2021**, *46*, 1151–1172. (In Chinese)
4. Liu, W.J.; Zheng, R.C. Characteristics and Movement of Ore-Forming Fluids in the Huayuan Lead-Zinc Deposit. *Miner. Depos.* **2000**, *19*, 173–181. (In Chinese)
5. Yang, S.X.; Yu, P.R.; Lao, K.T. Metallogenic regularity and prospecting direction of lead-zinc deposits in northwestern Hunan. *Land Resour. Her.* **2006**, *3*, 92–98. (In Chinese)
6. Hu, Y.S.; Ye, L.; Wei, C.; Li, Z.L.; Huang, Z.L.; Wang, H.Y. Trace elements in sphalerite from the Dadongla Zn-Pb deposit, western Hunan–eastern Guizhou Zn-Pb metallogenic belt, south China. *Acta Geol. Sin.* **2020**, *94*, 2152–2164. [CrossRef]
7. Tang, Z.Y.; Deng, F.; Li, K.; Duan, Q.F.; Zou, X.W.; Dai, P.Y. Stratigraphic characteristics of the Cambrian Qingxudong formation in relation to lead-zinc mineralization in western Hunan-eastern Guizhou area. *Geol. China* **2012**, *39*, 1034–1041. (In Chinese)
8. Mao, D.L. Geological Characteristics and Genesis of the Danaopo Pb–Zn Deposit in Huayuan County, Hunan Province. *Mod. Min.* **2016**, *32*, 90–94. (In Chinese)
9. Schneider, J.; Boni, M.; Lapponi, F.; Bechstadt, T. Carbonate Hosted Zinc-Lead Deposits in the Lower Cambrian of Hunan, South China: A Radiogenic (Pb, Sr) Isotope Study. *Econ. Geol.* **2002**, *97*, 1815–1827. [CrossRef]
10. Ye, L.; Pan, Z.P.; Li, C.Y.; Liu, T.G.; Xia, B. Isotopic Geochemical Characters in Niujiaotang Cd Rick Znic Deposit, Duyun, Guizhou. *Mineral. Petrol.* **2005**, *25*, 70–74. (In Chinese)
11. Liu, J.S.; Zou, X.W.; Tang, C.Y.; Cui, S.; Xia, J.; Gan, J.M.; Zhao, W.Q.; Jin, S.C. Preliminary Discussion on Relationship between Pb–Zn Deposits and Paleo-oil Reservoirs in Western Hunan and Eastern Guizhou Province. *Geol. Miner. Resour. South China* **2012**, *28*, 220–225. (In Chinese)
12. Cai, Y.X.; Yang, H.M.; Duan, R.C.; Lu, S.S.; Zhang, L.G.; Liu, C.P.; Qiu, X.F. Fluid Inclusions and S, Pb, C Isotope Geochemistry of Pb–Zn Deposits Hosted by Lower Cambrian in Western Hunan-Eastern Guizhou Area. *Geoscience* **2014**, *28*, 29–41. (In Chinese)
13. Li, K.; Duan, Q.F.; Zhao, S.R.; Tang, Z.Y. Material sources and ore-forming mechanism of the Huayuan Pb–Zn ore deposit in Hu'nan Province: Evidence from S, Pb, Sr isotopes of sulfides. *Geol. Bull. China* **2017**, *36*, 811–822. (In Chinese)
14. Yang, S.X.; Lao, K.T. A tentative discussion on genesis of lead-zinc deposits in northwest Hunan. *Miner. Depos.* **2007**, *26*, 330–340. (In Chinese)
15. Zhou, Y.; Duan, Q.F.; Tang, J.X.; Cao, L.; Li, F.; Huang, H.L.; Gan, J.M. The Large-Scale Low-Temperature Mineralization of Lead-Zinc Deposits in Western Hunan-Evidence from Fluid Inclusions. *Geol. Explor.* **2014**, *50*, 515–532. (In Chinese)
16. Yu, Y.S.; Guo, F.S.; Dai, Y.P.; Liu, A.S.; Zhou, Y. Ore genesis of Tangbian Pb–Zn deposit in Tongren, Guizhou: Evidence from ore-forming fluids and isotopes. *Miner. Depos.* **2017**, *36*, 330–344. (In Chinese)
17. Li, K.; Zhao, S.R.; Tang, Z.Y.; Duan, Q.F.; Li, J.W. Fluid Sources and Ore Genesis of the Pb–Zn Deposits of Huayuan Ore-Concentrated District, Northwest Hunan Province, China. *Earth Sci.* **2018**, *43*, 2449–2464. (In Chinese)
18. Xu, L.; Luo, C.G.; Wen, H.J.; Zhou, Z.B.; de Fourestier, J. The role of fluid mixing in the formation of vein-type Zn-Pb deposits in eastern Guizhou Province, SW China: Insights from elemental compositions and chlorine isotopes of fluid inclusions. *J. Asian Earth Sci.* **2022**, *239*, 105403. [CrossRef]
19. Xia, X.J.; Fu, S.Y. Mineralization Pattern of North Western Hunan Lead-zinc Mine. *Nonferrous Met.* **2010**, *62*, 35–38. (In Chinese)
20. Wei, H.T.; Shao, Y.J.; Xiong, Y.Q.; Liu, W.; Kong, H.; Li, Q.; Sui, Z.H. Metallogenic model of Huayuan Pb–Zn ore field in the western Hunan Province, South China. *J. Cent. South Univ.* **2017**, *48*, 2402–2413. (In Chinese)
21. Zhou, J.X.; An, Y.L.; Yang, Z.M.; Luo, K.; Sun, G.T. New discovery of extraordinary enrichment of selenium in Dingtoushan Pb–Zn deposit, Qinglong City, Guizhou Province, and its geological significance. *Miner. Depos.* **2020**, *39*, 568–578. (In Chinese)
22. Li, Z.F. A Preliminary Discussion on the Origin of Pb–Zn Ore Deposits in Western Hunan and Eastern Guizhou. *Guizhou Geol.* **1991**, *8*, 363–371. (In Chinese)
23. Li, K.; Wu, C.X.; Tang, C.Y.; Duan, Q.F.; Yu, Y.S. Carbon and oxygen isotopes of Pb–Zn ore deposits in western Hunan and eastern Guizhou provinces and their implications for the ore-forming process. *Geol. China* **2014**, *41*, 1608–1619. (In Chinese)
24. Wei, H.T.; Shao, Y.J.; Ye, Z.; Zhou, H.D. Geochemical characteristics of trace elements of sphalerite from Huayuan Pb–Zn ore field, western Hunan, China. *J. Chengdu Univ. Technol.* **2021**, *48*, 142–153. (In Chinese)
25. Yang, D.Z.; Zhou, J.X.; Luo, K.; Yu, J.; Zhou, Z.H. New discovery and research value of Zhulingou zinc deposit in Guiding, Guizhou. *Bull. Mineral. Petrol. Geochem.* **2020**, *39*, 344–345. (In Chinese)
26. Zhou, J.X.; Meng, Q.T.; Ren, H.Z.; Sun, G.T.; Zhang, Z.J.; An, Q.; Zhou, C.X. Discovery of super large paragenetic (associated) germanium deposit in Huangsi anticline, Guizhou. *Geotecton. Metallog.* **2020**, *44*, 1025–1026. (In Chinese)

27. Zhou, J.X.; Yang, D.Z.; Yu, J.; Luo, K.; Zhou, Z.H. Ge extremely enriched in the Zhulingou Zn deposit, Guiding City, Guizhou Province, China. *Geol. China* **2021**, *48*, 665–666. (In Chinese)
28. Zhou, J.X.; Luo, K.; Sun, G.T. Se extremely enriched in the Dingtoushan Pb–Zn deposit, Qinglong City, Guizhou Province, China. *Geol. China* **2021**, *48*, 339–340. (In Chinese)
29. Zhou, J.X.; Yang, D.Z.; Xiao, S.; An, Y.L.; Luo, K. Discovery and Significance of Thallium Supernormal Enrichment in Huodehong Lead—Zinc Deposit, Northeastern Yunnan. *Geotecton. Metallog.* **2021**, *45*, 427–429. (In Chinese)
30. Meng, Q.T.; Zhou, J.X.; Sun, G.T.; Zhao, Z.; An, Q.; Yang, X.Y.; Lu, M.D.; Xiao, K.; Xu, L. Geochemical characteristics and ore prospecting progress of the Banbianjie Zn deposit in Guiding City, Guizhou Province, China. *Acta Mineral. Sin.* **2021**, *23*, 616–619. (In Chinese)
31. Tu, G.Z.; Gao, Z.M.; Hu, R.Z.; Zhang, Q.; Li, C.Y.; Zhao, Z.H.; Zhang, B.G. *Dispersed Element Geochemistry and Mineralization Mechanism*; Geological Publishing House: Beijing, China, 2004; 421p. (In Chinese)
32. Zhai, M.G.; Wu, F.Y.; Hu, R.Z.; Jiang, S.Y.; Li, W.C.; Wang, R.C.; Wang, D.H.; Qi, T.; Qin, K.Z.; Wen, H.J. Critical metal mineral resources: Current research status and scientific issues. *Bull. Natl. Nat. Sci. Found. China* **2019**, *33*, 106–111. (In Chinese)
33. Wen, H.J.; Zhou, Z.B.; Zhu, C.W.; Luo, C.G.; Wang, D.Z.; Du, S.J.; Li, X.F.; Chen, M.H.; Li, H.Y. Critical scientific issues of super-enrichment of dispersed metals. *Acta Petrol. Sin.* **2019**, *35*, 3271–3291. (In Chinese)
34. Li, K.X.; Leng, C.B.; Ren, Z.; Liu, F.; Xu, D.R.; Ye, L.; Luo, T.Y. Progresses of researches on the dispersed elements associated with lead-zinc deposits. *Acta Mineral. Sin.* **2021**, *41*, 225–233. (In Chinese)
35. Hu, R.Z.; Su, W.C.; Qi, H.W.; Bi, X.W. Geochemistry, Occurrence and Mineralization of Germanium. *Bull. Mineral. Petrol. Geochem.* **2000**, *19*, 215–217. (In Chinese)
36. Zhang, Q.; Liu, Y.P.; Ye, L.; Shao, S.X. Study on Specialization of Dispersed Element Mineralization. *Bull. Mineral. Petrol. Geochem.* **2008**, *27*, 247–253. (In Chinese)
37. Wang, D.H.; Wang, R.J.; Li, J.K.; Zhao, Z.; Yu, Y.; Dai, J.J.; Chen, Z.H.; Li, D.X.; Qu, W.J.; Deng, M.C.; et al. The progress in the strategic research and survey of rare earth, rare metal and rare-scattered elements mineral resources. *Geol. China* **2013**, *40*, 361–370. (In Chinese)
38. Tao, Y.; Hu, R.Z.; Tang, Y.Y.; Ye, L.; Qi, H.W.; Fan, H.F. Types of dispersed elements bearing ore-deposits and their enrichment regularity in Southwest China. *Acta Geol. Sin.* **2019**, *93*, 1210–1230. (In Chinese)
39. Mao, J.W.; Yang, Z.X.; Xie, G.Q.; Yuan, S.D.; Zhou, Z.H. Critical minerals: International trends and thinking. *Miner. Depos.* **2019**, *38*, 689–698. (In Chinese)
40. Schwartz, M. Cadmium in zinc deposits: Economic geology of a polluting element. *Int. Geol. Rev.* **2000**, *42*, 445–469. [CrossRef]
41. Xue, C.J.; Chen, Y.C.; Yang, J.M.; Wang, D.H.; Yang, W.G.; Yang, Q.B. Jinding Pb–Zn Deposit: Geology and Geochemistry. *Miner. Depos.* **2002**, *21*, 270–277. (In Chinese)
42. Bradley, D.C.; Leach, D.L.; Symons, D.; Emsbo, P.; Premo, W.; Breit, G.; Sangster, D.F. Reply to discussion on "Tectonic controls of Mississippi Valley-type lead-zinc mineralization in orogenic forelands" by Kesler, S.E.; Christensen, J.T.; Hagni, R.D.; Heijlen, W.; Kyle, J.R.; Misra, K.C.; Muchez, P.; van derVoo, R, Mineralium Deposita. *Miner. Depos.* **2004**, *39*, 515–519. [CrossRef]
43. Si, R.J.; Gu, X.X.; Pang, X.C.; Fu, S.H.; Li, F.Y.; Zhang, M.; Li, Y.H.; Li, X.Y.; Li, J. Geochemical Character of Dispersed Element in Sphalerite from Fule Pb–Zn Polymetal Deposit, Yunnan Province. *Mineral. Petrol.* **2006**, *26*, 75–80. (In Chinese)
44. Zhou, J.X.; Huang, Z.L.; Zhou, G.F.; Li, X.B.; Ding, W.; Gu, J. The Occurrence States and Regularities of Dispersed Elements in Tianqiao Pb–Zn Ore Deposit, Guizhou Province, China. *Acta Mineral. Sin.* **2009**, *29*, 471–480. (In Chinese)
45. Zhou, J.X.; Huang, Z.L.; Zhou, M.F.; Li, X.B.; Jin, Z.G. Constraints of C-O-S-Pb isotope compositions and Rb-Sr isotopic age on the origin of the Tianqiao carbonate-hosted Pb–Zn deposit, SW China. *Ore Geol. Rev.* **2013**, *53*, 77–92. [CrossRef]
46. Hu, R.Z.; Chen, W.T.; Xu, D.R.; Zhou, M.F. Reviews and new metallogenic models of mineral deposits in South China: An introduction. *J. Asian Earth Sci.* **2017**, *137*, 1–8. [CrossRef]
47. Mandelbrot, B.B. Stochastic Models for the Earth's relief, the shape and the fractal dimension of the coastlines, and the number-area rule for islands. *Proc. Natl. Acad. Sci.* **1975**, *72*, 3825–3828. [CrossRef]
48. Zhang, J.; Wang, D.H.; Sun, B.S.; Chen, Z.H. Metallogenic Spatial Analysis Based on Fractal Theory: A Case Study of the Kangguertage Gold Ore Belt in Xinjiang. *Acta Geosci. Sinica.* **2009**, *30*, 58–64. (In Chinese)
49. Tan, K.X.; Hao, X.C.; Dai, T.G. Fractal Features of Fractures in China and Their Implication for Geotectonics. *Geotecton. Et Metallog.* **1998**, *22*, 17–20. (In Chinese)
50. Hu, X.Q.; Shi, Z.J.; Tian, Y.M.; Wang, C.C.; Cao, J.F. Multifractal feature and significance of Maokou Formation faults in the southeast of Sichuan. *J. Chengdu Univ. Technol.* **2014**, *41*, 476–482. (In Chinese)
51. Li, F.; Liu, G.S.; Zhou, Q.W.; Zhao, G.B. Application of fractal theory in the study of the relationship between fracture and mineral. *J. Hefei Univ. Technol.* **2016**, *39*, 701–706. (In Chinese)
52. Sun, T.; Li, H.; Wu, K.X.; Chen, L.K.; Liu, W.M.; Hu, Z.J. Fractal and Multifractal Characteristics of Regional Fractures in Tongling Metallogenic Area. *Nonferrous Met. Eng.* **2018**, *8*, 111–115. (In Chinese)
53. Chen, P.; Peng, S.Y.; Wang, P.F.; Chen, X.X.; Yang, T.; Liu, Y.J. Fractal and multifractal characteristics of geological faults in coal mine area and their control on outburst. *Coal Sci. Technol.* **2019**, *47*, 42–52. (In Chinese)
54. Cui, Z.L.; Liu, X.Y.; Zhou, J.X. Fractal characteristics of faults and its geological significance in Sichuan-Yunnan-Guizhou Pb–Zn metallogenic province, China. *Glob. Geol.* **2021**, *40*, 75–92. (In Chinese)

55. Cui, Z.L.; Zhou, J.X.; Luo, K. Fractal structure analysis and application prospect of Xingguo-Ningdu fluorite metallogenic belt in southern Jiangxi, China. *J. Jilin Univ.* **2022**, *52*, 1–18. [CrossRef]

56. Tan, K.X.; Liu, S.S.; Xie, Y.S. Multifractal analysis of ore deposits distribution in Alty, Xinjiang, China. *Geotecton. Et Metallog.* **2000**, *24*, 333–341. (In Chinese)

57. Cui, Z.L.; Kong, D.K. Fractal characteristic and range of reconnaissance in Xiong'er mountain ore-concentrated area and its vicinity, Henan Province. *Ind. Miner. Process.* **2021**, *50*, 1–7. (In Chinese)

58. Han, X.B.; Li, J.B.; Feng, Z.H.; Zhang, S.Y.; Liang, J.C.; Long, J.P. Spatial Fractal Characteristics of Gold-Silver Deposits in the Western and Northern Yunkai Uplift. *J. Guilin Univ. Technol.* **2010**, *30*, 15–20. (In Chinese)

59. Shi, G.D.; Jin, B.L.; Chen, B. Fractal study on correlation between distribution of ore spots and main traces in Anhui. *J. Guilin Univ. Technol.* **2020**, *40*, 271–277. (In Chinese)

60. Mokhtari, Z.; Sadeghi, B. Geochemical anomaly definition using multifractal modeling, validated by geological field observations: Siah Jangal area, SE Iran. *Geochemistry* **2021**, *81*, 125774. [CrossRef]

61. Cheng, Q.M.; Agterberg, F.P.; Bonham-Carter, G.F. A spatial analysis method for geochemical anomaly separation. *J. Geochem. Explor.* **1996**, *56*, 183–195. [CrossRef]

62. Parsa, M.; Sadeghi, M.; Grunsky, E. Innovative methods applied to processing and interpreting geochemical data. *J. Geochem. Explor.* **2022**, *237*, 106983. [CrossRef]

63. Xie, S.Y.; Bao, Z.Y. Application of multifractal to ore-forming potential evaluation. *J. Chengdu Univ. Technol.* **2004**, *31*, 28–33. (In Chinese)

64. Xie, S.; Cheng, Q.; Zhang, S.; Huang, K. Assessing microstructures of pyrrhotites in basalts by multifractal analysis. *Nonlinear Process. Geophys.* **2010**, *17*, 319–327. [CrossRef]

65. Zuo, R.G. Identifying geochemical anomalies associated with Cu and Pb–Zn skarn mineralization using principal component analysis and spectrum-area fractal modeling in the Gangdese Belt, Tibet (China). *J. Geochem. Explor.* **2011**, *111*, 13–22. [CrossRef]

66. Sun, X.; Zheng, Y.Y.; Wang, C.M.; Zhao, Z.Y.; Geng, X.B. Identifying geochemical anomalies associated with Sb-Au-Pb–Zn-Ag mineralization in North Himalaya, southern Tibet. *Ore Geol. Rev.* **2016**, *73*, 1–12. [CrossRef]

67. Zuo, R.G.; Wang, J. Fractal/multifractal modeling of geochemical data: A review. *J. Geochem. Explor.* **2016**, *164*, 33–41. [CrossRef]

68. Zhao, M.Y.; Xia, Q.L.; Wu, L.R.; Liang, Y.Q. Identification of multi-element geochemical anomalies for Cu–polymetallic deposits through staged factor analysis, improved fractal density and expected value function. *Nat. Resour. Res.* **2022**, *31*, 1867–1887. [CrossRef]

69. Lyu, C.; Cheng, Q.M.; Zuo, R.G.; Wang, W.P. Mapping Spatial Distribution Characteristics of Lineaments Extracted from Remote Sensing Image Using Fractal and Multifractal Models. *J. Earth Sci.* **2017**, *28*, 507–515. [CrossRef]

70. Cheng, Q.M. What are Mathematical Geosciences and its frontiers? *Earth Sci. Front.* **2021**, *28*, 6–25. (In Chinese)

71. Yin, F.G.; Xu, X.S.; Wan, F.; Chen, M. Characteristic of Seqence and Stratigraphical Division in Evolution of Upper Yangtze Region During Caledonian. *J. Stratigr.* **2002**, *26*, 315–319. (In Chinese)

72. Shu, L.S.; Zhou, X.M.; Deng, P.; Wang, B.; Jiang, S.Y.; Yu, J.H.; Zhao, X.X. Mesozoic tectonic evolution of the southeast China block: New insights from basin analysis. *J. Asian Earth Sci.* **2009**, *34*, 376–391. [CrossRef]

73. Yan, D.P.; Qiu, L.; Chen, F.; Li, L.; Zhao, L.; Yang, W.X.; Zhang, Y.X. Structural style and kinematics of the Mesozoic Xuefengshan intraplate orogenic belt. *Earth Sci. Front.* **2018**, *25*, 1–13. (In Chinese)

74. Cello, G. Fractal analysis of a Quaternary fault array in the central Apennines, Italy. *J. Struct. Geol.* **1997**, *19*, 945–953. [CrossRef]

75. Wang, W.; Liu, M.Y.; Shi, G.W.; Li, J.Q.; Wang, Y.Q. Fractal characteristics study of fractures in Zhaxikang ore-concentrated area of Longzi County, Tibet. *Geol. Miner. Resour. South China* **2016**, *32*, 358–365. (In Chinese)

76. Wu, C.J. Fractal characteristics of lineament in Fozichong lead-zinc district. *China Min. Mag.* **2017**, *26*, 233–236. (In Chinese)

77. Kruhl, J.H. *Fractals and Dynamic Systems in Geoscience*; Spring: New York, NY, USA, 1994; 421p.

78. Dong, F.Q. Fractal Characteristics of Fractures and Its Geological Significance in Gudui-Longzi Region in Southern Tibet. *Gold Sci. Technol.* **2012**, *20*, 41–45. (In Chinese)

79. Ding, S.J.; Zhai, Y.S. Fractal Study of Structural Traces in Jiaojia Gold Deposit, Jiaodong, China. *Earth Sci.* **2000**, *25*, 416–420. (In Chinese)

80. Zhou, Q.Y.; Tan, K.X.; Xie, Y.S. Fractal Character of Structural Control on Uranium Mineralization in South China. *J. Univ. South China* **2009**, *23*, 32–36. (In Chinese)

81. Xie, Y.S.; Yin, J.W.; Tan, K.X.; Tang, Z.P.; Duan, X.Z.; Hu, Y.; Wang, Z.Q.; Li, C.G.; Wang, Z.Z.; Feng, Z.G. Tectono-Magmatic Activization and Fractal Dynamics of Hydrothermal Uranium Ore Formation in South. *Geotecton. Metallog.* **2015**, *39*, 510–519. (In Chinese)

82. Lu, X.W.; Ma, D.S. Fractal dimensions of fracture systems in antimony metallogenic zones of central Hunan and their indicating significance for migration of ore forming fluids and location of ore deposits. *Miner. Depos.* **1999**, *18*, 168–174. (In Chinese)

83. Han, X.B.; Liang, J.C.; Feng, Z.H.; Zhang, G.L.; Chen, M.H. Fractal Features of Fractures and Their Relation to Silver-Gold Mineralization in Southeast Guangxi. *Guangxi Sci.* **2003**, *10*, 117–121. (In Chinese)

84. Wang, N.; Liu, Y.S.; Peng, N.; Wu, C.L.; Liu, N.Q.; Nie, B.F.; Yang, X.Y. Fractal characteristics of fault structures and their use for mapping ore-prospecting potential in the Qitianling area, southern Hunan province, China. *Acta Geol. Sin.* **2015**, *89*, 121–132.

85. He, H.; An, L.; Liu, W.; Yang, X.; Gao, Y.; Yang, L. Fractal characteristics of fault systems and their geological significance in the Hutouya poly-metallic ore field of Qimantage, east Kunlun, China. *Geol. J.* **2017**, *52*, 419–424. [CrossRef]

86. Mao, Z.L.; Peng, S.L.; Lai, J.Q.; Wang, L. Fractal studies of fracture and metallogenic prediction in the east area of Gejiu ore district. *Contrib. Geol. Miner. Resour. Res.* **2004**, *19*, 17–19. (In Chinese)
87. Sun, T.; Liao, Z.Z.; Wu, K.X.; Chen, L.K.; Liu, W.M.; Yi, C.Y. Fractal distribution characteristics and geological significances of fracture structure in Southern Jiangxi. *J. Jiangxi Univ. Sci. Technol.* **2017**, *38*, 48–54. (In Chinese)
88. Zhao, J.L.; Li, J.; Zhang, Z.; Yang, C.C.; Zhang, X.J. Multifractal characteristics of spatial distribution of the faults in west subsag of Bozhong Sag. *Oil Drill. Prod. Technol.* **2018**, *40*, 14–16. (In Chinese)
89. Shi, Z.J.; Luo, Z.T.; Peng, D.J.; He, Z.H. Multifractal characteristics of fault spatial distribution in Sichuan area. *Geoscience* **1995**, *9*, 467–474. (In Chinese)
90. Liao, J.F.; Feng, Z.H.; Luo, C.Q.; Kang, Z.Q. Fractal characteristics analysis of fractures in Shuiyanba ore field of Hezhou, Guangxi. *Miner. Depos.* **2012**, *31*, 459–464. (In Chinese)
91. Cui, Z.L.; Yao, Y.L.; Cheng, J.H.; Luo, K.; Zhou, J.X. Structural Fractal Texture Characteristics and Its Prospecting Significance for the Yadu-Mangdong Metallogenic Belt in NW Guizhou Province. *Geol. J. China Univ.* **2022**, *28*, 592–605. (In Chinese)
92. Lovejoy, S.; Schertzer, D. Multifractals, universality classes, and satellite and radar measurements of cloud and rain fields. *J. Geophys Res.* **1990**, *95*, 2021–2034. [CrossRef]
93. Stanley, H.E.; Meakin, P. Multifractal phenomena in physics and chemistry. *Nature* **1998**, *335*, 405–409. [CrossRef]
94. Cheng, Q.M.; Zhang, S.Y.; Zuo, R.G.; Chen, Z.J.; Xie, S.Y.; Xia, Q.L.; Xu, D.Y.; Yao, L.Q. Progress of multifractal filtering techniques and their applications in geochemical information extraction. *Earth Sci. Front.* **2009**, *16*, 158–198. (In Chinese)
95. Halsey, T.C.; Jensen, M.H.; Kadanoff, L.P.; Procaccia, I.I.; Shraiman, B.I. Fractal measures and their sing ularities, the characterization of strange sets. *Phys. Rev.* **1986**, *33*, 1141–1151. [CrossRef] [PubMed]
96. Cheng, Q. Multifractality and spatial statistics. *Comput. Geosci.* **1999**, *25*, 949–962. [CrossRef]
97. Xie, S.Y.; Bao, Z.Y. The Method of Moments and Its Application to The Study of Mineralization in Shaoguan District, North Guangdong, China. *J. Chang. Univ. Sci. Technol.* **2003**, *33*, 443–448. (In Chinese)
98. Grau, J.; Méndez, V.; Tarquis, A.M.; Díaz, M.C.; Saa, A. Comparison of gliding box and box-counting methods in soli image analysis. *Geoderma* **2006**, *134*, 349–359. [CrossRef]
99. Zhou, G.Z.; Wang, C.Z.; Liu, Y.Z.; Xiao, T.F. Application of Multifractal Theory in the Environmental Sciences. *Bull. Mineral. Petrol. Geochem.* **2013**, *33*, 107–113. (In Chinese)
100. Cao, H.Q.; Zhu, G.X.; Li, X.T.; Xia, W.F. Multi-fractal and its application in terrain character analysis. *J. Beijing Univ. Aeronaut. Astronaut.* **2004**, *30*, 1182–1185. (In Chinese)
101. Wen, H.M.; Xiao, C.X.; Li, R.; Yang, B.; Wang, H. On Multifractal Analysis Method in Well Logging Interpretation. *Well Logging Technol.* **2004**, *28*, 381–385. (In Chinese)
102. Wang, W.G. *Software System Design of Geochemical Data Processing—Fractal Volume*; China University of Geosciences: Beijing, China, 2006; 64p. (In Chinese)
103. Liu, X.M. *Fractal Software System Used in Tungsten Forecast of Dajishan*; China University of Geosciences: Beijing, China, 2009; 66p. (In Chinese)
104. Shi, J.F.; Wang, C.N. Fractal analysis of gold deposits in China: Implication for giant deposit exploration. *Earth Sci.* **1998**, *23*, 616–619. (In Chinese)
105. Li, C.J.; Jiang, X.L.; Xu, Y.L.; Ma, T.H. Fractal Study of Mesozoic Hydrothermal Deposits in Zhejiang Province. *Chin. J. Geol.* **1996**, *31*, 55–64. (In Chinese)
106. Li, C.J.; Ma, T.H.; Zhu, X.S. *Fractal, Chaos and ANN in Mineral Exploration*; Geological Publishing House: Beijing, China, 1999; 140p. (In Chinese)
107. Fry, H. Random point distributions and strain measurement in rocks. *Tecomophysics* **1979**, *60*, 89–105. [CrossRef]
108. Stone, D.; Schwerdtner, W.M. Total strain within a major mylonite zone, southern Canadian Shield. *J. Struct. Geol.* **1981**, *3*, 193–194.
109. Ramsay, J.G.; Huber, M.I. *The Techniques of Modern Structural Geology, Volume 1: Strain Analysis*; Academic Press: London, UK, 1983.
110. Lacassin, R.; Driessche, J. Finite strain determination of gneiss: Application of Fry's method to porphyroid in the southern Massif Central (France). *J. Struct. Geol.* **1983**, *5*, 245–253. [CrossRef]
111. Vearncombe, J.; Vearncombe, S. The spatial distribution of mineralization: Applications of Fry analysis. *Econ. Geol.* **1999**, *94*, 475–486. [CrossRef]
112. Austin, J.R.; Blenkinsop, T.G. Local to regional scale structural controls on mineralisation and the importance of a major lineament in the eastern Mount Isa Inlier, Australia: Review and analysis with autocorrelation and weights of evidence. *Ore Geol. Rev.* **2009**, *35*, 298–316. [CrossRef]
113. Parsa, M.; Maghoudi, A.; Yousefi, M. Spatial analyses of exploration evidence data to model skarn-type copper prospectivity in the Varzaghan district, NW Iran. *Ore Geol. Rev.* **2018**, *92*, 97–112. [CrossRef]

Article

3D Quantitative Metallogenic Prediction of Indium-Rich Ore Bodies in the Dulong Sn-Zn Polymetallic Deposit, Yunnan Province, SW China

Fuju Jia [1], Zhihong Su [1], Hongliang Nian [2], Yongfeng Yan [1,*], Guangshu Yang [1], Jianyu Yang [3], Xianwen Shi [1], Shanzhi Li [2], Lingxiao Li [2], Fuzhou Sun [2] and Ceting Yang [1]

1   College of Land Resources and Engineering, Kunming University of Science and Technology, Kunming 650093, China
2   No. 317 Geological Party of Yunnan Nonferrous Geological Bureau, Qujing 655000, China
3   Hualian Zinc & Indium Co., Ltd., Maguan 663701, China
*   Correspondence: 11301017@kust.edu.cn

Abstract: The southwestern South China Block is one of the most important Sn polymetallic ore districts in the world, of which the Dulong Sn-Zn polymetallic deposit, closely related to Late Cretaceous granitic magmatism, contains 0.4 Mt Sn, 5.0 Mt Zn, 0.2 Mt Pb, and 7 Kt In, and is one of the largest Sn-Zn polymetallic deposits in this region. In this paper, on the basis of a 3D model of ore bodies established by the cut-off grade of the main ore-forming elements, the In grades were estimated by the ordinary Kriging method and the In-rich cells were extracted. The 3D models of strata, faults, granites, and granite porphyries in the mining area were established and assigned the attributes to the cells, which built buffer zones representing the influence space of the geological factors. The weight of evidence and artificial neural network methods were used to quantitatively evaluate the contribution of each geological factor to mineralization. The results show that the Neoproterozoic Xinzhai Formation (Pt_3x), fault (F_1), and Silurian granites (S_3L) have considerable control effects on the occurrence of In-rich ore bodies. The metallogenic predictions according to the spatial coupling relationship of each geological factor in 3D space were carried out, and then the 3D-space-prospecting target areas of In-rich ore bodies were delineated. In addition, the early geological maps and data information of the mining area were comprehensively integrated in 3D space. The feasibility of 3D quantitative metallogenic prediction based on the deposit model was explored by comparing the two methods, and then, the 3D-space prospecting target area was delineated. The ROC curve evaluation shows that the results of two methods have indicative value for prospecting. The modeling results may support its use for future deep prospecting and exploitation of the Dulong and other similar deposits.

Keywords: 3D deposit model; weight of evidence method; neural network method; metallogenic prediction; the Dulong Sn-Zn polymetallic deposit; SW China

Citation: Jia, F.; Su, Z.; Nian, H.; Yan, Y.; Yang, G.; Yang, J.; Shi, X.; Li, S.; Li, L.; Sun, F.; et al. 3D Quantitative Metallogenic Prediction of Indium-Rich Ore Bodies in the Dulong Sn-Zn Polymetallic Deposit, Yunnan Province, SW China. *Minerals* **2022**, *12*, 1591. https://doi.org/10.3390/min12121591

Academic Editor: Behnam Sadeghi

Received: 17 October 2022
Accepted: 9 December 2022
Published: 12 December 2022

Publisher's Note: MDPI stays neutral with regard to jurisdictional claims in published maps and institutional affiliations.

1. Introduction

Indium (In) is widely used in the aerospace, radio, electronic, and medical industries and is defined as a "critical mineral" by many economically developed countries and regions around the world. The types of skarn and massive sulfide deposits are the main sources of indium, which account for 29% and 28% of global indium resources, respectively, followed by epithermal and sedimentary Pb-Zn deposits, which account for 19.9% and 18.0% of the global indium resources, respectively [1,2]. The representative skarn In-rich deposits in the world include Dulong and Dachang in China, Ayawilca in Peru, and Tellerhauser and Pohla-Globenstein in Germany. Massive sulfide In-rich deposits include Kidd Creek, Geco/Manitouwadge, and Heath Steele in Canada; and Gaiskoye, Podolskoye, and Sibaiskoye in Russia. Indium mainly forms in Zn and Cu sulfide ores. The main In-rich sulfide is sphalerite, which accounts for 95% of global indium resources [3].

The Dulong Sn-Zn polymetallic deposit is located in the outer contact zone of the southwest side of the Cretaceous granite. The mining area is about 8 km long from north to south and 1.5 km wide from east to west. The Dulong mining area has proven reserves of 0.4 Mt of Sn, 5 Mt of Zn, 0.2 Mt of Pb, 7 Kt of In, and 3 Kt of Cd, and it is also rich in Ag, Cu, Ga, and Ge. The development and utilization values of the mineral resources are high, and the mining area has considerable prospecting potential [4]. Indium exists in the Dulong Sn-Zn polymetallic deposit as associated elements of polymetallic sulfide ores, and its main carrier mineral is marmatite, with a small amount in chalcopyrite. The In-rich ore bodies occur as stratified or stratiform-like in skarn, and the ore-bearing skarn is mainly composed of diopside, tremolite, chlorite, epidote, and actinolite, whose occurrence regularity are as follows: (a) The skarn and In-rich ore bodies are stratified, stratiform-like or lenticular restricted in the Neoproterozoic Xinzhai Formation and are consistent with the stratigraphic occurrence; (b) Occurs in the contact zone between marble and clastic rocks and often forms multilayered skarn and In-rich ore bodies in the area where these two types of rocks interact; (c) Occurs in the upper and lower plates of the fault (F_1); (d) The upper part of the Cretaceous granite ridge uplift is the site of concentrated occurrence of the skarn and In-rich ore bodies. At present, the average drilling depth of the mine area is about 312 m, and the risks of deep exploration are gradually increasing. Therefore, a comprehensive analysis of the existing geological data and an exploration of the valuable prospecting space in the deep and peripheral parts of the mine area are necessary in order to improve the success rate of prospecting.

The deposit's 3D geological model mainly includes models of ore bodies, strata, faults, and magmatic rocks. The 3D ore body model is widely used to resource reserve estimations [5–13]. 3D geological models can intuitively show the spatial forms of geological bodies and reflect the spatial position relationships between geological bodies, with varying degree of success [14–18]. Traditional 2D metallogenic prediction methods include the weight of evidence, information value, neural network, logistic regression, and fractal analysis methods. These 2D metallogenic prediction methods have played an important role in regional metallogenic prediction, but are difficult in one given deposit [19–22]. Based on the establishment of the deposit model, this study reveals the control effects of strata, faults, and granites on In-rich ore bodies using 3D weight of evidence and 3D neural network methods. Additionally, a 3D quantitative metallogenic prediction was carried out to explore the mining area's deep and peripheral prospecting targets. The results can provide clues for the same types of ore deposits.

2. Geological Background and Deposit Model

2.1. Geological Background

The Dulong Sn-Zn polymetallic deposit is located in the joint part of the Cathaysia, Yangtze, Indochina, and North Viet blocks (Figure 1a). The main outcrop strata are Neoproterozoic, Cambrian, and Devonian. The Neoproterozoic Xinzhai Formation is the main ore-bearing stratum of the Dulong Sn-Zn polymetallic deposit (Figures 1b and 2a). The upper part of the Neoproterozoic Xinzhai Formation is composed of gray-green quartz-mica schist, calcite marble, and skarn lenticles. The middle part is a composite lithological section that composes of quartz-mica schist, marble, skarn, granulite, and a small amount of gneiss with frequent changes in the lithofacies, complex rock assemblage, and skarn geological bodies appearing in groups and zones. This is the main occurrence horizon of the mining area's Sn-Zn industrial ores. The lower part consists of marble lens and dark gray biotite-plagioclase gneiss, biotite plagioclase hornblende gneiss, and granite gneiss, in addition to plagioclase granulite with a small amount of garnet skarn and siliceous marble lenticle.

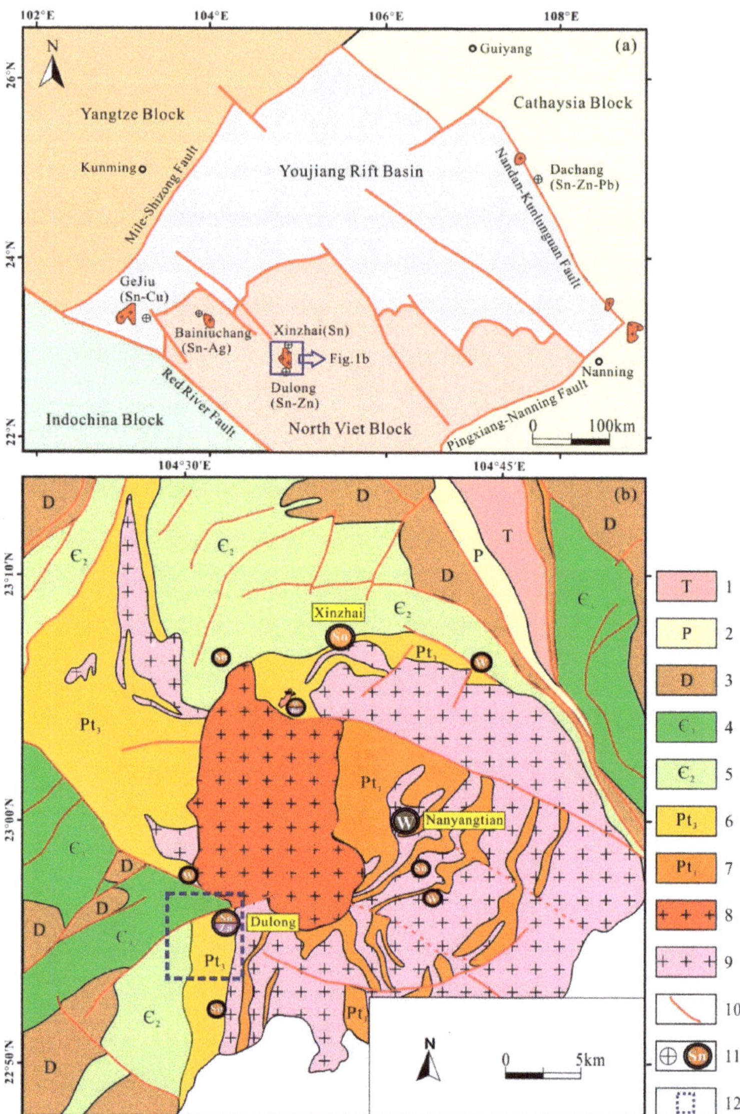

Figure 1. Geotectonic location and regional geological map (modified using Ref. [23]). (**a**) The geotectonic location of the Dulong Sn-Zn polymetallic deposit; (**b**) The regional geological map of the Dulong Sn-Zn polymetallic deposit. 1—Triassic mudstone and tuff; 2—Permian siliceous rocks and mudstones; 3—Devonian carbonate rocks; 4—Upper Cambrian dolomite and limestone; 5—Middle Cambrian dolomite and phyllite; 6—Upper Proterozoic marble and schist; 7—Lower Proterozoic schist and granulite; 8—Cretaceous granite; 9—Silurian gneiss granite; 10—fault; 11—deposit; 12—3D geological modeling area.

The well-developed south-north strike faults, which mainly interlayer dislocation faults with multistage activity characteristics, were observed with the same occurrence as strata in the mining area. Among them, the largest is the F_0 fault that comprises the eastern boundary of the ore bodies. Moreover, the F_1 fault is located in the middle of the mining area, and its hanging wall and footwall have ore bodies. The F_2 fault is located parallel to F_0 and F_1 in the mining area's western section. The F_3 fault strike is NE-SW, staggered to the north-south fault, and has the characteristics of late activity (Figure 2b).

The Cretaceous granites have an outcrop area of about 153 km^2. The granites can be roughly divided into three sub-stages according to their evolutionary sequences [24]. The first sub-stage is gray-white porphyry-bearing medium-coarse-grained dimica granite, accounting for 2/3 of the total outcrop area of the magmatic rock. The second sub-stage is gray-white medium-fine-grained dimica granite, which intrudes into the first sub-stage pluton in the shape of a rock cluster. The third sub-stage is gray-white granite porphyry, which is interspersed in the early granite and metamorphic rock series in the shape of rock branches and veins. The Silurian granite is medium-fine-grained, mostly metamorphic to (eyeball-shaped) granite gneiss, and the zircon U-Pb ages are 427–436 Ma [25].

A series of dome-shaped metamorphic rocks are exposed around the Cretaceous granite, which is called the "Laojunshan metamorphic core complex" [26,27], and the metamorphic strata are mainly composed of the Neoproterozoic Xinzhai and Lower-Middle Cambrian Formations, while the Upper Cambrian, Lower Ordovician, and Devonian Formations also suffered from mild metamorphism. Metamorphism is dominated by regional metamorphism, and strong migmatization occurs in the late stages.

2.2. Deposit Geology

The Dulong deposit is located in the southwest outer contact zone of the Laojunshan granites. The stratigraphic trend is north-south, dips to the west, and the dip angle is 10°–35°. The strata exposed in the mine area include the Middle Cambrian Longha Formation, Tianpeng Formation and Neoproterozoic Xinzhai Formation, which are subject to moderate metamorphism, and the Xinzhai Formation is the ore-bearing stratum of the Sn-Zn polymetallic ore body. Normal faults are very well-developed in the mine area, mainly as interstratified glide. The typical faults are F_0, F_1, F_2, and F_3, where F_0 is the eastern boundary of the ore body output; F_1 has ore body output in both upper and lower plates, which is closely related to the spatial location of the ore bodies; and F_2 is the boundary between the ore-bearing Xinzhai Formation and the Tianpeng Formation. The exposed magmatic rocks in the mine area mainly include Silurian granite, Cretaceous granite, and Cretaceous granite porphyry. Silurian granite is located in the lower plate of fault F_0, Cretaceous granite is exposed in the northern part of the mine area and the burial depth gradually increases to the south, Cretaceous granite porphyry intrudes into the aforementioned strata and granite in the form of veins, and there is no ore body output inside the granite and granite porphyry. The alterations of the surrounding rocks in the mine area mainly include marbleization, skarnization, silicification, etc. The types of skarns mainly include chlorite skarn, epidote skarn, actinolite skarn, diopside skarn, tremolite skarn, clinotetrahedrite skarn, and garnet skarn, among which chlorite skarn and actinolite skarn have the best ore-bearing properties.

The ore bodies occur stratified, stratiform-like, or lenticular restricted in the Neoproterozoic Xinzhai Formation and are consistent with the stratigraphic occurrence. Metal minerals mainly include marmatite, pyrite, chalcopyrite, cassiterite, pyrrhotite, and magnetite. Gangue minerals include diopside, chlorite, epidote, actinolite, clinozoisite, garnet, tremolite, quartz, and calcite. Indium enters marmatite by coupled substitution as (Cu^+ or Ag^+, In^{3+})$\leftrightarrow$($2Zn^{2+}$) and is mainly contained homogeneously in the marmatite [28].

2.3. Deposit Model

The Dulong Sn-Zn polymetallic deposit modeling area is 5.0 km wide from east to west, 5.7 km long from north to south, and the surface elevation is 650–1768 m. The minimum modeling elevation is -1000 m. The boundaries of the strata, magmatic rocks, and ore bodies according to the geological survey and prospecting engineering data of the mining area were determined using 3Dmine software, and the model of each geological factor according to the occurrence of these boundaries was appropriately extrapolated

(Figure 2a,b). The modeling strata include the Middle Cambrian Longha, Middle Cambrian Tianpeng, and Neoproterozoic Xinzhai Formations. Our modeled magmatic rocks are Cretaceous granite (K_1H), Cretaceous granite porphyry (K_2K), and Silurian granite (S_3L). Furthermore, our modeled faults are the F_0, F_1, and F_2 NS-strike faults, and the F_3 NE-SW-strike fault. The ore bodies' model with boundary grades of Zn 1.00%, Sn 0.15%, and Cu 0.20% was established. It has been proven that if the cell division is too small, it not only causes the geological phenomenon to be artificially divided, but also brings inconvenience to the huge data processing work; if the cell division is too large, it makes the morphological distribution of ore-bearing units unreliable. Taking into account the strata, structure, granite, and ore-body occurrence and scale in the deposit model, as well as the data processing capability of the computer, and taking a 20 m × 20 m × 20 m cube as the unit block, the ore deposit modeling area was represented by 7,994,700 cells, of which 19,101 were ore-bearing cells (Figure 2a,b).

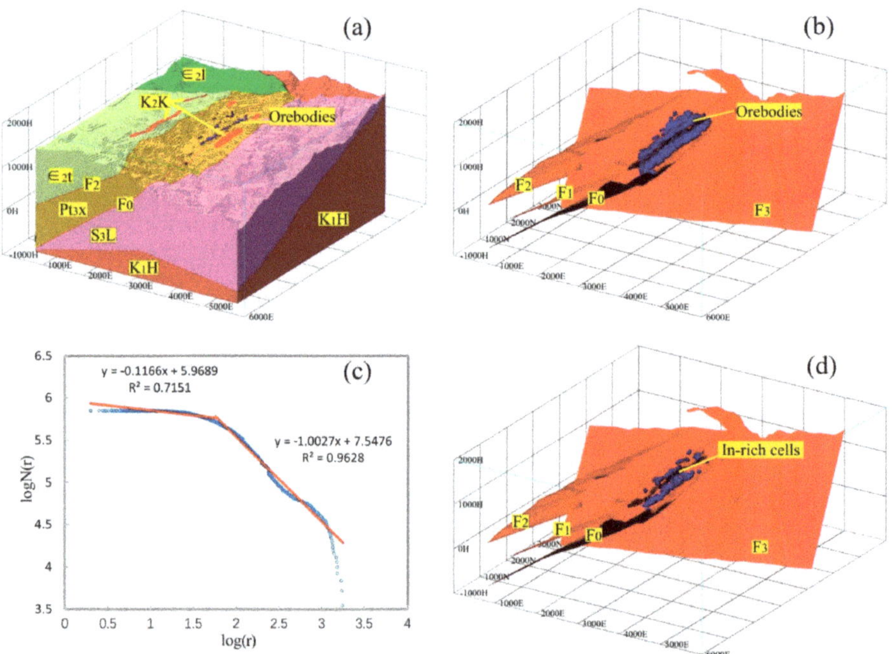

Figure 2. Geology and In-rich ore bodies model in Dulong mining area. (**a**) Geological model of mining area; (**b**) Faults and ore bodies model; (**c**) Indium-grade log–log plot of ore bodies; (**d**) Faults and In-rich ore bodies model. $\in_2l$—Middle Cambrian Longha Formation; $\in_2t$—Middle Cambrian Tianpeng Formation; Pt_3x—Neoproterozoic Xinzhai Formation; K_1H—Cretaceous granite; K_2K—Cretaceous granite porphyry; S_3L—Silurian granite; F_0-F_3—Fault number.

The 3D spatial positioning for In-grade test data from drilling was carried out, and the polymetallic ore body model by the ordinary Kriging method was estimated. Thus, the In-grade attributes corresponding to 10,320 ore-bearing cells were obtained. In these cells, the maximum, minimum, and mean values were 1780 g/t, 2 g/t, and 69 g/t, respectively. The In-grade log–log plot of the ore bodies was obtained by fractal analysis (Figure 2c) and calculated the grade boundary of the In-rich cells to be 60 g/t. The 3752 cells with indium grades extracted were greater than 60 g/t, forming an In-rich ore model (Figure 2d).

3. Prediction Method and Ore Control Factors

3.1. Weight of Evidence Method

Bonham Carter et al. (1988) and Agterberg (1989) first proposed the weight of evidence method for multivariate statistics and the fusion of discrete data using probability and Bayesian theory [19]. The posterior probability of determining the favorable area for mineralization is obtained through the superposition and composite analysis of several kinds of mineralization-related geological information. Each metallogenic factor is regarded as an evidence factor of the metallogenic prospect, and the weight value of each evidence factor represents its contribution to the metallogenic prediction [29]. The evidence weight analysis needs to verify the conditional independence of different evidence and then calculate the posterior probability of each basic unit divided into the study area. The value represents the metallogenic probability, and the area where the posterior probability is greater than the critical value is the metallogenic prospect area.

The modeling area is evenly divided into T units with the same volume, the number of ore-bearing cells in the model is D, and P is the probability of any cell containing ore.

The formula of prior probability is:

$$P_{pri} = P(D) = \frac{D}{T} \tag{1}$$

The priori probability (O_{pri}) is:

$$O_{pri} = O(D) = \frac{P(D)}{1 - P(D)} = \frac{D}{T - D} \tag{2}$$

where O_{pri} is used to calculate the ratio of the prior probability of ore-bearing to non-ore-bearing blocks in the attributes of each metallogenic geological factor.

The weight of any geological factor's binary image is defined as:

$$W^+ = \ln \frac{P(B/D)}{P(B/\overline{D})} = \ln \frac{N(B \cap D)/N(D)}{(N(B) - N(B \cap D))/(N(T) - N(D))} \tag{3}$$

$$W^- = \ln \frac{P(\overline{B}/D)}{P(\overline{B}/\overline{D})} = \ln \frac{(N(D) - N(B \cap D))/N(D)}{(N(T) - N(B) - N(D) + N(B \cap D))/(N(T) - N(D))} \tag{4}$$

In the formula, W^+ represents that the weight value of each metallogenic geological factor's attribute exists within the prediction area, and W^- represents that the weight value of each metallogenic geological factor's attribute does not exist within the prediction area. If the weight value is 0, it means that the data in the prediction area is missing. B and $\overline{B}$ represent the presence or absence of each metallogenic geological factor's attributes in the layer.

The correlation between geological factors and known ore-bearing units is represented by contrast C:

$$C = W^+ - W^- \tag{5}$$

The value of C being greater than 0 indicates that the attribute of the geological factor is favorable for mineralization. This means it can participate in quantitative mineralization prediction. A C value equal to 0 indicates that the attribute of this geological factor has no guiding significance for metallogenic prediction, and a C value of less than 0 indicates that the attribute of the geological factor is not conducive to mineralization and should be discarded.

If the n evidence factors are conditionally independent from the ore-bearing unit distribution, the logarithm of the posterior probability is:

$$\ln\left(O\left(D \middle| B_1{}^k \cap B_2{}^k \cap B_3{}^k \cdots B_n{}^k\right)\right) = \sum_{j=1}^{n} W_j{}^k + \ln(O(D)) \tag{6}$$

The posterior probability O_{pos} is expressed as:

$$O_{pos} = \exp\left\{ \ln\left(O_{pri}\right) + \sum_{j=1}^{n} W_j^k \right\} \tag{7}$$

$$W_j^k = \left\{ \begin{array}{ll} W^+ & \text{evidence factor} \\ W^- & \text{no evidence factor} \\ 0 & \text{missing data} \end{array} \right. \tag{8}$$

The posterior probability is expressed as:

$$P_{pos} = \frac{O_{pos}}{\left(1 + O_{pos}\right)} \tag{9}$$

The posterior probability of each favorable ore-forming factor was obtained through the above calculation. The posterior probability value is between 0 and 1, and the higher the value, the greater the possibility of bearing ore [30,31].

3.2. Artificial Neural Network Method

An artificial neural network (ANN) is a mathematical model composed of highly nonlinear and linear operations and is established by simulating the thinking mode and organizational form of the human brain. ANNs can automatically simulate the natural relationships between variables when they are used to make prediction, carry out global optimization searches, reduce manual interventions, and improve the accuracy of predictions [32,33]. Moreover, ANNs can be divided into feedforward and feedback neural networks, as well as competitive learning networks. At present, the most widely used multilayer perceptron and BP networks are feedforward neural networks.

The multilayer perceptron provided by the SPSS19 software is composed of several perceptron layers and adjustable weight connections, generally including an input layer, one or more hidden layers, and an occurrence layer. The input layer is used to store predictive variables, the hidden layer contains nodes or cells that cannot be observed, and the output layer contains output variables (Figure 3).

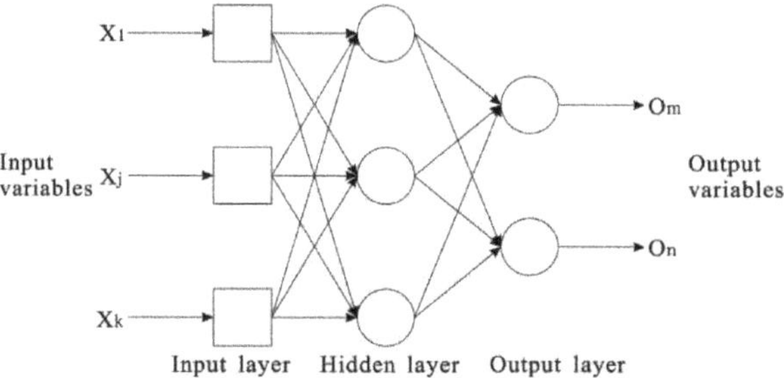

Figure 3. Feedforward architecture of artificial neural networks (modified using Ref. [21]).

3.3. Ore-Controlling Factors and Prediction Variable Selection

The Neoproterozoic Xinzhai Formation (Pt_3x), faults (F_0 and F_1), Cretaceous granite porphyries (K_2K), Cretaceous granites (K_1H), and Silurian granites (S_3L) are geological factors closely related to the ore bodies in the Dulong mining area. The search is from the Xinzhai Formation floor to the interior of the strata, forming the distance attribute (Pt_3x distance). The Xinzhai Formation distance attribute ranges from 0 to 350 m and contains all 3752 In-rich cells (Figure 4a). The search is from the fault plane to both sides to determine the distance attributes (F_0 and F_1 distances, Figure 4b,c). The F_0 distance value contains all In-rich cells in the range from 0 to 500 m, and the F_1 distance value contains all In-rich cells in the range from 0 to 250 m. The search is from the granite porphyry vein to the periphery to establish the distance attribute (K_2K distance, Figure 4d). The K_2K distance value ranges from 0 to 750 m, which includes all indium-rich cells. The search is from the granite roof to the outside of the contact zone to determine the distance attributes (K_1H and S_3L distances) from the peripheral centroid point to the granite roof (Figure 4e,f). The K_1H distance ranges from 0 to 650 m, and the S_3L distance ranges from 0 to 500 m, including all In-rich cells. The 3D buffer zones were established according to the ore-bearing properties of the peripheral space of each geological factor to represent the influenced space of the strata, structure, and granite on mineralization (Figure 4).

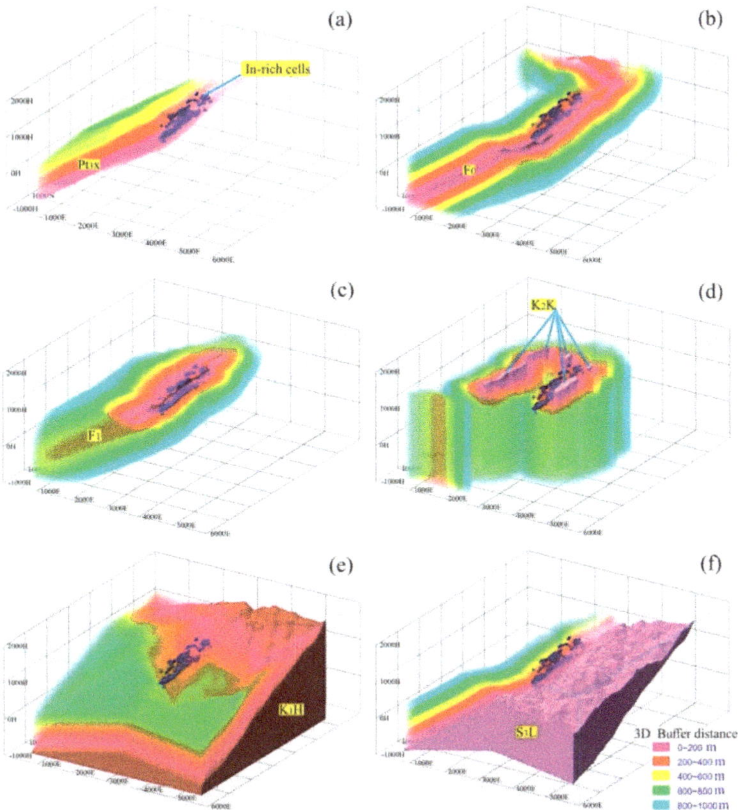

Figure 4. Ore-controlling geological factors and 3D buffer zone. (**a**) Buffer zone of Neoproterozoic Xinzhai Formation (Pt₃x); (**b**,**c**) Buffer zone of faults (F₀ and F₁); (**d**–**f**) Buffer zone of granite porphyries and granites (K_2K, K_1H, and S_3L).

4. 3D Quantitative Metallogenic Prediction

4.1. Metallogenic Prediction by Weight of Evidence Method

The positive W^+ and negative W^- weights in the zoning intervals of each geological factor were calculated according to the buffer zone of each geological variable, with 50 m as the basic bandwidth (Table 1).

Table 1. Statistical table of geological variable weight scores.

Geological Variables	Distance Interval	Number of Ore Blocks in Zone	Number of no Ore Blocks in Zone	Number of Ore Blocks Outside Zone	Number of no Ore Blocks Outside Zone	W^+	W^-	Contrast C	Sort
Xinzhai formation (Pt_3x)	0~50	102	64,363	3650	7,926,585	1.2165	−0.0195	1.2359	39
	50~100	802	66,550	2950	7,924,398	3.2452	−0.2321	3.4773	6
	100~150	1355	64,918	2397	7,926,030	3.7945	−0.4399	4.2344	1
	150~200	681	66,048	3071	7,924,900	3.0892	−0.1920	3.2812	9
	200~250	557	67,496	3195	7,923,452	2.8665	−0.1522	3.0187	11
	250~300	249	70,513	3503	7,920,435	2.0177	−0.0598	2.0775	21
	300~350	6	73,400	3746	7,917,548	−1.7481	0.0076	−1.7558	58
Fault (F_0)	0~50	30	94,831	3722	7,896,117	−0.3949	0.0039	−0.3988	54
	50~100	141	110,668	3611	7,880,280	0.9982	−0.0244	1.0226	41
	100~150	274	113,660	3478	7,877,288	1.6359	−0.0615	1.6974	30
	150~200	770	119,640	2982	7,871,308	2.6179	−0.2146	2.8325	13
	200~250	837	125,076	2915	7,865,872	2.6569	−0.2366	2.8936	12
	250~300	589	132,552	3163	7,858,396	2.2475	−0.1540	2.4015	18
	300~350	334	138,231	3418	7,852,717	1.6382	−0.0758	1.7140	29
	350~400	445	145,115	3307	7,845,833	1.8766	−0.1079	1.9845	23
	400~450	303	150,811	3449	7,840,137	1.4537	−0.0652	1.5189	34
	450~500	29	158,081	3723	7,832,867	-0.9398	0.0122	−0.9520	57
Fault (F_1)	0~50	886	82,661	2866	7,908,287	3.1280	−0.2590	3.3870	7
	50~100	952	93,397	2800	7,897,551	3.0777	−0.2809	3.3586	8
	100~150	1170	99,859	2582	7,891,089	3.2170	−0.3611	3.5782	4
	150~200	664	107,693	3088	7,883,255	2.5750	−0.1812	2.7562	16
	200~250	80	114,691	3672	7,876,257	0.3958	−0.0071	0.4029	48
Cretaceous granite porphyry (K_2K)	0~50	82	218,801	3670	7,772,147	−0.2254	0.0057	−0.2311	53
	50~100	346	163,915	3406	7,827,033	1.5031	−0.0760	1.5791	32
	100~150	475	172,750	3277	7,818,198	1.7675	−0.1135	1.8810	24
	150~200	385	198,484	3367	7,792,464	1.4186	−0.0831	1.5017	36
	200~250	333	221,363	3419	7,769,585	1.1644	−0.0648	1.2292	40
	250~300	523	236,707	3229	7,754,241	1.5488	−0.1200	1.6688	31
	300~350	450	230,613	3302	7,760,335	1.4245	−0.0985	1.5230	33
	350~400	263	224,860	3489	7,766,088	0.9127	−0.0441	0.9568	42
	400~450	182	226,482	3570	7,764,466	0.5374	−0.0210	0.5583	45
	450~500	157	229,881	3595	7,761,067	0.3747	−0.0136	0.3883	49
	500~550	190	229,624	3562	7,761,324	0.5666	−0.0228	0.5894	44
	550~600	171	233,344	3581	7,757,604	0.4452	−0.0170	0.4622	47
	600~650	94	233,387	3658	7,757,561	−0.1534	0.0043	−0.1577	52
Cretaceous granite (K_1H)	0~50	43	177,989	3709	7,812,959	−0.6645	0.0110	−0.6755	55
	50~100	108	186,994	3644	7,803,954	0.2071	−0.0055	0.2126	51
	100~150	413	173,086	3339	7,817,862	1.6257	−0.0947	1.7204	28
	150~200	317	162,447	3435	7,828,501	1.4246	−0.0677	1.4923	37
	200~250	136	153,947	3616	7,837,001	0.6321	−0.0175	0.6495	43
	250~300	246	147,643	3506	7,843,305	1.2666	−0.0492	1.3157	38
	300~350	386	142,937	3366	7,848,011	1.7495	−0.0905	1.8400	26
	350~400	382	137,890	3370	7,853,058	1.7750	−0.0900	1.8650	25
	400~450	527	132,028	3225	7,858,920	2.1402	−0.1347	2.2749	20
	450~500	550	124,884	3202	7,866,064	2.2386	−0.1428	2.3813	19
	500~550	391	118,220	3361	7,872,728	1.9522	−0.0951	2.0473	22
	550~600	229	112,180	3523	7,878,768	1.4696	−0.0488	1.5185	35
	600~650	24	106,574	3728	7,884,374	−0.7348	0.0070	−0.7418	56

Table 1. *Cont.*

Geological Variables	Distance Interval	Number of Ore Blocks in Zone	Number of no Ore Blocks in Zone	Number of Ore Blocks Outside Zone	Number of no Ore Blocks Outside Zone	W^+	W^-	Contrast C	Sort
	0~50	32	40,709	3720	7,950,239	0.5153	−0.0035	0.5188	46
	50~100	130	46,549	3622	7,944,399	1.7830	−0.0294	1.8125	27
	100~150	263	44,562	3489	7,946,386	2.5313	−0.0671	2.5984	17
Silurian granite (S$_3$L)	150~200	753	43,859	2999	7,947,089	3.5991	−0.2185	3.8176	3
	200~250	846	43,842	2906	7,947,106	3.7159	−0.2500	3.9660	2
	250~300	604	43,981	3148	7,946,967	3.3758	−0.1700	3.5458	5
	300~350	319	44,444	3433	7,946,504	2.7270	−0.0833	2.8103	15
	350~400	451	44,394	3301	7,946,554	3.0744	−0.1225	3.1969	10
	400~450	323	44,459	3429	7,946,489	2.7391	−0.0844	2.8235	14
	450~500	30	44,382	3722	7,946,566	0.3644	−0.0025	0.3668	50

The contrast C value represents the strength of correlation between ore-controlling geological factors and mineralization [34]. The top 10 geological factors of contrast C values in each zone are mainly the Xinzhai Formation (Pt$_3$x), Silurian granites (S$_3$L), and fault F$_1$. The ore bodies are strictly limited in the Xinzhai Formation (Pt$_3$x), and the C value is high, 100–200 m away from the stratum floor, which is conducive to mineralization. The metamorphic zone that is 150–400 m away from the Silurian granite roof (S$_3$L) has a high C value, which is conducive to mineralization. The C value of the spatial range 0~150 m away from the F$_1$ fault plane is high, which is conducive to mineralization. The positive weight W^+ value of each geological factor's zoning interval with a C value was assigned greater than 0 to the cells in the zoning interval and calculated the posterior probability P_{pos} value of each cell.

The ROC curve can be used to determine the best prediction boundary of the binary classification and to evaluate the prediction ability. The optimal prediction boundary by the "Youden index", that is, sensitivity-(1-specificity), was determined, and the boundary value corresponding to the maximum value of the index value is the optimal prediction boundary.

A total of 3,086,482 posterior probability P_{pos} values were obtained through the calculations, the minimum, maximum, and mean values were 0.0011, 0.9999, and 0.1035, respectively, the variance was 0.0756, and the standard deviation was 0.2749. After taking the P_{pos} value of the cell as the test variable and determining whether it belonged to the In-rich cells as the state variable, our ROC curve analysis showed that the best prediction boundary of the P_{pos} value for the In-rich ore bodies was 0.820. We obtained a total of 247,555 cells (P_{pos} > 0.820) according to the prediction boundary, of which 3665 cells belonged to In-rich cells. The high value of the posterior probability P_{pos} of the prediction space was mainly distributed in the In-rich cells and the southwest of the mining area. The deep space in the south and west of the mining area can be used as a prospecting target space (Figure 5).

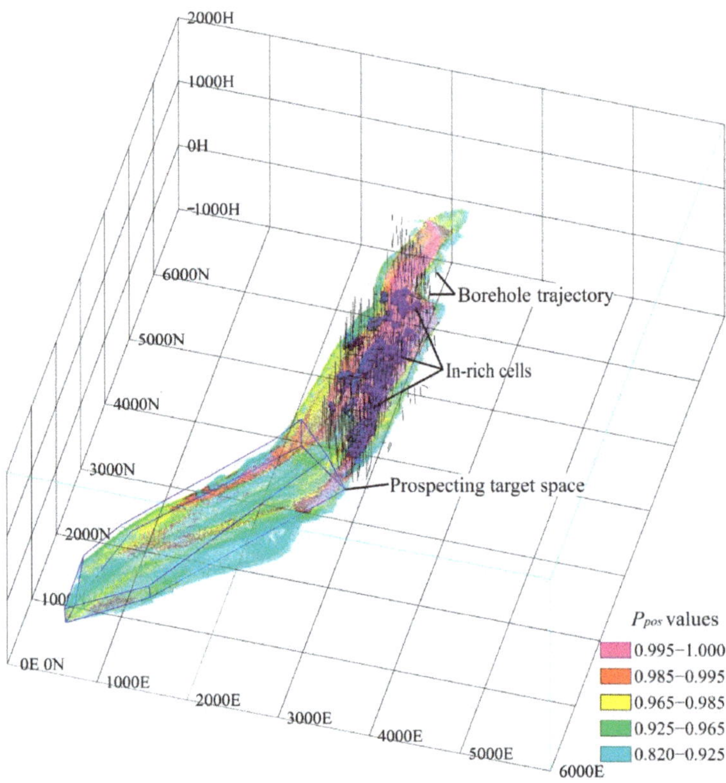

Figure 5. Metallogenic prediction map of weight of evidence method ($P_{pos} > 0.820$).

4.2. Metallogenic Prediction by Artificial Neural Network Method

The In-rich attribute of the cell was taken as the dependent variable in the SPSS software, and the distance attribute from the centroid point was taken to the floor surface of the stratum, the fault plane, the granite porphyry, and the granite roof as the covariate. Additionally, the multilayer perceptron was used to predict the mineralization by the artificial neural network method.

The software evaluated the importance of each geological variable when performing the artificial neural network metallogenic prediction (Figure 6). The results show that the faults (F_1, 0.349), the Xinzhai Formation (Pt_3x, 0.286), and Silurian granites (S_3L, 0.199), had a greater impact on the metallogenic prediction.

Because the In-rich attributes of the cells are the categorical dependent variables (1 and 0 represent the In-rich and non-In-rich cells, respectively), the prediction result is the pseudo-probability of the In-rich cell (the In-rich attribute is 1). A total of 379,647 pseudo-probability values were obtained through the calculations, with the minimum, maximum and mean values of 0.000, 0.392, and 0.102, respectively, a variance of 0.000, and a standard deviation of 0.016. Using ROC curve analysis and taking the pseudo-probability as the test variable to determine whether it belonged to the In-rich cell as the state variable suggested that the optimal prediction boundary of the pseudo-probability for the In-rich ore bodies was 0.007. A total of 215,723 unit blocks (pseudo-probability > 0.007) were obtained according to the prediction boundary, among which 3287 unit blocks belonged to indium-rich unit blocks. The high pseudo-probability of In-rich ore bodies in the prediction model was mainly distributed in the In-rich and the southwest cells of the mining area.

The deep space in the southwest of the mining area can be used as a prospecting target space (Figure 7).

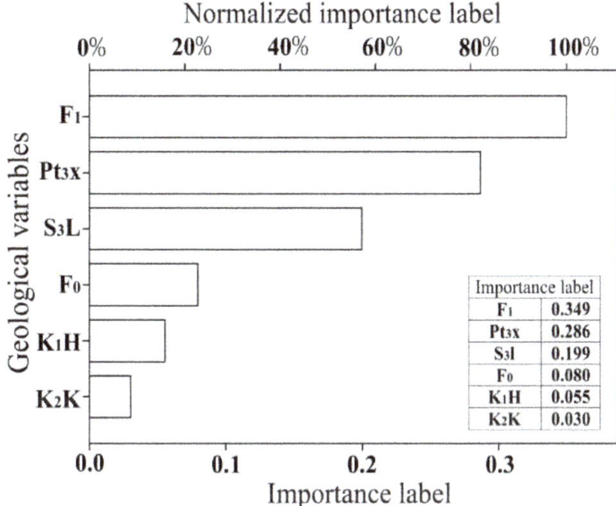

Figure 6. Evaluation results of geological variables by artificial neural network method.

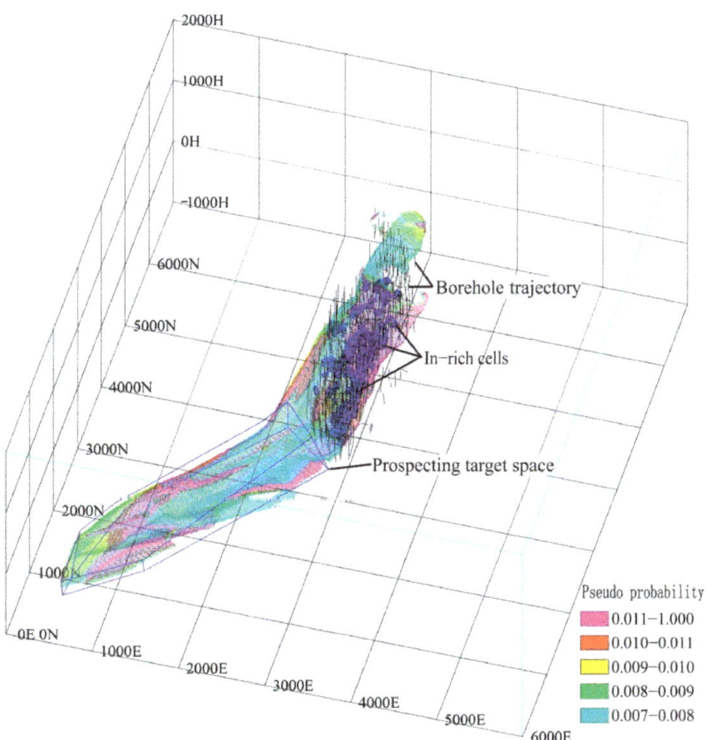

Figure 7. Metallogenic prediction map of artificial neural network method (prediction probability of indium-rich cell > 0.007).

5. Discussion

5.1. Considerable Ore-Controlling Geological Factor

The weight of evidence method reflects the importance of geological factors in a given zoning interval through the contrast C value. The artificial neural network method takes the geological factors involved in the prediction as the dependent variable and evaluates its importance. The Neoproterozoic Xinzhai Formation (Pt$_3$x), fault F_1, and Silurian granite (S$_3$L) are of great significance for metallogenic predictions based on the evaluation results of this study regarding the importance of the two methods to geological factors.

The main ore bodies of the Dulong Sn-Zn polymetallic deposit are parallel to the strata layers of the Neoproterozoic Xinzhai Formation (Pt$_3$x), and they are prone to skarnization and the formation of industrial ore bodies in the contact zone between carbonate and clastic rocks. Therefore, the control effects of the fixed horizon and lithofacies combination space are conducive to the selective metasomatism and mineralization of ore-bearing hydrothermal fluids.

The hanging wall and footwall of the fault (F_1) are the main spaces for ore-body occurrence. Some studies have believed that the fault is an ore-conducting and ore-hosting structure. Therefore, the F_1 fault plane and the surrounding space are favorable for metallization.

The results of the evidence weight and artificial neural network methods show that the Silurian granite (S$_3$L) is promising for prospecting, which may be related to the composite of various geological interfaces in the mining area's Cretaceous granite roof. The interface surface is also the floor of the Xinzhai Formation and the plane of fault (F_0). Above, the control effect of the Xinzhai Formation on mineralization was described. The upper part of the fault (F_0) is the ore bodies' occurrence space, and the lower part has no ore. The fault plane acts as a geochemical barrier to the ore-forming fluids. The space above the Cretaceous granite roof is favorable for mineralization.

In addition, the cassiterite U-Pb ages of the deposit are 82.0–96.6 Ma [35], which are close to the ages of Cretaceous granites and granite porphyries (K_1H and K_2K). Hence, the Cretaceous granites play a considerable role in mineralization. However, there are fewer occurrences of ore bodies in contact zones of the Cretaceous granites and granite porphyries (K_1H and K_2K). For example, the zoning statistical results of the evidence weight method show that the contrast C value of the zoning interval of Cretaceous granite (K_1H) 0–300 m from the granite roof is not high (Table 1), while the most favorable space for mineralization reflected by the contrast is the zoning interval of 400–550 m from the granite roof. Therefore, the evidence and neural network methods mainly consider the spatial position relationship between the geological factors and the ore bodies. The prospecting target areas delineated by these two methods reflect the spatial coupling of various geological factors.

5.2. Comparison of Forecasting Methods

The area under the ROC curve (AUC) is a considerable evaluation index, and its probability value is between 0.00 and 1.00. It can intuitively and quantitatively evaluate the quality of the prediction model by combining it with the shape of the ROC curve. An AUC = 0.50 indicates that the classifier is similar to a random guess with no predictive value, while AUC = 1.0 indicates a perfect classifier [36].

According to the prediction results, the weight of evidence and artificial neural network methods were gathered, the posterior probability P_{pos} and pseudo-probability values are the test variables, and whether it belongs to the In-rich cell determines its use as the state variable to draw the ROC curve (Figure 8). The results show that the AUC values of the evidence weight and artificial neural network methods are 0.805 and 0.730, respectively. However, the weight of evidence method's prediction capability is better than that of the artificial neural network method. It is speculated that the weight of evidence method carried out detailed zoning for each geological factor's buffer zone and excluded the influence of low-contrast zoning on the metallogenic prediction. The prospecting space predicted by the two methods was basically the same (Figures 6 and 7).

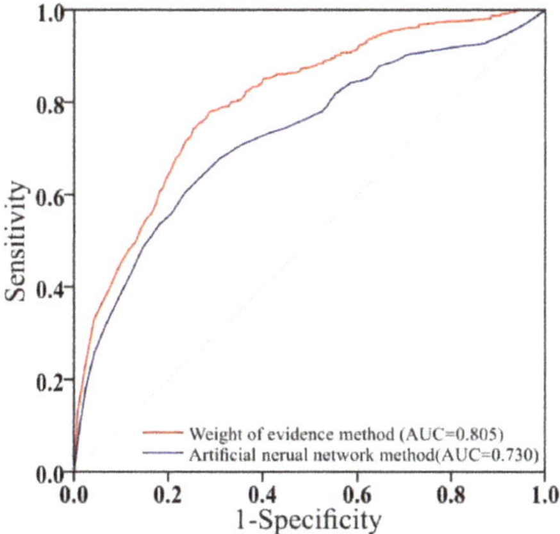

Figure 8. ROC curve graph about prediction results.

5.3. Prospecting Target Area

The geological variables such as strata, structures, and granites in the evaluation range were included by using the weight of evidence and artificial neural network methods. The units with prospecting potential are mainly distributed in the southwest of the mining area through calculation. Therefore, this area was delineated as the prospecting target area of In-rich ore bodies (Figure 9).

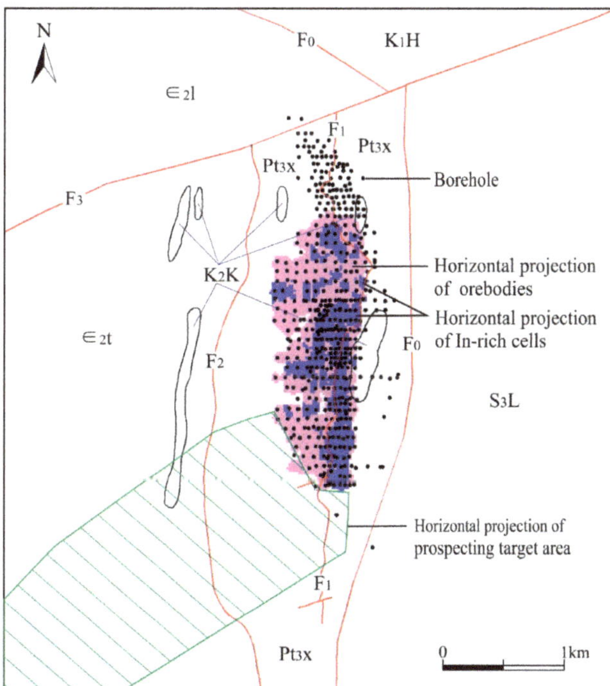

Figure 9. Metallogenic prediction map of indium-rich ore bodies. $\in_2 l$—Middle Cambrian Longha formation; $\in_2 t$—Middle Cambrian Tianpeng formation; $Pt_3 x$—Neoproterozoic Xinzhai formation; $K_1 H$—Cretaceous granite; $K_2 K$—Cretaceous granite porphyry; $S_3 L$—Silurian granite.

6. Conclusions

The results from evaluating ore-controlling geological factors by the weight of evidence and artificial neural network methods show that the Neoproterozoic Xinzhai Formation (Pt_3x), fault (F_1), and Silurian granites (S_3L) have considerable control effects on the occurrence of In-rich ore bodies. The results of the ROC curve evaluation show that the prediction space delineated by the weight of evidence (AUC = 0.805) and artificial neural network methods (AUC = 0.730) has an indicative value for prospecting. The two target delineation methods reflect the spatial coupling of ore-controlling factors. The prospecting target areas obtained for In-rich ore bodies will be informative for future prospecting work in the Dulong mining area.

Author Contributions: Methodology, F.J., Z.S., G.Y., X.S. and C.Y.; formal analysis, F.J., Z.S., Y.Y., G.Y., X.S. and C.Y.; investigation, H.N., J.Y., S.L., L.L. and F.S.; resources, H.N., J.Y., S.L., L.L. and F.S.; data curation, F.J., Z.S., H.N., Y.Y. and G.Y.; writing—original draft preparation, F.J.; funding acquisition, Y.Y. All authors have read and agreed to the published version of the manuscript.

Funding: The research was funded by the National Natural Science Foundation of China (No. 42072094, No. 42162012) and Development Research Center of China Geological Survey (Grant No: DD20190166-2).

Data Availability Statement: Not applicable.

Acknowledgments: We would like to thank Jiaxi Zhou of Yunnan University for his valuable suggestions.

Conflicts of Interest: The authors declare no conflict of interest.

References

1. Werner, T.T.; Mudd, G.M.; Jowitt, S.M. The world's by-product and critical metal resources part III: A global assessment of indium. *Ore Geol. Rev.* **2017**, *86*, 939–956. [CrossRef]
2. Li, X.F.; Xu, J.; Zhu, Y.T.; Lv, Y.H. Critical minerals of indium: Major ore types and scientific issues. *Acta Petrol. Sin.* **2019**, *35*, 3292–3302.
3. Lerouge, C.; Gloaguen, E.; Wille, G.; Bailly, L. Distribution of In and other rare metals in cassiterite and associated minerals in Sn ± W ore deposits of the western Variscan Belt. *Eur. J. Mineral.* **2017**, *29*, 739–753. [CrossRef]
4. Li, T.J.; Zhou, L.; Zhao, Y.K.; Zhu, G.S.; Li, H.L. History and present situation of mineral resources exploitation and utilization in dulong mine. *Acta Mineral. Sin.* **2016**, *36*, 463–470. [CrossRef]
5. Carranza, E.J.M. Controls on mineral deposit occurrence inferred from analysis of their spatial pattern and spatial association with geological features. *Ore Geol. Rev.* **2009**, *35*, 383–400. [CrossRef]
6. Chen, J.P.; Shi, R.; Chen, Z.P.; Wang, L.M.; Sun, Y. 3D positional and quantitative prediction of the Xiaoqinling gold ore belt in Tongguan, Shaanxi, China. *Acta Geol. Sin.* **2012**, *86*, 653–660. (In English) [CrossRef]
7. Payne, C.E.; Cunningham, F.; Peters, K.J.; Nielsen, S.; Puccioni, E.; Wildman, C.; Partington, G.A. From 2D to 3D: Prospectivity modelling in the Taupo Volcanic Zone, New Zealand. *Ore Geol. Rev.* **2014**, *71*, 558–577. [CrossRef]
8. Li, X.H.; Yuan, F.; Zhang, M.M.; Jia, C.; Jowitt, S.M.; Ord, A.; Zheng, T.K.; Hu, X.Y.; Li, Y. Three-dimensional mineral prospectivity modeling for targeting of concealed mineralization within the Zhonggu iron orefield, Ningwu Basin, China. *Ore Geol. Rev.* **2015**, *71*, 633–654. [CrossRef]
9. Nielsen, S.H.H.; Cunningham, F.; Hay, R.; Partington, G.; Stokes, M. 3D prospectivity modelling of orogenic gold in the Marymia Inlier, Western Australia. *Ore Geol. Rev.* **2015**, *71*, 578–591. [CrossRef]
10. Wang, G.W.; Li, R.X.; Carranza, E.J.M.; Zhang, S.T.; Yan, C.H.; Zhu, Y.Y.; Qu, J.N.; Hong, D.M.; Song, Y.W.; Han, J.W.; et al. 3D geological modeling for prediction of subsurface Mo targets in the Luanchuan district, China. *Ore Geol. Rev.* **2015**, *71*, 592–610. [CrossRef]
11. Xiao, K.Y.; Li, N.; Porwal, A.; Holden, E.J.; Bagas, L.; Lu, Y.J. Gis-based 3D prospectivity mapping: A case study of Jiama copper-polymetallic deposit in Tibet, China. *Ore Geol. Rev.* **2015**, *71*, 611–632. [CrossRef]
12. Mao, X.C.; Zhang, B.; Deng, H.; Zou, Y.H.; Chen, J. Three-dimensional morphological analysis method for geologic bodies and its parallel implementation. *Comput. Geosci.* **2016**, *96*, 11–22. [CrossRef]
13. Hu, X.Y.; Yuan, F.; Li, X.H.; Jowitt, S.M.; Jia, C.; Zhang, M.M.; Zhou, T.F. 3D characteristic analysis-based targeting of concealed Kiruna-type Fe oxide-apatite mineralization within the Yangzhuang deposit of the Zhonggu orefield, southern Ningwu volcanic basin, middle-lower Yangtze River metallogenic Belt, China. *Ore Geol. Rev.* **2018**, *92*, 240–256. [CrossRef]
14. Li, R.X.; Wang, G.W.; Carranza, E.J.M. GeoCube: A 3D mineral resources quantitative prediction and assessment system. *Comput. Geosci.* **2016**, *89*, 161–173. [CrossRef]

15. Yang, F.; Wang, G.W.; Santosh, M.; Li, R.X.; Tang, L.; Cao, H.W.; Guo, N.N.; Liu, C. Delineation of potential exploration targets based on 3d geological modeling: A case study from the laoangou pb-zn-ag polymetallic ore deposit, china. *Ore Geol. Rev.* **2017**, *89*, 228–252. [CrossRef]

16. Mao, X.C.; Ren, J.; Liu, Z.K.; Chen, J.; Tang, L.; Deng, H.; Bayless, R.C.; Yang, B.; Wang, M.J.; Liu, C.M. Three-dimensional prospectivity modeling of the Jiaojia-type gold deposit, Jiaodong peninsula, eastern China: A case study of the Dayingezhuang deposit. *J. Geochem. Explor.* **2019**, *203*, 27–44. [CrossRef]

17. Zhang, M.M.; Zhou, G.Y.; Shen, L.; Zhao, W.G.; Liao, B.S.; Yuan, F.; Li, X.H.; Hu, X.Y.; Wang, C.B. Comparison of 3D prospectivity modeling methods for Fe-Cu skarn deposits: A case study of the Zhuchong Fe-Cu deposit in the Yueshan orefield (Anhui), eastern China. *Ore Geol. Rev.* **2019**, *114*, 103126. [CrossRef]

18. Mohammadpour, M.; Bahroudi, A.; Abedi, M. Three dimensional mineral prospectivity modeling by evidential belief functions, a case study from kahang porphyry cu deposit. *J. Afr. Earth Sci.* **2021**, *174*, 104098. [CrossRef]

19. Bonham-Carter, G.F.; Agterberg, F.P.; Wright, D.F. Weights of evidence modeling: A new approach to mapping mineral potential. *Stat. Appl. Earth Sci.* **1990**, *89*, 171–183.

20. Agterberg, F.P.; Bonham-Carter, G.F.; Cheng, Q.M.; Wright, D.F. Weights of evidence modeling and weighted logistic regression for mineral potential mapping. *Comput. Geol.* **1993**, *25*, 13–32. [CrossRef]

21. Brown, W.M.; Gedeon, T.D.; Groves, D.I.; Barnes, R.G. Artificial neural networks: A new method for mineral prospectivity mapping. *J. Aust. Earth Sci.* **2000**, *47*, 757–770. [CrossRef]

22. Porwal, A.; Gonzalez-Alvarez, I.; Markwitz, V.; McCuaig, T.C.; Mamuse, A. Weights-of-evidence and logistic regression modeling of magmatic nickel sulfide prospectivity in the Yilgarn Craton, Western Australia. *Ore Geol. Rev.* **2010**, *38*, 184–196. [CrossRef]

23. Yang, G.S.; Wang, K.; Yan, Y.F.; Jia, F.J.; Li, P.Y.; Mao, Z.B.; Zhou, Y. Genesis of the ore-bearing skarns in Laojunshan Sn-W-Zn-In polymetallic ore district, southeastern Yunnan Province, China. *Acta Petrol. Sin.* **2019**, *35*, 3333–3354. [CrossRef]

24. Peng, T.P.; Fan, W.M.; Zhao, G.C.; Peng, B.X.; Xia, X.P.; Mao, Y.S. Petrogenesis of the Early Paleozoic strongly peraluminous granites in the western South China Block and its tectonic implications. *J. Asinan Earth Sci.* **2015**, *98*, 399–420. [CrossRef]

25. Feng, J.R.; Mao, J.W.; Pei, R.F. Ages and geochemistry of Laojunshan granites in southeastern Yunnan, China: Implications for W-Sn polymetallic ore deposits. *Mineral. Petrol.* **2013**, *107*, 573–589. [CrossRef]

26. Yan, D.P.; Zhou, M.F.; Wang, Y.; Wang, C.L.; Zhao, T.P. Structural styles and chronological evidences from Dulong-Song Chay tectonic dome: Earlier spreading of south china sea basin due to late mesozoic to early cenozoic extension of south china block. *Earth Sci.* **2005**, *30*, 402–412.

27. Yan, D.P.; Zhou, M.F.; Wang, C.Y.; Xia, B. Structural and geochronological constraints on the tectonic evolution of the Dulong-Song Chay tectonic dome in Yunnan province, SW China. *J. Asian Earth Sci.* **2006**, *28*, 332–353. [CrossRef]

28. Murakami, H.; Ishihara, S. Trace elements of Indium-bearing sphalerite from tin-polymetallic deposits in Bolivia, China and Japan: A femto-second LA-ICPMS study. *Ore Geol. Rev.* **2013**, *53*, 223–243. [CrossRef]

29. Liu, S.; Xue, L.; Qie, R.; Zhang, X.; Meng, Q. An application of GIS-based weights of evidence for gold prospecting in the northwest of Heilongjiang Province. *J. Jilin Univ. (Earth Sci. Ed.)* **2007**, *37*, 889–894. [CrossRef]

30. Schaeben, H. A mathematical view of weights-of-evidence, conditional independence, and logistic regression in terms of Markov random fields. *Math. Geosci.* **2004**, *46*, 691–709. [CrossRef]

31. Lindsay, M.D.; Betts, P.G.; Ailleres, L. Data fusion and porphyry copper prospectivity models, southeastern Arizona. *Ore Geol. Rev.* **2014**, *61*, 120–140. [CrossRef]

32. Shao, Y.J.; He, H.; Zhang, Y.Z.; Liang, E.Y.; Ding, Z.W.; Chen, X.L.; Liu, Z.F. Metallogenic prediction of Xiangxi gold deposit based on BP neural networks. *J. Cent. South Univ. Sci. Technol.* **2007**, *12*, 38-06. [CrossRef]

33. Li, S.M.; Yao, S.Z.; Zhou, Z.G. Research on quantitative prediction of mineral resources. *Contrib. Geol. Miner. Resour. Res.* **2007**, *3*, 22-01. [CrossRef]

34. Bonham-Carter, G.F.; Agterberg, F.P.; Wright, D.F. Integration of geological datasets for gold exploration in Nova Scotia. *Digit. Geol. Geogr. Inf. Syst.* **1988**, *10*, 15–23. [CrossRef]

35. Wang, X.J.; Liu, Y.P.; Miao, Y.L.; Bao, T.; Ye, L.; Zhang, Q. In-situ LA-MC-ICP-MS cassiterite U-Pb dating of Dulong Sn-Zn polymetallic deposit and its significance. *Acta Petrol. Sin.* **2014**, *30*, 867–876.

36. Lee, S.; Dan, N.T. Probabilistic landslide susceptibility mapping in the Lai Chau Province of Vietnam: Focus on the relationship between tectonic fractures and landslides. *Environ. Geol.* **2005**, *48*, 778–787. [CrossRef]

 *minerals*

Article

The Origin of Carbonate Components in Carbonate Hosted Pb-Zn Deposit in the Sichuan-Yunnan-Guizhou Pb-Zn Metallogenic Province and Southwest China: Take Lekai Pb-Zn Deposit as an Example

Zhiwei He [1], Bo Li [2,*], Xinfu Wang [2], Xianguo Xiao [3], Xin Wan [2] and Qingxi Wei [4]

1   College of Earth Science, Chengdu University of Technology, Chengdu 610000, China
2   Faculty of Land and Resource Engineering, Kunming University of Science and Technology, Kunming 650093, China
3   Non-Ferrous Metals and Nuclear Industry Geological Exploration Bureau of Guizhou, Guiyang 550000, China
4   Kunming Prospecting Design Institute of China Nonferrous Metals Industry Co., Ltd., Kunming 650051, China
*   Correspondence: libo8105@kust.edu.cn

Citation: He, Z.; Li, B.; Wang, X.; Xiao, X.; Wan, X.; Wei, Q. The Origin of Carbonate Components in Carbonate Hosted Pb-Zn Deposit in the Sichuan-Yunnan-Guizhou Pb-Zn Metallogenic Province and Southwest China: Take Lekai Pb-Zn Deposit as an Example. *Minerals* **2022**, *12*, 1615. https://doi.org/10.3390/min12121615

Academic Editor: Maria Boni

Received: 14 November 2022
Accepted: 13 December 2022
Published: 15 December 2022

Publisher's Note: MDPI stays neutral with regard to jurisdictional claims in published maps and institutional affiliations.

Abstract: The Lekai lead–zinc (Pb-Zn) deposit is located in the northwest of the Sichuan–Yunnan–Guizhou (SYG) Pb-Zn metallogenic province, southwest China. Even now, the source of the metallogenic fluid of Pb-Zn deposits in the SYG Pb-Zn metallogenic province has not been recognized. Based on traditional lithography, rare earth elements (REEs), and carbon–oxygen (C–O) isotopes, this work uses the magnesium (Mg) isotopes of hydrothermal carbonate to discuss the fluid source of the Lekai Pb-Zn deposit and discusses the fractionation mechaism of Mg isotopes during Pb-Zn mineralization. The REE distribution patterns of hydrothermal calcite/dolomite are similar to that of Devonian sedimentary carbonate rocks, which are all present steep right-dip type, indicating that sedimentary carbonate rocks may be serve as the main source units of ore-forming fluids. The C–O isotopic results of hydrothermal dolomite/calcite and the $\delta^{13}C_{PDB}$–$\delta^{18}O_{SMOW}$ diagram show that dolomite formation is closely related to the dissolution of marine carbonate rocks, and calcite may be affected to some extent by basement fluid. The Mg isotopic composition of dolomite/calcite ranges from −3.853‰ to −1.358‰, which is obviously lighter than that of chondrites, mantle, or seawater and close to that of sedimentary carbonate rock. It shows that the source of the Mg element in metallogenic fluid of Lekai Pb-Zn deposit may be sedimentary carbonate rock rather than mantle, chondrites, or seawater. In addition, the mineral phase controls the Mg isotope fractionation of dolomite/calcite in the Lekai Pb-Zn deposit. Based on the geological, mineralogical, and hydrothermal calcite/dolomite REE, C–O isotope, and Mg isotope values, this work holds that the mineralization of the Lekai Pb-Zn deposit is mainly caused by basin fluids, influenced by the basement fluids; the participation of basement fluids affects the scale and grade of the deposit.

Keywords: Sichuan–Yunnan–Guizhou Pb-Zn metallogenic province; Lekai Pb-Zn deposit; calcite/dolomite REE; C–O–Mg isotope; metallogenic fluid source

1. Introduction

Located in the southwest margin of Yangtze Block, the Sichuan–Yunnan–Guizhou (SYG) Pb-Zn metallogenic province is one of the main production bases of Pb-Zn–Ag–Ge and other metal elements in China and also an important part of the giant south China Mesozoic low-temperature metallogenic domain [1]. The Pb-Zn deposits in this region are mainly hosted in Sinian to Permian carbonates, which are obviously controlled by the fracture and generally present the characteristics of one orebody occurring in multiple strata. The northeast Yunnan metallogenic belt in the metallogenic province is mainly controlled by the northeast (NE)–trending tectonic belt, producing the super large Pb-Zn deposits (i.e., Huize and Lemachang) and large Pb-Zn deposits (i.e., Maozu and Maoping) [2,3]. Sinian, Cambrian, Devonian, and Carboniferous are the main ore host sequences for the Pb-Zn orebodies in northeast Yunnan. The metallogenic belt in northwestern Guizhou is mainly controlled by the northwest (NW)–trending tectonic belt, producing one super large Pb-Zn deposit (i.e., Zhugongtang) and a large number of small to medium-sized Pb-Zn deposits. Carboniferous and Permian are the main ore host sequences for the Pb-Zn orebodies in northwestern Guizhou. Compared with the northeast Yunnan metallogenic belt, the ore host sequences of the northwestern Guizhou metallogenic belt are relatively newer, the deposit scale is smaller, the average grade is lower, and the types of Ag–Ge–Cd–In–Ga and other metal elements are fewer. The Lekai Pb-Zn deposit is located in the northwest Guizhou metallogenic belt, but it has a similar metallogenic geological setting to the northeast Yunnan metallogenic belt (Figure 1). This study of the Lekai Pb-Zn deposit is conducive to the comparative study of metallogenic materials and fluid sources of the northwest Guizhou and northeast Yunnan metallogenic belts.

Recently, a variety of geochemical methods have been widely used to determine the properties and sources of ore-forming fluid in the SYG Pb-Zn metallogenic province, including the petrography of fluid inclusions in quartz, calcite, and sphalerite [2,4]; H–O–Sr isotopic composition of calcite [5,6]; C–O isotope and rare earth element (REE) analysis of calcite; and S–Pb isotope analysis of sulfides [1,7–10]. Concurrently, accompanied by the development of high–precision testing instruments, such as NanoSIMS and laser ablation inductively coupled plasma mass spectroscopy, the application of microscale in situ testing technology tends to be mature, such as the measurement of sulfide in situ trace elements and S–Pb isotopes, which greatly improves the testing accuracy [5,11,12]. At present, the nontraditional stable isotope geochemistry of Fe, Cu, Zn, Cd, and Mg has made great progress and shown great application potential in tracing the source of ore-forming materials [13–19]. In particular, the Zn-Cd isotopes of sphalerite provide more direct evidence for the source of metal elements in Pb-Zn deposits [16,20,21]. However, there is no consensus on the contributions of sedimentary rocks, basement rocks, or magmatic rocks to metallogenic materials.

The information contained in hydrothermal carbonate rocks can well reflect the source and physicochemical characteristics of metallogenic fluids, and this information has been widely used in research on ore deposits. Mg is the core element of hydrothermal carbonate rocks, and isotopic testing technology (multicollector-inductively coupled plasma mass spectroscopy [MC–ICP–MS]) has led to the establishment of the isotope geochemical system [22–26], which has gradually become a new means to study carbonate rocks. Mg isotope has unique advantages in revealing the process of epigenetic geological processes, dolomite genesis, crust–mantle material circulation, and the exploration of fluid properties and sources of low-temperature deposits [27–29].

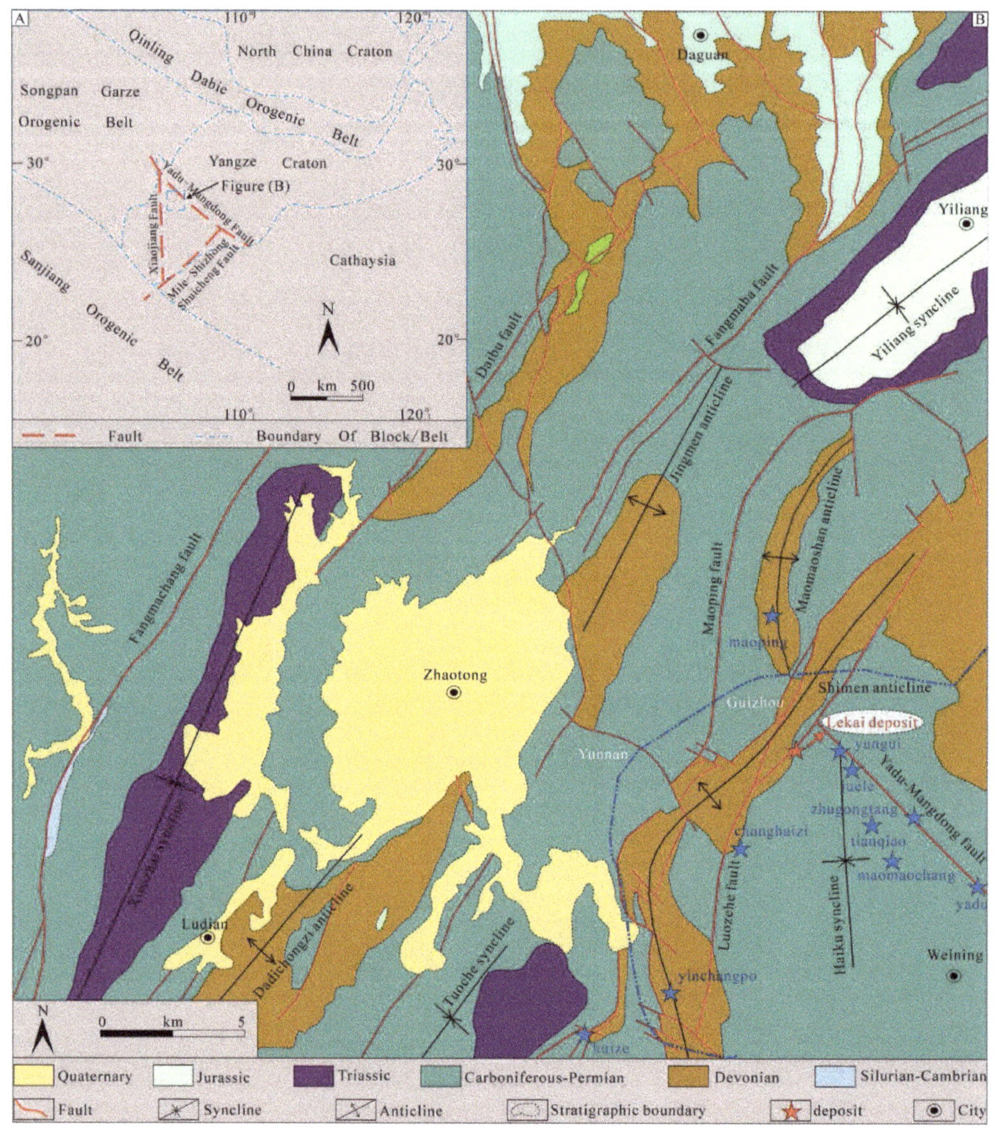

Figure 1. (**A**) Regional geological framework of south China [7]; (**B**) Regional geological map of Lekai Pb-Zn deposit in SYG Pb-Zn metallogenic province [30].

The Mg isotopic composition of main reservoirs and the mechanism of Mg isotopic behavior in geological processes provide a theoretical basis for the tracing of metallogenic fluids. At present, there has been a certain degree of data accumulation in the study of Mg isotopes [13–15,23,25,29,31–39]. These data have revealed that the Earth's materials, mantle peridotites, and basalts were relatively heavy Mg isotopic compositions (peridotites had δ^{26}Mg values ranging from $-0.48‰$ to $0.06‰$, average: $-0.23 \pm 0.19‰$; basalts had δ^{26}Mg values ranging from $-0.09‰$ to $0.46‰$, average: $-0.24 \pm 0.12‰$), which are relatively homogeneous and similar to those of chondrites (δ^{26}Mg values ranging from $-0.35‰$ to $-0.20‰$, average: $-0.28 \pm 0.06‰$) [23,34]. Compared with mantle rocks, the Mg isotopic composition of seawater is relatively light but also relatively homogeneous (δ^{26}Mg values range from $-0.87‰$ to $-0.75‰$, average: $-0.83 \pm 0.07‰$) [23,33,34]. The Mg isotopic composition of carbonate rocks is the lightest, and its variation is the largest (δ^{26}Mg $= -4.84‰$ to $-1.00‰$, average: $-3.09 \pm 2.66‰$) [14,23,36–38]. The δ^{26}Mg values of dolomite and limestone range from $-2.29‰$ to $-1.09‰$ and $-4.47‰$ to $-2.43‰$, respectively [29,31–39]. The δ^{26}Mg values of sedimentary rocks (excluding carbonate rocks) mainly range from $-0.94‰$ to $0.92‰$, with an average of $-0.06 \pm 0.60‰$ [25]. Therefore, there are significant differences in the Mg isotopic composition and distribution range among each of Earth's reservoirs. In particular, the greatest differences in Mg isotopic composition are between mantle and sedimentary rocks. This is the important basis of Mg isotopes in tracing the source of metallogenic materials and restricting mineralization.

This study selects the hydrothermal carbonate minerals (i.e., dolomite and calcite) of the Lekai Pb-Zn deposit as the main objects, which are based on systematic petrography and geology. Detailed REE and C–O isotopic compositions of dolomite and calcite were studied. The Mg isotopic data of hydrothermal carbonate rocks are experimentally measured, and the REE, C, O, and Mg isotopic compositions of hydrothermal carbonate rocks in the deposit are comprehensively analyzed using these data. The REE and C, O, and Mg isotopic compositions of the hydrothermal carbonate rocks in the Lekai Pb-Zn deposit are analyzed. The sources of metallogenic material and fractionation factors of Mg isotopes in the mineralization process of the Lekai Pb-Zn deposit are identified.

2. Geological Background and Deposit Geology

The Yangtze Block is mainly composed of basement metamorphic rocks, marine/continental sedimentary rocks, and igneous rocks. The basement metamorphic rocks are a set of dioritic, granitic mixed gneiss, migmatitic, and other deep metamorphic rocks of the Paleoproterozoic Kangding group and a set of shallow to medium metamorphic rocks of the Mesoproterozoic Kunyang/Huili Formation. The caprock sequence is mainly composed of Sinian to Permian marine facies and Mesozoic to Cenozoic continental sedimentary rocks. The igneous rocks are mainly dominated by the Emeishan basalt and homologous diabase in the Late Permian [12,40]. The tectonic deformation of the Yangtze Block is dominated by deep faults and folds. It has experienced the Hercynian, Indosinian, and Yanshanian geological evolution stages and was affected by the Himalayan orogeny events. These tectonic events mainly controlled the sedimentation, magmatism, and mineralization in this region [9,38,41]. The Lekai Pb-Zn deposit is mainly controlled by a series of faults and folds formed by the late Indosinian tectonic event.

The Lekai Pb-Zn deposit is the southern extension of the Huize–Yiniang–Niujie oblique strike-slip fault–fold belt in the northeast Yunnan metallogenic belt and controlled by the NE-trending Luozehe fault–fold belt (Figure 1). The folds mainly developed in the NE-trending Shimen anticline and a series of secondary folds. The broken parts of the secondary folds on the SE wing of Shimen anticline are favorable for mineralization. The faults are mainly presented in the NE– and NW –trending, with "ξ–type" (xi-type) and "λ–type" (lambda–type) structural styles. The NE–trending faults are closely related to mineralization and control the distribution of orebodies. The NW– trending faults are smaller in scale and are post–mineralization faults (Figure 2). The Pb-Zn orebodies present layered, stratoid, and lenticular shapes and occur in the folds and faults intersections within the coarse-grained and altered dolomite in the Devonian Wangchengpo formation (D_3w). There are four Pb-Zn orebodies (mineralized bodies) have been identified, which all show the obvious characteristics of slow widening, fast narrowing, expansion, and contraction in plane and section views (Figures 3 and 4A). The Lekai deposit mainly experienced the following two metallogenic formation periods: a hydrothermal metallogenic period and a supergene oxidation period. Ore minerals (i.e., sphalerite, galena, and pyrite and gangue minerals (i.e., dolomite and calcite) are mainly developed in the hydrothermal mineralization period. Ore structures consist of brecciated (Figure 4B), veined (Figure 4C,D), and disseminated (Figure 4D–F) types, and textures consist of hypidiomorphic–idiomorphic grain (Figure 5A), metasomatic (Figure 5B,C), codissolved (Figure 5E), interstitial (Figure 5D,F), and crushed (Figure 5F) morphologies. The hydrothermal mineralization period can be divided into the following three stages: (i) pyrite + dolomite + sphalerite; (ii) galena + pyrite + sphalerite + calcite; (iii) galena + pyrite. Cerusite, smithsonite, and limonite are the main minerals in the supergene oxidation period (Figure 6).

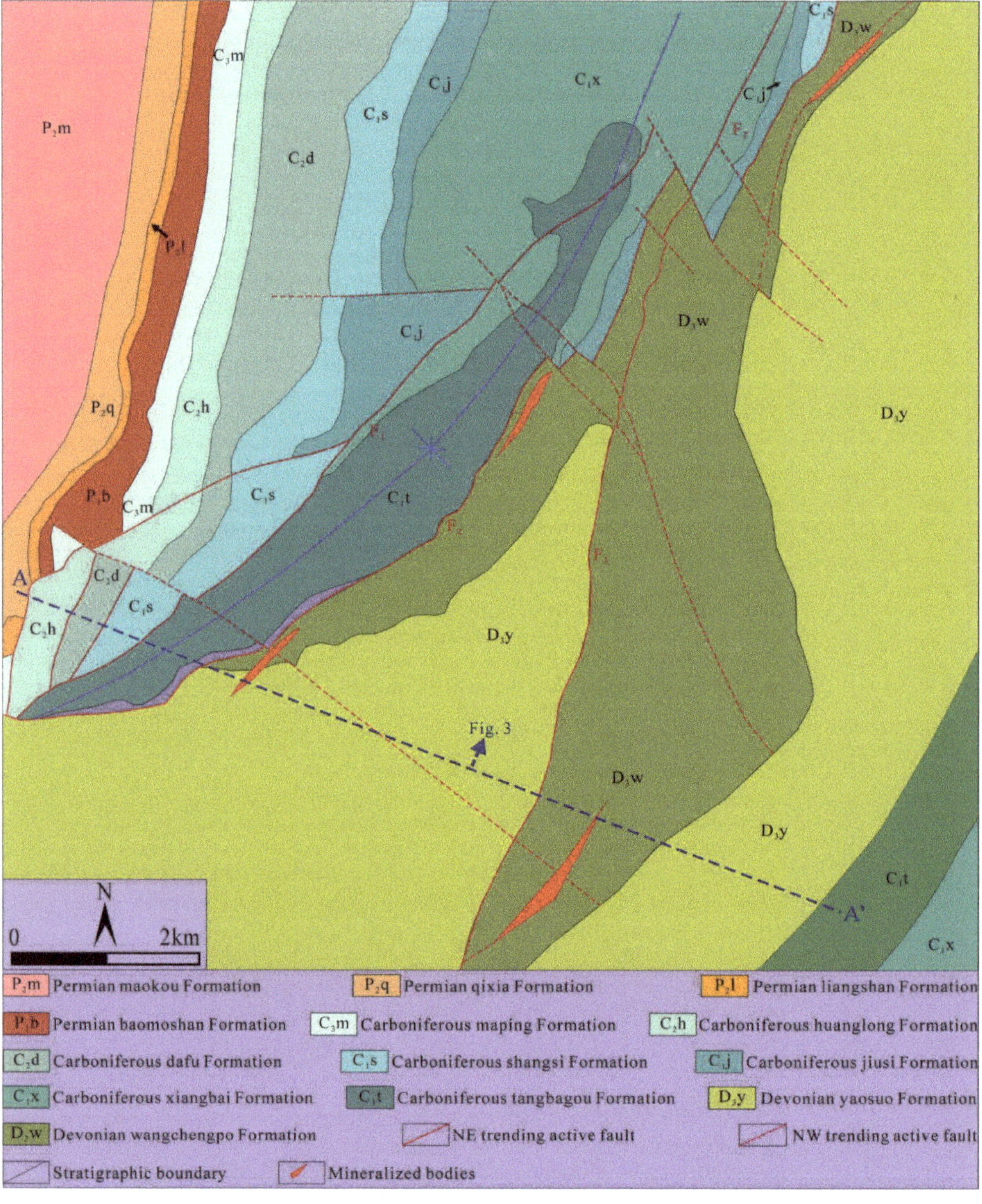

Figure 2. Geological map of Lekai Pb-Zn deposit in SYG Pb-Zn metallogenic province [42].

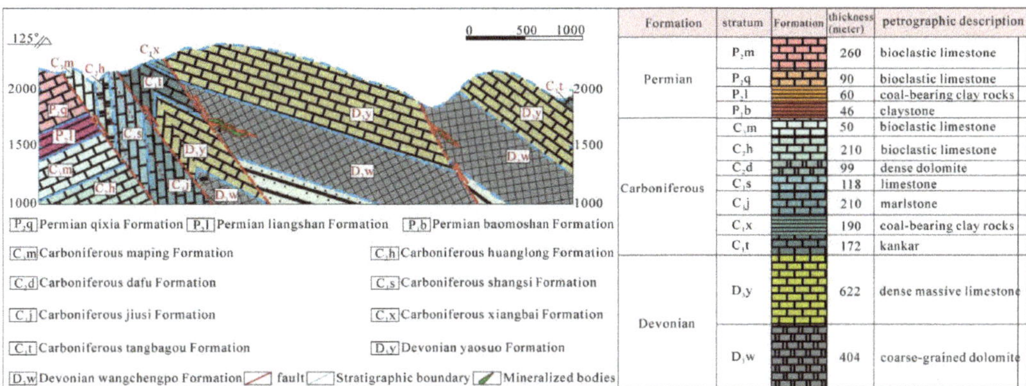

Formation	stratum	Formation	thickness (meter)	petrographic description
Permian	P₂m		260	bioclastic limestone
	P₂q		90	bioclastic limestone
	P₁l		60	coal-bearing clay rocks
	P₁b		46	claystone
Carboniferous	C₃m		50	bioclastic limestone
	C₂h		210	bioclastic limestone
	C₂d		99	dense dolomite
	C₁s		118	limestone
	C₁j		210	marlstone
	C₁x		190	coal-bearing clay rocks
	C₁t		172	kankar
Devonian	D₃y		622	dense massive limestone
	D₁w		404	coarse-grained dolomite

Figure 3. Cross-section map of Lekai Pb-Zn deposit in SYG Pb-Zn metallogenic province [30].

Figure 4. The Pb-Zn orebody and ore structure of Lekai Pb-Zn deposit. (**A**) Stratoid Pb-Zn orebodies occur within interlayer fracture zone; (**B**) Brecciated Pb-Zn ore, Galena occurs in brecciated dolomite as an agglomerate; (**C**) Disseminated Pb-Zn ore, Veined dolomite and pyrite occur in altered dolomite; (**D**) Cloddy Pb-Zn ore, Agglomerate and veined galena occur in dolomite; (**E**) Disseminated Pb-Zn ore, Agglomerate sphalerite and calcite occur in dolomite; (**F**) Cloddy Pb-Zn ore, Sphalerite, galena, and calcite are filled within the dolomite pores. Sp = sphalerite; Py = pyrite; Gn = galena; Cal = calcite; Dol = dolomite.

Figure 5. Mineral assemblages and sequence of Lekai Pb-Zn deposit. (**A**) Sphalerite, galena, and pyrite are hypidiomorphic-idiomorphic grain; Pyrite is surrounded by calcite, sphalerite, and galena form codissolved texture; (**B**) Veined galena metasomatized dolomite; (**C**) Sphalerite, galena, and pyrite are granularly cemented by calcite; (**D**) The later granular pyrite developed in veined calcite and metasomatized earlier formed sphalerite; (**E**) Pyrite, sphalerite, and galena show a codissolved texture; (**F**) A small amount of sphalerite is filled within the fracture zones of crushed pyrite. Sp = sphalerite; Py = pyrite; Gn = galena; Cal = calcite; Dol = dolomite.

	Lekai deposit			
Metallogenic stages	Hydrothermal mineralization Stage			Supergene Stage
Stages	i	ii	iii	
Mineral assemblage	Py+Dol	Sp+Gn+Py+Cal	Gn+Py+Cal	
Pyrite				
Sphalerite				
Galena				
Calcite				
Dolomite				
Limonite				
Cerussite				
Smithsonite				

More ———— Less ————

Figure 6. Mineral paragenesis of the Lekai Pb-Zn deposit [30].

3. Sampling and Methods

The samples were mainly collected from adit LD11 of the Lekai Pb-Zn deposit. We take the method of continuous block in adit for sampling. The top and bottom segment of the Pb-Zn orebody is in contact with the altered dolomite, which is mainly disseminated ore, and the center of the orebody is brecciated and cloddy ore. The representative and fresh hydrothermal carbonate samples were collected based on the field's detailed geological cataloging and observations. Firstly, the samples were crushed to 40–60 mesh. Hydrothermal calcite and dolomite with purities greater than 99% were selected under the binocular microscope, after which the samples were ultrasonically cleaned and repeatedly purified. The pure single-grain samples were pounded to 200 mesh with an agate mortar for REE, C, O, and Mg isotope analyses. REE and C–O isotope analyses were performed at the State

Key Laboratory of Ore Deposit Geochemistry, Institute of Geochemistry, Chinese Academy Sciences, and Mg isotope analyses were undertaken at Aoshi analytical testingCo., Ltd, Guangzhou, China.

The REE contents of the calcite/dolomite samples were measured via inductively coupled plasma mass spectroscopy (ICP-MS; ELAN DRC-e four-stage bar inductively coupled plasma mass spectrometer; PerkinElmer, Woodbridge, ON, Canada); the analysis uncertainty was less than 5%. Bulk C–O isotope analyses were conducted using a Finnigan MAT-253 mass spectrometer (Thermo Fisher Scientific, Waltham, MA, USA). Calcite/dolomite samples were reacted with 100% H_3PO_4 to produce CO_2. The analytical precision rates calculated from replicate analyses of unknown samples were better than 0.2‰ (2σ) and 1‰ (2σ) for $\delta^{13}C$ and $\delta^{18}O$, respectively. The $\delta^{13}C$ and $\delta^{18}O$ values were reported relative to the Vienna Pee Dee Belemnite (V-PDB) standard and Standard Mean Ocean Water (SMOW), respectively.

Hydrothermal calcite/dolomite samples were prepared via alkali fusion and digestion using acids, followed by separation of Mg via ion exchange AG 50W-8X (Bio-Rad, Hercules, CA, USA) and measurement via MC-ICP-MS (NEPTUNE PLUS) (Thermo Fisher Scientific, Shanghai, China) for Mg isotopes (ratio). Si was added to all analytical solutions (purified Mg fractions, standards, and blanks) as a spike internal standard to correct mass bias and ensure the best precision of Mg isotope values. Delta values for Mg were calculated against IRMM-3704 CRM.

4. Results

4.1. REE Contents

Hydrothermal calcite/dolomite is characterized by an increase in total REE (excluding Y, ΣREE) concentrations from 2.45 to 29.11 ppm, with an average of 6.05 ppm (n = 13) (Table 1, Figure 7A,B). The concentrations of light (LREEs) and heavy REEs (HREEs) range from 2.02 to 25.02 ppm (average: 5.18 ppm, n = 13) and 0.31 to 4.09 ppm (average: 0.87 ppm, n = 13), respectively. The LREE/HREE ratios of calcite/dolomite samples were between 4.33 and 8.06 (average: 5.93, n = 13); the $(La/Yb)_N$ values range from 6.17 to 13.32 and showed LREE enrichment patterns. The differences between LREE and HREE concentrations were obvious, and the chondrite-normalized REE patterns were consistently steep right-sloping types (Figure 7A). The Eu and Ce show moderate negative anomalies, with δEu and δCe values ranging from 0.56 to 0.69 and 0.63 to 0.87, respectively.

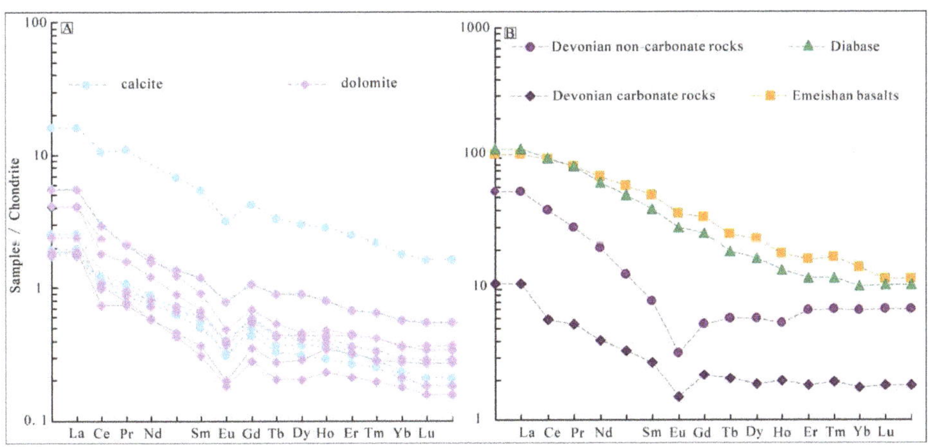

Figure 7. (**A**) The rare earth elements distribution of calcite and dolomite in Lekai Pb-Zn deposits; (**B**) The rare earth elements distribution of carbonate rocks and non-carbonate rocks in Devonian, Emeishan basalts, and Diabase in SYG Pb-Zn metallogenic province [43].

Table 1. Analysis results of REE contents of hydrothermal calcite/dolomite samples from the Lekai Pb-Zn deposit.

Sample Minerals	LK635	LK600	LD01R24	LD01R3	LD01R2	LD01R1	Average	LK122–3	LK122	LD03R8	LD03R6	LD01R3	LD01R2	LD01R1	Average
			Calcite (ppm)							Dolomite (ppm)					
La	0.72	5.89	0.93	0.64	0.67	2.04	1.82	1.51	1.54	0.89	0.69	0.86	0.93	1.47	1.13
Ce	1.09	10.10	1.17	0.95	1.03	2.85	2.87	2.24	1.73	0.98	0.71	0.97	1.01	2.01	1.38
Pr	0.13	1.49	0.15	0.12	0.13	0.29	0.39	0.29	0.22	0.11	0.10	0.14	0.14	0.28	0.18
Nd	0.56	6.00	0.62	0.52	0.58	1.11	1.57	1.20	0.86	0.42	0.42	0.56	0.55	1.11	0.73
Sm	0.12	1.26	0.13	0.14	0.15	0.28	0.35	0.21	0.16	0.07	0.09	0.12	0.11	0.21	0.14
Eu	0.03	0.28	0.03	0.04	0.03	0.07	0.08	0.04	0.03	0.02	0.02	0.03	0.02	0.05	0.03
Gd	0.13	1.30	0.15	0.18	0.18	0.33	0.38	0.21	0.17	0.09	0.11	0.14	0.13	0.23	0.15
Tb	0.02	0.19	0.02	0.03	0.03	0.05	0.06	0.03	0.03	0.01	0.02	0.02	0.02	0.04	0.02
Dy	0.12	1.16	0.14	0.16	0.17	0.34	0.35	0.18	0.17	0.08	0.11	0.14	0.14	0.23	0.15
Ho	0.03	0.24	0.03	0.03	0.04	0.07	0.07	0.04	0.04	0.02	0.03	0.03	0.03	0.05	0.03
Er	0.07	0.62	0.08	0.08	0.09	0.17	0.19	0.11	0.11	0.05	0.08	0.09	0.09	0.15	0.1
Tm	0.01	0.08	0.01	0.01	0.01	0.02	0.02	0.02	0.02	0.01	0.01	0.01	0.01	0.02	0.01
Yb	0.05	0.44	0.06	0.05	0.07	0.14	0.14	0.09	0.09	0.05	0.07	0.08	0.07	0.13	0.08
Lu	0.01	0.06	0.01	0.01	0.01	0.02	0.02	0.01	0.01	0.01	0.01	0.01	0.01	0.02	0.01
Y	1.51	9.31	2.02	1.90	1.95	3.36	3.34	2.41	2.25	1.26	1.96	2.01	1.97	2.66	2.07
ΣREE	3.08	29.11	3.52	2.95	3.19	7.78	8.27	6.19	5.16	2.79	2.45	3.18	3.26	5.99	4.15
LREE	2.65	25.02	3.02	2.41	2.59	6.64	7.06	5.50	4.53	2.48	2.02	2.66	2.75	5.12	3.58
HREE	0.43	4.09	0.50	0.55	0.60	1.14	1.22	0.69	0.63	0.31	0.44	0.52	0.51	0.87	0.57
LREE/HREE	6.15	6.12	6.05	4.40	4.33	5.80	5.48	7.93	7.22	8.06	4.64	5.12	5.40	5.87	6.32
La_N/Yb_N	9.49	9.07	10.92	8.12	6.17	9.71	8.91	11.34	11.31	13.32	6.73	7.55	8.56	7.64	9.49
δEu	0.69	0.66	0.60	0.67	0.63	0.69	0.66	0.62	0.62	0.62	0.56	0.61	0.62	0.67	0.62
δCe	0.82	0.80	0.74	0.80	0.82	0.87	0.81	0.79	0.70	0.74	0.63	0.67	0.67	0.74	0.71
Y/Ho	60.4	38.63	66.30	57.58	54.17	49.41	54.42	58.78	59.21	63.00	66.5	64.84	61.56	50.19	60.58
La/Ho	28.64	24.44	30.52	19.30	18.50	30.00	25.23	36.83	40.53	44.35	23.40	27.74	28.91	27.74	32.78
Tb/La	0.03	0.03	0.02	0.04	0.04	0.03	0.03	0.02	0.02	0.01	0.02	0.02	0.02	0.02	0.02
Sm/Nd	0.20	0.21	0.20	0.28	0.26	0.25	0.23	0.18	0.18	0.17	0.20	0.21	0.19	0.19	0.19

4.2. C and O Isotopes

The C and O isotopic compositions of calcite/dolomite samples separated from the sulfide ore are listed in Table 2. The $\delta^{13}C_{PDB}$ and $\delta^{18}O_{SMOW}$ values of calcite ranging from $-6.02‰$ to $2.40‰$ (average: $-3.26‰$, $n = 6$) and $14.72‰$ to $20.14‰$ (average: $16.19‰$, $n = 6$), respectively. The $\delta^{13}C_{PDB}$ and $\delta^{18}O_{SMOW}$ values of dolomite ranging from $-0.33‰$ to $2.29‰$ (average: $0.93‰$, $n = 7$) and $19.06‰$ to $26.16‰$ (average: $22.27‰$, $n = 7$), respectively.

Table 2. C-O isotopic composition of hydrothermal calcite/dolomite in Lekai Pb-Zn deposit.

Sample Number	Mineral	$\delta^{13}C$ (‰V_{PDB})	Std.ev	$\delta^{18}O$ (‰V_{PDB})	Std.ev	$\delta^{18}O$ (‰$_{SMOW}$)
LD01R1	Calcite	-3.57	0.02	-14.99	0.02	15.41
LD01R2	Calcite	-3.78	0.05	-15.32	0.10	15.07
LD01R3	Calcite	-3.59	0.04	-15.66	0.03	14.72
LD01R24	Calcite	-5.00	0.05	-13.92	0.09	16.51
LK600	Calcite	2.40	0.07	-10.40	0.10	20.14
LK635	Calcite	-6.02	0.08	-15.11	0.08	15.28
Average		-3.26	0.05	-14.23	0.07	16.19
LD01R1	Dolomite	0.53	0.05	-9.86	0.06	20.70
LD01R2	Dolomite	1.52	0.02	-6.78	0.03	23.87
LD01R3	Dolomite	1.61	0.03	-8.57	0.03	22.03
LD03R6	Dolomite	1.01	0.05	-8.77	0.07	21.82
LD03R8	Dolomite	-0.33	0.05	-11.45	0.08	19.06
LK122	Dolomite	2.29	0.04	-4.56	0.03	26.16
LK122-3	Dolomite	-0.11	0.02	-8.34	0.03	22.26
Average		0.93	0.04	-8.33	0.05	22.27

4.3. Mg Isotopes

The Mg isotopic compositions of calcite and dolomite samples are given in Table 3. The $\delta^{26}Mg$ values of seven hydrothermal carbonates (calcite and dolomite) range from $-3.853‰$ to $-1.358‰$, with an average of $-2.433‰$. Three calcite and four dolomite samples had $\delta^{26}Mg$ values ranging from $-3.853‰$ to $-3.483‰$ (average: $-3.613‰$) and $-1.751‰$ to $-1.358‰$ (average: $-1.548‰$), respectively.

Table 3. Mg isotopic composition of hydrothermal calcite/dolomite in Lekai Pb-Zn deposit.

Sample Number	Sample Name	δ^{25}Mg	Std.ev	δ^{26}Mg	Std.ev
LD01R2	Calcite	−1.736	0.059	−3.503	0.085
LD01R24	Calcite	−1.905	0.061	−3.853	0.078
LK600	Calcite	−1.746	0.071	−3.483	0.093
Average	Calcite	−1.796	0.064	−3.613	0.085
LD01R1	Dolomite	−0.701	0.041	−1.429	0.069
LD01R2	Dolomite	−0.801	0.047	−1.655	0.074
LD03R8	Dolomite	−0.659	0.051	−1.358	0.062
LK122	Dolomite	−0.859	0.040	−1.751	0.060
Average	Dolomite	−0.755	0.045	−1.548	0.066

5. Discussion

5.1. Genetic Relationship of Hydrothermal Carbonate Rocks and the Nature of Fluids

REE geochemistry is a useful tool in investigating hydrothermal mineralization and understanding the genesis of carbonate rocks in different geological environments [42]. Calcite and dolomite are the main carbonate minerals, which were closely associated with galena, pyrite, and sphalerite in the Lekai Pb-Zn deposit. They have similar REE distribution patterns and show consistent ΣREE, Eu, and Ce anomalies, etc.; in addition, they show constant Y/Ho values and are roughly horizontally distributed on the Y/Ho–La/Ho diagram (Figure 8A), indicating that dolomite and calcite may have crystallized at approximately the same period and originated from the same fluid system.

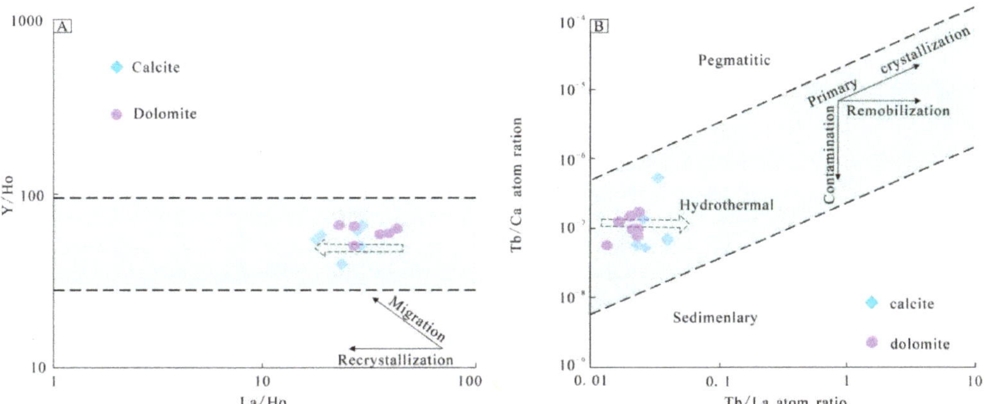

Figure 8. Plots of (**A**)Y/Ho versus La/Ho ratios and (**B**) Tb/Ca versus Tb/La ratios for the Lekei calcite and dolomite. (The original base map of A and B after [44,45], respectively).

In fact, there are slight differences in LREE/HREE fractionation between dolomite and calcite, although they both show LREE-enriched patterns with constant LREE/HREE ratios. The average LREE/HREE of dolomite (6.32, n = 6) is slightly higher than that of calcite (5.48, n = 7). Morgan et al. (1980) believed that the REE pattern in hydrothermal minerals is mainly controlled by the ionic radius of cations and that the LREEs are easier to incorporate into carbonate rock crystal lattices than the HREEs because the differences in ionic radii between LREE^{3+} and Ca^{2+} are smaller than these between HREE^{3+} and Ca^{2+}. Therefore, carbonate rocks should have LREE-enriched patterns as found in this study. LREE/HREE differentiation of hydrothermal carbonates is mainly controlled by physicochemical conditions, which control REE leaching and fluid migration in source rocks [46]. Bau and Möller (1992) [47] suggested that the leaching and fluid migration of REEs may take place in two different states—adsorption and complexation. Specifically, with the increase of pH value and the decrease of temperature, complexation is more common than adsorption, and a partition mode rich in LREEs is generated in adsorption conditions (such as low pH and high temperature), or ligand-rich fluids can form LREE-poor partition patterns through the complexation process [48–50] because HREEs can form more stable complexes with ligands (CO_3^{2-} and OH^-). Therefore, the fluids in the early stage tend to be rich in LREEs, while HREEs tend to precipitate preferentially in the late stage. We think that dolomite with higher LREE contents may crystallize earlier than calcite. In the process of fluid evolution and carbonate precipitation, not only is there LREE/HREE differentiation, but the Tb/La and Sm/Nd ratios also have differences. Constantopoulos (1988) [50] and Chesley et al. (1991) [51] proposed that the ratios of Tb/La and Sm/Nd of early carbonate rocks are lower than those of late carbonate rocks. The Tb/La ratio of dolomite in the Lekai deposit ranges from 0.01 to 0.02, with an average of 0.02 ($n = 7$), which is slightly lower than that of calcite (0.02–0.04, average: 0.03, $n = 6$); the Sm/Nd ratio of dolomite is between 0.17 and 0.20 (average: 0.19, $n = 7$), which is slightly lower than the Sm/Nd ratio of calcite (0.20–0.28, average: 0.24, $n = 6$). This also supports the view that calcite crystallization may be slightly later than dolomite crystallization.

The dolomite and calcite in the Lekai deposit show moderate negative Eu and Ce anomalies, which indicate that low oxygen fugacity and low-temperature environments occurred during their deposition [52,53]. Considering that there are only a few differences between dolomite and calcite in REE compositions, Tb/La and Sm/Nd ratios, and δEu and δCe values, we preliminarily suspect that they are products of different stages in the evolution of homologous fluids. Dolomite formed earlier and was more influenced by the surrounding rock of Devonian, while calcite formed later, showing more characteristics of the ore-forming fluids.

5.2. Sources of Metallogenic Fluids

5.2.1. REE and C–O Isotopic Constraints

Y and Ho usually show similar geochemical behavior, so the Y/Ho ratio is an important parameter to trace fluid processes [42]. The Y/Ho ratios of hydrothermal dolomite and calcite range from 50.19 to 66.50 and 38.63 to 60.40, respectively, indicating the existence of hydrothermal sources (for which the Y/Ho ratios are approximately 20–110 [44]). The La/Ho ratio of the Devonian surrounding rock (average: 33.79, $n = 2$), dolomite (average: 32.78, $n = 7$), and calcite (average: 25.23, $n = 6$) in the study area gradually decrease, indicating that dolomite and calcite may be hydrothermal carbonates formed by the recrystallization of Devonian rocks in the host rock. Conversely, the REE patterns of dolomite and calcite in the Lekai mining district are consistent with those of Devonian rocks (carbonate and noncarbonate rocks); however, magmatic rocks (Emeishan basalt and diabase) show completely different REE patterns, indicating that dolomite and calcite may have a genetic relationship with the Devonian rocks rather than the Emeishan basalt and diabase. In addition, the Tb/Ca–Tb/La diagram (Figure 8B) established by Möller et al. (1976) [45] can distinguish the relationships among dolomite, calcite, and other calcium-bearing minerals and sedimentary, hydrothermal, and magmatic rocks. In particular, the Tb/Ca ratio may be a good indicator of the formation environment of carbonate rocks [54]. All dolomite and calcite in the Lekai mining district fall into the hydrothermal area, and there are no data within the magmatic area, suggesting their genesis may not be related to the magmatic fluid. The ratios of La/Ho and Tb/La are the criteria for the degree of crystallization differentiation of carbonate rocks [54]. The Y/Ho–La/Ho and Tb/Ca–Tb/La diagrams show the changing trends of La/Ho and Tb/La ratios, which better indicate the crystallization differentiation trends of dolomite and calcite. As expected, dolomite crystallizes earlier than calcite, which may indicate the inheritance from sedimentary Devonian rocks to dolomite and calcite.

The δ^{13}C values of dolomite and calcite from the Lekai hydrothermal solution range from -0.33‰ to 2.29‰ (average: 0.93‰, $n = 7$) and -6.02‰ to 2.40‰ (average: -3.26‰, $n = 6$), respectively, which are significantly higher than those of organic carbon in sediments (-30.0‰ $< \delta^{13}C_{PDB} < -10.0$‰ [55]), slightly higher than those of mantle-derived magma ($-8.0‰ < \delta^{13}C_{PDB} < -4.0‰$, [56,57]), and basically corresponding within the range of marine carbonate ($-4.0‰ < \delta^{13}C_{PDB} < 4.0‰$, [58]). It is suggested that the carbon of hydrothermal dolomite and calcite may be derived from marine carbonate rocks. The $\delta^{18}O_{SMOW}$ values of hydrothermal dolomite range from 20.60‰ to 26.10‰ (average: 22.60‰, $n = 7$), which are significantly higher than that of the common mantle-derived magma ($6.0‰ < \delta^{18}O_{SMOW} < 10.0‰$, [56,57]), and are completely distributed within the marine carbonate rocks ($20.0‰ < \delta^{18}O_{SMOW} < 30.0‰$, [58]). The $\delta^{18}O_{SMOW}$ values of calcite range from 14.30‰ to 20.30‰ (average: 16.10‰, $n = 6$) and are distributed between the common mantle-source magma and marine carbonate rocks and obviously lower than those of marine carbonate rock. Although the dissolution of carbonate rocks will lead to almost constant $\delta^{13}C$ values and reduced $\delta^{18}O$ values [59], the $\delta^{18}O_{SMOW}$ value of dolomite does not change significantly, which suggests that the oxygen isotopes of dolomite may be inherited from marine carbonate rocks, whereas the oxygen isotopes of calcite may be influenced by other fluids.

The $\delta^{13}C_{PDB}$–$\delta^{18}O_{SHOW}$ diagram (Figure 9A) shows that the C–O isotopic composition of dolomite and calcite have obvious differences. The C–O isotopic values decrease from dolomite to calcite, suggesting that calcite may crystallize later than dolomite. Calcite is obviously affected by other fluids and does not retain the characteristics of marine carbonate rocks as dolomite does, which is consistent with the REE analyses. Studies have found that the mineralization of Pb-Zn deposits in northwest Guizhou is a mixture of two kinds of fluids, namely fluid rich in metal elements that originated from the basement and fluid rich in sulfate that originated from the stratum [60]. Therefore, we primarily think the basement fluid is the main factor affecting the O isotopic values of hydrothermal calcite.

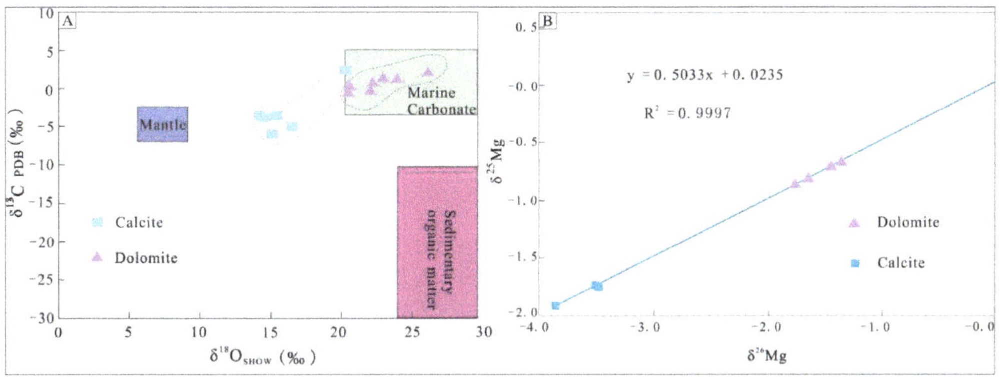

Figure 9. (**A**) Diagram of $\delta^{13}C$ vs. $\delta^{18}O$ of calcite and dolomite in Lekai Pb-Zn deposit (Mantle [56]; Marine Carbonate rocks [58]; Sedimentary organic matters [57]); (**B**) Diagram of $\delta^{26}Mg$ vs. $\delta^{25}Mg$ of calcite and dolomite in Lekai Pb-Zn deposit.

5.2.2. Mg Isotopic Constraints

The Mg isotopic measurements of the Lekai Pb-Zn deposit are all located on the mass fractionation line with a slope of 0.50 (within experimental error), and the fitting degree is very high (R^2 = 0.9997), which indicates that the interference of homoectopic elements in the mass spectrometry measurement process can be ignored. However, the $\delta^{26}Mg$–$\delta^{25}Mg$ diagrams of hydrothermal dolomite and calcite samples are distributed in different intervals (Figure 9B), indicating that there is mass fractionation between hydrothermal dolomite and calcite.

It is found that chondrites, mantle peridotites, and basalts have relatively homogeneous Mg isotopic compositions (Figure 10A) [61], suggesting that the mass balance fractionation of Mg isotopes is very small in processes of high-temperature magmatism, such as crystallization differentiation and partial melting [61,62]. In other words, the

Mg isotopic composition of carbonatites, which are genetically related to mantle-derived igneous rocks, should be similar to that of mantle-derived igneous rocks. For example, the δ^{26}Mg values of dolomite (igneous carbonatite dyke) in the Bayan Obo deposit range from $-0.94‰$ to $-0.10‰$, with an average of $-0.50‰$ [63]; and the δ^{26}Mg values of the H8 dolomite range from $-1.18‰$ to $0.56‰$ (average: $-0.42‰$) and are closely related to those of mantle rocks [64] (Figure 10B). In addition, some studies show that Mg isotopic compositions of different ages and types of sedimentary carbonate rocks have no obvious correlations beyond similar distribution intervals [24,64]. For example, ancient (Paleoproterozoic to Triassic) and modern (Cenozoic) dolostones have similar δ^{26}Mg compositions, ranging from $-3.25‰$ to $-0.45‰$ and $-3.46‰$ to $-0.38‰$, respectively [26,39]. Therefore, carbonatites can be well distinguished from sedimentary carbonate rocks. The Mg isotopic compositions of hydrothermal carbonate rocks (dolomite and calcite) in the Lekai Pb-Zn deposit range from $-3.853‰$ to $-1.358‰$, which are significantly lighter than those of chondrites, mantle rock, and seawater (Figure 10B), and close to those of sedimentary carbonate rocks. These data show that the source area of Mg in the metallogenic fluid of the Lekai Pb-Zn deposit may be sedimentary carbonate rocks, and has little relationship with mantle, sedimentary rock, or seawater.

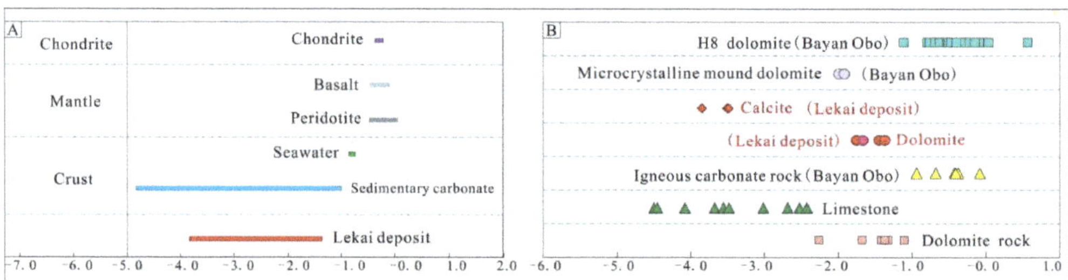

Figure 10. (**A**) Magnesium isotopic composition of different reservoirs of the Earth (Chondrite [13]; Mantle [34]; Crust [23,33]; (**B**) Magnesium isotopic composition of Lekai Pb-Zn deposit and Bayan Obo deposit (Bayan Obo deposit [63]).

The Mg isotopic compositions of hydrothermal dolomite and calcite in the Lekai Pb-Zn deposit are distributed in the sedimentary carbonate rocks interval. However, there is still a certain fraction (Figure 9B) wherein the average δ^{26}Mg value of hydrothermal dolomite is $-1.548‰$, whereas that of calcite is $-3.613‰$. It is found that Mg isotope fractionation of carbonate minerals is mainly affected by a combination of factors, including the following: the growth rate of carbonate minerals [28,65]; temperature [66]; the existing form of Mg in the aqueous solution [67]; biological processes [68,69]; and the type of carbonate minerals (different minerals have different Mg–O bond lengths) [23,36]. The REE characteristics of dolomite and calcite in the Lekai hydrothermal system show that they originated from the same fluid system, but the dolomite crystallized earlier than calcite, which indicates that the Mg isotopic fractionations of dolomite and calcite are not caused by the existence of Mg in the hydrothermal fluid. In addition, the Mg isotopic compositions of hydrothermal dolomite and calcite are relatively concentrated, suggesting that the influence of mineral growth rate is not significant. The biological effects first need to meet the temperature conditions for bacterial life and that of other organisms (<100–120 °C, [70]). The homogenization temperature of hydrothermal mineral fluid inclusions of Pb-Zn deposits in northwest Guizhou is between 160 °C and 260 °C [4]; therefore, the effect of biological action on Mg isotope fractionation can be eliminated. Some experimental and field data have shown that Mg isotopic fractionation during inorganic precipitation of carbonate minerals is positively correlated with the temperature at medium and low temperatures [43,66], but the influence of temperature on Mg isotopic fractionation is relatively weak, so it is not the main controlling factor [14,35,71]. However, this is not consistent with the fact that dolomite with a little earlier crystallization should have a slightly heavier Mg isotopic composition than calcite δ^{26}Mg. Mg isotopic fractionation is obviously beyond the influence range of temperature. Liu et al. (2010) [72] found that mineral phase is the main controlling factor of Mg isotopic fractionation in carbonate minerals, that is, the bond length or bond energy strength of the corresponding chemical bonds formed by Mg in different mineral phases determines the degree of Mg isotope fractionation. Generally, the bond length is determined by the coordination number of the cation (Mg), followed by the coordination number of the anion (generally O). The lower the coordination number, the shorter the bond length, and the stronger the bond energy, the more favorable it is for the enrichment of heavy Mg isotopes [14,72]. It is found that Mg–Ca substitution leads to strong deformation of the Mg ion lattice during dolomitization, resulting in the reduction of the coordination number and formation of stronger Mg–O bond energy. The concentrations of Mg and Ca in carbonate rocks also affect the Mg–O bond energy. The ratio of Mg to Ca in dolomite is more conducive to the enrichment of ^{26}Mg [73]. Therefore, dolomite has a heavier Mg isotopic composition than calcite. In addition, the relevant experimental data also show that carbonate minerals tend to be enriched in light Mg isotopes, and calcite has a greater fractionation coefficient than dolomite [14,36,66,74,75]. This is consistent with the results of this study. Therefore, we primarily think that the mineral phase controls the Mg isotope fractionation of dolomite and calcite in the Lekai Pb-Zn deposit.

5.3. Mineralization

Studies have confirmed that the mineralization of Pb-Zn deposits in the SYG Pb-Zn metallogenic province is a mixture of two kinds of fluids, such as acidic fluid rich in metal elements that originated from the basement is fed into the overlying sedimentary strata by deep faults and mixed with alkaline fluid rich in sulfate in the strata. As a result, thermochemical sulfate reduction (TSR) produces a large amount of S^{2-}, which combines with metal cations such as Pb^{2+}, Zn^{2+}, and Fe^{2+} in the metal fluid for mineralization [76] (Figure 11). The participation of basement fluid directly affects the scale and grade of the orebody. From west to east, the deep lithospheric Anninghe, Ganluo–Xiaojiang, and Yadu–Mangdong Faults are distributed in the SYG Pb-Zn metallogenic province. Our results show that the Pb-Zn deposits are distributed along these three faults in a lenticular linear manner. The Ganluo–Xiaojiang Fault in the central part has the highest mineralization intensity and the best continuity in the distribution of deposits, whereas the Anninghe and Yadu–Mangdong Faults on the west and east sides, respectively, are relatively inferior in mineralization intensity and scale.

The metallogenic belt of northeast Yunnan in the ore concentration area is mainly controlled by the Ganluo–Xiaojiang Fault, and the basement metamorphic rocks have developed in this area. The ore host strata of the Pb-Zn deposits are mainly Sinian, Devonian, and other older strata, which are greatly affected by the basement metallic fluid and form large-scale and high-grade Pb-Zn orebodies easily, such as the Huize, Maoping, and other super large Pb-Zn deposits and the Maozu, Fule, and other large Pb-Zn deposits. In particular, the amount of Pb + Zn in the Huize Pb-Zn deposit exceeds 5 million tons, and that in the Maoping Pb-Zn deposit exceeds 3 million tons [2,3]. However, the Pb-Zn deposits in the northwest Guizhou metallogenic belt are mainly controlled by the Yadu–Mangdong Fault at the eastern boundary of the SYG Pb-Zn metallogenic province. The basement metamorphic rocks are not developed. The ore host strata are mainly Carboniferous, Permian, and other strata, which are less affected by the metal-bearing fluids in the basement. It is not easy to form large-scale and high-grade Pb-Zn orebodies in this area. There are few metal elements, such as Ag, Ge, Cd, and Ga. The Lekai Pb-Zn deposit is located at the intersection of the northern part of the Yadu–Mangdong Fault and the secondary Luozehe Fault of the Xiaojiang Fault. The ore-bearing strata are Devonian carbonate rocks with good structural and lithological conditions. However, the deposit is small in scale and low grade. The main factor is that the basement fluid is not involved in the deposit, and the basement fluid is the main source area of metal elements.

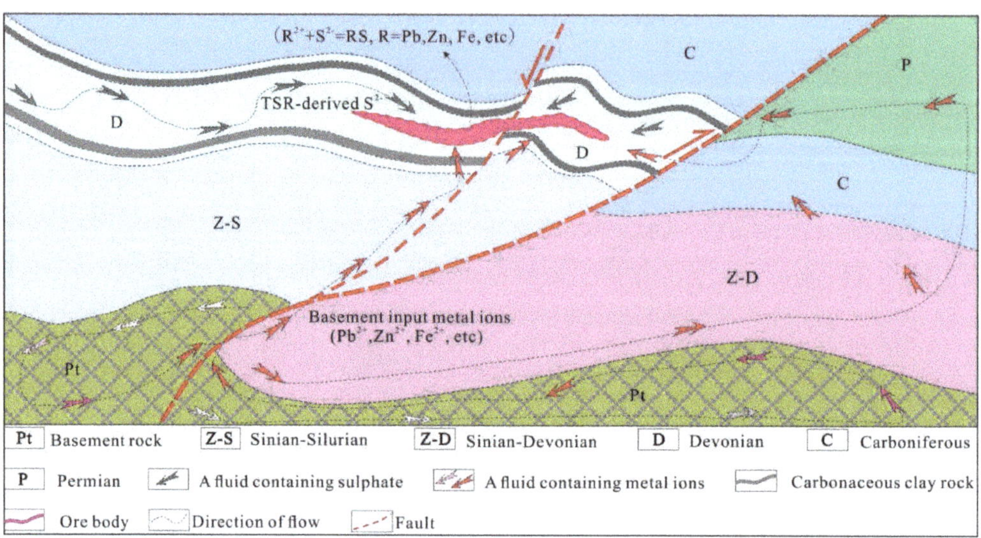

Figure 11. Metallogenic model map of Lekai Pb-Zn deposit [76].

6. Conclusions

Based on the in-depth analysis of the geological characteristics of the Lekai Pb-Zn deposit in the SYG Pb-Zn metallogenic province, and the systematic study of REE and C–O–Mg isotopic geochemistry of hydrothermal calcite/dolomite, the following four points are obtained:

(I)   The mineralization of the Lekai Pb-Zn deposit is mainly metasomatism and filling and controlled by faults and lithology. The orebody is stratoid and lenticular and develops veined, massive, brecciated, and disseminated structures, showing obvious epigenetic metallogenic characteristics.

(II) The REE characteristics of hydrothermal calcite/dolomite in the Lekai Pb-Zn deposit show that the metallogenic materials are provided by the carbonate rocks, and the basin fluid is the main metallogenic fluid. The formation environment of the Pb-Zn deposit has low oxygen fugacity and low temperature. The C and O isotopic compositions of calcite/dolomite indicate that the metallogenic process is mainly influenced by the basement fluid, followed by basin fluid.

(III) Mg isotopic analysis of hydrothermal calcite/dolomite in the Lekai Pb-Zn deposit shows that the source of metallogenic fluid may be sedimentary carbonate rocks, rather than the mantle, chondrites, or seawater. The Mg isotopic fractionation of calcite/dolomite is controlled by the mineral phase.

(IV) The mineralization of Pb-Zn deposits in the SYG Pb-Zn metallogenic province may be the result of two fluids mixing (basement fluid and basin fluid). The participation of basement fluid directly affects the scale and grade of the orebody. The Lekai Pb-Zn deposit is obviously less affected by the basement fluid and shown small in deposit scale and grade.

Author Contributions: Conceptualization, Z.H. and B.L.; methodology, Z.H.; validation, Z.H., X.W. (Xinfu Wang) and X.X.; formal analysis, Z.H. and X.X. investigation, B.L. and X.W. (Xin Wan); data curation, Q.W.; writing—original draft preparation, Z.H; writing—review and editing, Z.H., X.W. (Xinfu Wang) and B.L.; visualization, B.L. and X.W. (Xinfu Wang); supervision, B.L. and X.X.; funding acquisition, B.L. All authors have read and agreed to the published version of the manuscript.

Funding: This research was financially supported by the National Natural Science Foundation of China (No. 41862007), the Key Disciplines Construction of Kunming University of Science and

Technology (No. 14078384), and the Yunnan Ten Thousand Talents Plan Young & Elite Talents Project (No. YNWR-QNBJ-2018-093).

Data Availability Statement: The data used to support this study are included within the article.

Acknowledgments: We sincerely thank the Editorial Board members and anonymous reviewers for their constructive comments.

Conflicts of Interest: The authors declare that they have no known competing financial interest or personal relationships that could have appeared to influence the work reported in this paper.

References

1. Hu, R.Z.; Fu, S.; Huang, Y.; Zhou, M.; Fu, S.; Zhao, C.; Wang, Y.; Bi, X.; Xiao, J. The giant South China Mesozoic low-temperature metallogenic province: Reviews and a new geodynamic model. *J. Asian Earth Sci.* **2017**, *137*, 9–34. [CrossRef]
2. Han, R.S.; Zou, H.J.; Hu, B.; Hu, Y.Z.; Xue, C.D. Features of fluid inclusions and sources of ore-forming fluid in the Maoping carbonate-hosted Zn-Pb-(Ag-Ge) deposit, Yunnan, China. *Acta Petrol. Sin.* **2007**, *23*, 2109–2118.
3. Han, R.S.; Hu, Y.Z.; Wang, X.K.; Hou, B.H.; Huang, Z.L.; Chen, J.; Wang, F.; Wu, P.; Li, B.; Wang, H.J.; et al. Mineraliztion model of rich Ge-Ag-bearing Zn-Pb polymetallic deposit concentrated district in Northeastern Yunnan, China. *Acta Geol. Sin. Ed.* **2012**, *86*, 280–294.
4. Zhu, L.Y.; Su, W.C.; Shen, N.P.; Dong, W.D.; Cai, J.L.; Zhang, Z.W.; Zhao, H.; Xie, P. Fluid inclusion and sulfur isotopic northwestern Guizhou, China. *Acta Petrol. Sin.* **2016**, *32*, 3431–3440.
5. Yuan, B.; Mao, J.W.; Yan, X.H.; Wu, Y.; Zhang, F.; Zhao, L.L. Sources of metallogenic materials and metallogenic mechanism of Daliangzi Ore Field in Sichuan Province: Constraints from geochemistry of S, C, H, O, Sr isotope and trace element in sphalerite. *Actor Petrol. Sircica* **2014**, *30*, 209–220.
6. Qiu, W.L.; Han, R.S. Study on hydrogen and oxygen isotopic characteristics of Zhaotong lead-zinc deposit. *Acta Geol. Sin.* **2015**, *89*, 173–174.
7. Zhou, J.X.; Huang, Z.L.; Zhou, M.; Li, X.; Jin, Z. Constraints of C-O-S-Pb isotope compositions and Rb-Sr isotopic age on the origin of the Tianqiao carbonate-hosted Pb-Zn deposit, southwest China. *Ore Geol. Rev.* **2013**, *53*, 77–92. [CrossRef]
8. Zhou, J.X.; Huang, Z.L.; Bao, G. Geological and sulfur-lead-strontium isotopic studies of the Shaojiwan Pb-Zn deposit, southwest China: Implications for the origin of hydrothermal fluids. *J. Geochemical. Explor.* **2013**, *128*, 51–61. [CrossRef]
9. Zhou, J.X.; Huang, Z.L.; Gao, J.G.; Yan, Z.F. Geological and C-O-S-Pb-Sr isotopic constraints on the origin of the Qingshan carbonate-hosted Pb-Zn deposit, Southwest China. *Ore Geol. Rev.* **2013**, *55*, 904–916. [CrossRef]
10. Zhou, J.X.; Huang, Z.L.; Bao, G.; Gao, J.G. Sources and thermo-chemical sulfate reduction for reduced sulfur in the hydrothermal fluids, southeastern SYG Pb-Zn metallogenic province, southwest China. *J. Asian Earth Sci.* **2013**, *24*, 759–771.
11. Ye, L.; Gao, W.; Yang, Y.L.; Liu, T.G.; Peng, S.S. Trace elements in sphalerite in Laochang Pb-Zn polymetallic deposit, Lancang, Yunnan Province. *Acta Petrologica. Sin.* **2012**, *28*, 1362–1372.
12. Zhou, J.X.; Luo, K.; Wang, X.C.; Simon, A.W.; Wu, T.; Huang, Z.L.; Cui, Y.L.; Zhao, X.Z. Ore genesis of the Fule Pb-Zn deposit and its relationship with the Emeishan Large Igneous Province: Evidence from mineralogy, bulk C-O-S and in situ S-Pb isotopes. *Gondwana Res.* **2018**, *54*, 161–179. [CrossRef]
13. Galy, A.; Young, E.D.; Ash, R.D.; O'Nions, R.K. The formation of chondrules at high gas pressures in the solar nebula. *Science* **2000**, *290*, 1751–1753. [CrossRef] [PubMed]
14. Galy, A.; Matthews, M.B.; Halicz, L.; O'Nions, R.K. Mg isotopic composition of carbonate: Insight from speleothem formation. *Earth Planet. Sci. Lett.* **2002**, *201*, 105–110. [CrossRef]
15. Galy, A.; Yoffo, O.; Janney, P.E.; Williams, R.W.; Cloquet, C.; Alard, O.; Halicz, L.; Wadhwa, M.; Hutcheon, I.D.; Ramon, E.; et al. Magnesium isotope heterogeneity of the isotopic standard SRM980 and new reference materials for magnesium-isotope-ratio measurements. *J. Anal. At. Spectrom.* **2003**, *18*, 1352–1356. [CrossRef]
16. Wilkinson, J.J.; Weiss, D.J.; Mason, T.F.D.; Coles, B.J. zinc isotope variation in hydrothermal systems: Preliminary evidence from the Irish Midlands ore field. Economic. *Geology* **2005**, *100*, 583–590. [CrossRef]
17. Fujii, T.; Moynier, F.; Pons, M.L.; Albarède, F. The origin of Zn isotope fractionation in sulfides. *Geochim. Cosmochim. Acta* **2011**, *75*, 7632–7643. [CrossRef]
18. Tang, S.H.; Zhu, X.K.; Li, J.; Yan, B.; Li, S.Z.; Li, Z.H.; Wang, Y.; Sun, J. New Standard Solutions for Measurement of Iron, Copper and zinc Isotopic Compositions by Multi-collector Inductively Coupled Plasma-Mass Spectrometry. *Rock Miner. Anal.* **2016**, *35*, 127–133.
19. Duan, J.; Tang, J.; Lin, B. zinc and lead isotope signatures of the Zhaxikang Pb-Zn deposit, South Tibet: Implications for the source of the mineralizing metals. *Ore Geol. Rev.* **2016**, *78*, 58–68. [CrossRef]
20. Gagnevin, D.; Boyce, A.J.; Barrie, C.D.; Menuge, J.F.; Blakeman, R.J. Zn, Fe and S isotope fractionation in a large hydrothermal system. *Geochim. Cosmochim. Acta* **2012**, *88*, 183–198. [CrossRef]
21. Zhou, J.X.; Xiang, Z.Z.; Zhou, M.F.; Feng, Y.X.; Luo, K.; Huang, Z.L.; Wu, T. The giant Upper Yangtze Pb-Zn province in southwest China: Reviews, new advances and a new genetic model. *J. Asian Earth Sci.* **2018**, *154*, 280–315. [CrossRef]
22. Galy, A.; Belshaw, N.S.; Halicz, L.; O'Nions, R.K. High-Precision measurement of magnesium isotopes by multiples ollector inductively coupled plasma mass spectrometry. *Int. J. Mass Spectrom.* **2001**, *208*, 89–98. [CrossRef]

23. Young, E.D.; Galy, A. The isotope geochemistry and cosmochemistry of magnesium. *Rev. Mineral. Geochem.* **2004**, *55*, 197–230. [CrossRef]

24. Ge, L.; Jiang, S.Y. Recent advances in research on magnesium isotope geochemistry. *Acta Petrol. Mineral.* **2008**, *27*, 367–374.

25. Ke, S.; Liu, S.A.; Li, W.H.; Yang, W.; Teng, F.Z. Advances and application in magnesium isotope geochemistry. *Acta Petrol. Sin.* **2011**, *27*, 383–397.

26. Teng, F.Z. Magnesium isotope geochemistry. *Rev. Mineral. Geochem.* **2017**, *82*, 219–287. [CrossRef]

27. Azmy, K.; Lavoie, D.; Wang, Z.R.; Brand, U.; Al-Aasm, I.; Jackson, S.; Girard, I. Magnesium-isotope and REE compositions of Lower Ordovician carbonates from eastern Laurentia: Implications for the origin of dolomites and limestones. *Chem. Geol.* **2013**, *356*, 64–75. [CrossRef]

28. Mavromatis, V.; Meister, P.; Oelkers, E.H. Using stable Mg isotopes to distinguish dolomite formation mechanisms: A case study from the Peru Margin. *Chem. Geol.* **2014**, *385*, 84–91. [CrossRef]

29. Geske, A.; Goldstein, R.H.; Mavromatis, V.; Richter, D.K.; Buhl, D.; Kluge, T.; John, C.M.; Immenhauser, A. The magnesium isotope ($\delta26Mg$) signature of dolomites. *Geochim. Gosmochimica Acta* **2015**, *149*, 131–151. [CrossRef]

30. Wan, X.; Han, R.S.; Li, B.; Xiao, X.G.; He, Z.W.; Wang, J.T.; Wei, Q.X. Tectono-geochemistry and deep prospecting prediction in the Lekai lead-zinc deposit, NW Guizhou Province, China. *Geol. China* **2020**. Available online: http://kns.cnki.net/kcms/detail/11.1167.P20200602.1143.011.html (accessed on 2 June 2020).

31. Young, E.D.; Galy, A.; Nagahara, H. Kinetic and equilibrium mass-dependent isotope fractionation laws in nature and their geochemical and cosmochemical significance. *Geochim. Cosmochim. Acta* **2002**, *66*, 1095–1104. [CrossRef]

32. Young, E.D.; Tonui, E.; Manning, C.E.; Schauble, E.; Macris, C.A. Spinel-olivine magnesium isotope thermometry in the mantle and implications for the Mg isotopic composition of Earth. *Earth Planet. Sci. Lett.* **2009**, *288*, 524–533. [CrossRef]

33. Chang, V.T.C.; Makishima, A.; Belshaw, N.S.; Keith O'Nions, R. Purification of Mg from low-Mg biogenic carbonates for isotope ratio determination using multiple collector ICP-MS. *J. Anal. At. Spectrom.* **2003**, *18*, 296–301. [CrossRef]

34. Pearson, N.J.; Griffin, W.L.; Alard, O.; O'Reilly, S.Y. The isotopic composition of magnesium in mantle olivine: Records of depletion and metasomatism. *Chem. Geol.* **2006**, *226*, 115–133. [CrossRef]

35. Immenhauser, A.; Bauhl, D.; Richter, D.; Niedermayr, A.; Riechelmann, D.; Dietzel, M.; Schulte, U. Magnesium-isotope fractionation during low-Mg calcite precipitation in a limestone cave-Field study and experiments. *Geochim. Cosmochim. Acta* **2010**, *74*, 4346–4364. [CrossRef]

36. Wang, Z.R.; Hu, P.; Gaetani, G.; Liu, C.; Saenger, C.; Cohen, A.; Hart, S. Experimental calibration of Mg isotope fractionation between aragonite and seawater. *Geochim. Cosmochim. Acta* **2013**, *102*, 113–123. [CrossRef]

37. Lavoie, D.; Jackson, S.; Girard, I. Magnesium isotopes in high-temperature saddle dolomite cements in the lower Paleozoic of Canada. *Sediment. Geol.* **2014**, *305*, 58–68. [CrossRef]

38. Yoshimura, T.; Tanimize, M.; Inoue, M.; Inoue, M.; Suzuki, A.; Iwasaki, N.; Kawahata, H. Mg isotope fractionation in biogenic carbonates of deep-sea coral, benthic foraminifera, and hermatypic coral. *Anal. Bioanal. Chem.* **2011**, *401*, 2755–2769. [CrossRef]

39. Huang, K.J.; Shen, B.; Lang, X.G.; Tang, W.; Peng, Y.; Ke, S.; Kaufman, A.; Ma, H.R.; Li, F.B. Magnesium isotopic composition of the Mesoproterozoic dolostones: Implications for Mg isotopic systematics of marine carbonates. *Geochim. Gosmochimica Acta* **2015**, *614*, 333–351. [CrossRef]

40. Guan, S.P.; Li, Z.X. Lead-sulfur isotope study of carbonate-hosted Pb-Zn deposits at the eastern margin of the kangdian axis. *Geol. Geochem.* **1999**, *27*, 45–54.

41. Zhang, C.Q.; Wu, Y.; Hou, L.; Mao, J.W. Geodynamic setting of mineralization of Mississippi Valley-type deposits in world-class SYG Zn-Pb triangle, southwest China: Implications from age-dating studies in the past decade and the Sm-Nd age of the Jinshachang deposit. *J. Asian Earth Sci.* **2015**, *103*, 103–114. [CrossRef]

42. Schwinn, G.; Markl, G. REE systematics in hydrothermal fluorite. *Chem. Geol.* **2005**, *216*, 225–248. [CrossRef]

43. Li, W.; Chakraborty, S.; Beard, B.L.; Romanek, C.S.; Johnson, C.M. Magnesium isotope fractionation during precipitation of inorganic calcite under laboratory conditions. *Earth Planet. Sci. Lett.* **2012**, *333*, 304–316. [CrossRef]

44. Bau, M.; Dulski, P. Comparative study of yttrium and rare-earth element behavior in fluorine-rich hydrothermal fluids. *Contrib. Mineral. Petrol.* **1995**, *119*, 213–223. [CrossRef]

45. Möller, P.; Parekh, P.P.; Schneider, H.J. The application of Tb/Ca–Tb/La abundance ratios to problems of fluorspar genesis. *Miner. Depos.* **1976**, *11*, 111–116. [CrossRef]

46. Michard, A. Rare earth element systematics in hydrothermal fluids. *Geochim. Cosmochim. Acta* **1989**, *53*, 745–750. [CrossRef]

47. Bau, M.; Möller, P. Rare earth element fractionation in metamorphogenic hydrothermal calcite, magnesite and siderite. *Miner. Petrol.* **1992**, *45*, 231–246. [CrossRef]

48. Bau, M. Rare-earth element mobility during hydrothermal and metamorphic fluid-rock interaction and the significance of the oxidation state of europium. *Chem. Geol.* **1991**, *93*, 219–230. [CrossRef]

49. Subías, I.; Fernández-Nieto, C. Hydrothermal events in the Valle de Tena (Spanish Western Pyrenees) as evidenced by fluid inclusions and trace-element distribution from fluorite deposits. *Chem. Geol* **1995**, *124*, 267–282. [CrossRef]

50. Constantopoulos, J. Fluid inclusions and rare-earth element geochemistry of fluorite from south-central Idaho. *Econ. Geol.* **1988**, *88*, 626. [CrossRef]

51. Chesley, J.T.; Halliday, A.N.; Scrivener, R.C. Samarium-Neodymium Direct of Fluorite. *Science* **1991**, *252*, 949–951. [CrossRef] [PubMed]

52. Xu, C.; Taylor, R.N.; Li, W.; Kynicky, J.; Chakhmouradian, A.R.; Song, W. Comparison of fluorite geochemistry from REE deposits in the Panxi region and Bayan Obo, China. *J. South. Hemisph. Earth Syst. Sci.* **2012**, *57*, 76–89. [CrossRef]

53. Pei, Q.M.; Zhang, S.T.; Santosh, M.; Cao, H.W.; Zhang, W.; Hu, X.K.; Wang, L. Geochronology, geochemistry, fluid inclusion and C, O and Hf isotope compositions of the Shuitou fluorite deposit, Inner Mongolia, China. *Ore Geol. Rev.* **2017**, *83*, 174–190. [CrossRef]

54. Möller, P.; Morteani, G. On the chemical fractionation of REE during the formation of Ca-minerals and its application to problems of the genesis of ore deposits. In *The Significance of Trace Elements in Solving Petrogenetic Problems*; Augustithis, S., Ed.; Theophrastus Publications: Athens, Greece, 1983; pp. 747–791.

55. Liu, J.M.; Liu, J.J. Basin fulid genetic model of sediment-hosted micro-disseminsted gold deposits in the gold-triangle area between Guizhou, Guangxi and Yunnan. *Acta Mineral. Sin.* **1997**, *17*, 448–456.

56. Taylor, H.P.; Frechen, J.; Degens, E.T. Oxygen and carbon isotope studies of carbonatites from the Laacher See District, West Germany and the Alnö District, southwesteden. *Geochim. Cosmochim. Acta* **1967**, *31*, 407–430. [CrossRef]

57. Hoefs, J. *Stable Isotope Geochemistry*, 4th ed.; Springer: Berlin, Germany, 1997; pp. 65–168.

58. Veizer, J.; Hoefs, J. The nature of O18/O16 and C13/C12 secular trends in sedimentary carbonate rocks. *Geochim. Et Cosmochim. Acts* **1976**, *40*, 1387–1395. [CrossRef]

59. Zhou, J.X.; Huang, Z.L.; Lv, Z.C.; Zhu, X.K.; Gao, J.G.; Mirnejad, H. Geology, isotope geochemistry and ore genesis of the Shanshulin carbonate-hosted Pb-Zn deposit, southwest China. *Ore Geol. Rev.* **2014**, *63*, 209–225. [CrossRef]

60. He, Z.W.; Li, Z.Q.; Li, B.; Chen, J.; Xiang, Z.P.; Wang, X.F.; Du, L.J.; Huang, Z.L. Ore genesis of the Yadu carbonate-hosted Pb-Zn deposit in Southwest China: Evidence from rare earth elements and C, O, S, Pb, and Zn isotopes. *Ore Geol. Rev.* **2021**, *131*, 104039. [CrossRef]

61. Teng, F.Z.; Li, W.Y.; Ke, S.; Marty, B.; Dauphas, N.; Huang, S.C.; Wu, F.Y.; Pourmand, A. Magnesium isotopic composition of the Earth and chondrites. *Geochim. Cosmochim. Acta* **2010**, *74*, 4150–4166. [CrossRef]

62. Teng, F.Z.; Wadhwa, M.; Helz, R.T. Investigation of magnesium isotope fractionation during basalt differentiation: Implications for a chondritic composition of the terrestrial mantle. *Earth Planet. Sci. Lett.* **2007**, *261*, 84–92. [CrossRef]

63. Sun, J. *The Origin of the Bayan Obo Ore Deposit, Inner Mongolia, China: The Iron and Magnesium Isotope Constraints*; China University of Geosciences: Beijing, China, 2013; pp. 1–115.

64. Ning, M.; Huang, K.J.; Shen, B. Applications and advances of the magnesium isotope on the 'dolomite problem'. *Acta Petrol. Sin.* **2018**, *34*, 3690–3708.

65. Mavromatis, V.; Gautier, Q.; Bosc, O.; Schott, J. Kinetics of Mg partition and Mg stable isotope fractionation during its incorporation in calcite. *Geochim. Cosmochim. Acta* **2013**, *114*, 188–203. [CrossRef]

66. Li, W.Q.; Beard, B.L.; Li, C.X.; Xu, H.F.; Johnson, C.M. Experimental calibration of Mg isotope fractionation between dolomite and aqueous solution and its geological implications. *Geochim. Et Cosmochim. Acta* **2015**, *157*, 164–181. [CrossRef]

67. Schott, J.; Mavromatis, V.; Fujii, T.; Pearce, C.R.; Oelkers, E.H. The control of carbonate mineral Mg isotope composition by aqueous speciation: Theoretical and experimental modeling. *Chem. Geol.* **2016**, *445*, 120–134. [CrossRef]

68. Pogge von Strandmann, P.A.E.; Burton, K.W.; James, R.H.; Calsteren, P.; Gislason, S.R.; Sigfússon, B. The influence of weathering processes on riverine magnesium isotopes in a basaltic terrain. *Earth Planet. Sci. Lett.* **2008**, *276*, 187–197. [CrossRef]

69. Hippler, D.; Buhl, D.; Witbaard, R.; Richter, D.K.; Immenhauser, A. Towards a better understanding of magnesium-isotope ratios from marine skeletal carbonates. *Geochim. Cosmochim. Acta* **2009**, *73*, 6134–6146. [CrossRef]

70. Basuki, N.I.; Taylor, B.E.; Spooner, E.T.C. Sulfur isotope evidence for thermo-chemical reduction of dissolved sulfate in Mississippi valley type zinc-lead mineralization, Bongara area, northern Peru. *Economic. Geol.* **2008**, *103*, 183–799. [CrossRef]

71. Tang, B.; Wang, J.T.; Fu, Y. Magnesium Isotope Composition of Different Geological Reservoirs and Controlling Factors of Magnesium Isotope Fractionation in the Formation of Carbonate Minerals-A Summary of Previous Results. *Rock Miner. Anal.* **2020**, *39*, 162–173.

72. Liu, S.A.; Teng, F.Z.; He, Y.S.; Ke, S.; Li, S.G. Investigation of magnesium isotope fractionation during granite differentiation: Implication for Mg isotopic composition of the continental crust. *Earth Planet. Sci. Lett.* **2010**, *297*, 646–654. [CrossRef]

73. Pinilla, C.; Blanchard, M.; Balan, E.; Natarajan, S.K.; Vuilleumier, R.; Mauri, F. Equilibrium magnesium isotope fractionation between aqueous Mg^{2+} and carbonate minerals: Insights from path integral molecular dynamics. *Geochim. Cosmochim. Acta* **2015**, *163*, 126–139. [CrossRef]

74. Saulnier, S.; Rollion-Bard, C.; Vigier, N.; Chaussidon, M. Mg isotope fractionation during calcite precipitation: An experimental study. *Geochim. Cosmochim. Acta* **2012**, *91*, 75–91. [CrossRef]

75. Mavromatis, V.; Pearce, C.R.; Shirokova, L.S.; Bundeleva, I.A.; Pokrovsky, O.S.; Benezeth, P.; Oelkers, E.H. Magnesium isotope fractionation during hydrous magnesium carbonate precipitation with and without cyanobacteria. *Geochim. Cosmochim. Acta* **2012**, *76*, 161–174. [CrossRef]

76. Xiao, X.G.; Li, B.; He, Z.W.; Wang, J.T.; Wei, Q.X.; Wan, X. Sources of metallogenic materials and genesis of Lekai lead-zinc deposit in northwestern Guizhou Province: Evidence from S and Pb isotopes. *Miner. Depos.* **2022**, *41*, 806–822. (In Chinese with English Abstract)

Article

The Origin of the Caiyuanzi Pb–Zn Deposit in SE Yunnan Province, China: Constraints from In Situ S and Pb Isotopes

Yongguo Jiang [1,2], Yinliang Cui [1,2], Hongliang Nian [2], Changhua Yang [2], Yahui Zhang [3,*], Mingyong Liu [2], Heng Xu [2], Jinjun Cai [2] and Hesong Liu [3]

[1] Faculty of Land Resources Engineering, Kunming University of Science and Technology, Kunming 650093, China
[2] Yunnan Nonferrous Geological Bureau, Kunming 650051, China
[3] Key Laboratory of Critical Minerals Metallogeny in Universities of Yunnan Province, School of Earth Sciences, Yunnan University, Kunming 650500, China
[*] Correspondence: yahuizh@126.com or zhangyahui@ynu.edu.cn

Abstract: Located at the intersection of the Tethys and Pacific Rim metallogenic belts, the Laojunshan polymetallic metallogenic province in SE Yunnan Province hosts many large-scale W–Sn and Sn–Zn polymetallic deposits. The newly discovered Caiyuanzi medium-sized Pb–Zn deposit is located in the northern part of this province and has eight sulfide ore bodies. All the ore bodies occur in the siliceous rocks of the Lower Devonian Pojiao Formation (D_1p). The ore bodies are conformable with stratigraphy and controlled by a lithologic horizon. The sulfide ores have banded or laminated structures. The ore minerals are mainly pyrite, chalcopyrite, sphalerite, and galena. In this study, in situ sulfur and lead isotopes were used to constrain the origin of the Caiyuanzi Pb–Zn deposit. The results show that the in situ $\delta^{34}S$ values of pyrite, chalcopyrite, and sphalerite range from 0.1‰ to 6.0‰, with an average of 4.7‰. This $\delta^{34}S$ signature reflects the mixing between magmatic-derived and reduced seawater sulfate sulfur. The in situ Pb isotopes characteristics of pyrite, galena, and sphalerite suggest that the sulfur and lead of ore minerals come from the upper crust. Integrating the data obtained from the studies including regional geology, ore geology, and S–Pb isotope geochemistry, we proposed that the Caiyuanzi Pb–Zn deposit is a hydrothermal deposit formed by sedimentary exhalative and magmatic hydrothermal superimposition.

Keywords: in situ S and Pb isotopes; the source of ore-forming elements; ore genesis; Caiyuanzi Pb–Zn deposit

Citation: Jiang, Y.; Cui, Y.; Nian, H.; Yang, C.; Zhang, Y.; Liu, M.; Xu, H.; Cai, J.; Liu, H. The Origin of the Caiyuanzi Pb–Zn Deposit in SE Yunnan Province, China: Constraints from In Situ S and Pb Isotopes. *Minerals* **2023**, *13*, 238. https://doi.org/10.3390/min13020238

Academic Editor: Mariko Nagashima

Received: 29 December 2022
Revised: 3 February 2023
Accepted: 5 February 2023
Published: 8 February 2023

1. Introduction

The Laojunshan polymetallic metallogenic province in SE Yunnan Province is located at the intersection of the Tethys and Pacific Rim metallogenic belts. This province hosts many large-scale W–Sn and Sn–Zn polymetallic deposits, such as the Dulong super-large Sn–Zn polymetallic deposit, Xinzhai large-scale Sn polymetallic deposit, and the large-scale Nanyangtian W–Sn deposit. In recent years, one large-scale (Hongshiyan Pb–Zn) and two medium-sized (Gaji Pb–Zn-Cu polymetallic and Caiyuanzi Pb–Zn) deposits have been discovered in the northern part of the province. The Pb–Zn deposits have a total Pb, Zn, and Cu metal resource of nearly 1.2 million tons, indicating that this province has good prospecting potential for these metals.

At present, the genesis of the Pb–Zn deposits is controversial [1–6], but theories include sedimentary exhalative (SEDEX) [1,2,6], and magmatic hydrothermal origins [3–5]. The main reason for the diversity in genetic views is the lack of understanding of the source of ore-forming materials. In this study, the in situ S and Pb isotopes of sphalerite, galena, pyrite, and chalcopyrite are used to trace the source of metallogenic elements and to discuss the ore genesis of the Caiyuanzi Pb–Zn deposit.

2. Geological Setting

2.1. Regional Geology

The Laojunshan metallogenic province is located at the junction of the Cathaysian, Yangtze, and Indochina blocks (Figure 1a), and in the northern part of the Song Chay metamorphosed dome (Figure 1b). The sedimentary environment in this province is complex and diverse and has experienced multiple periods of large-scale magmatic intrusion [7–11].

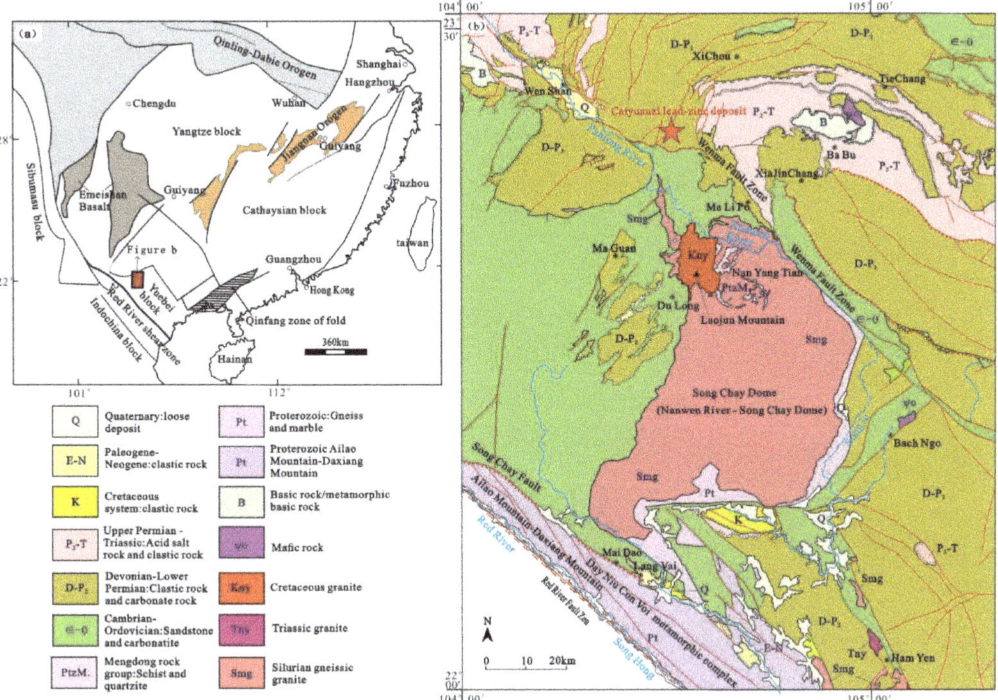

Figure 1. Geotectonic location map of the study area ((**a**), modified according to [9]) and the regional geological map ((**b**), quoted from [10]).

Since the Cambrian, this province has experienced repeated transgression and regression, ending in late Triassic marine sedimentation. The exposed strata are Cambrian, Devonian, and Permian, which show a trend of decreasing metamorphism. The Lower Cambrian is mainly sandy argillaceous slate and schist. The abundance of carbonate rocks gradually increases in the upper Middle Cambrian; the Lower Devonian is sandy argillaceous slate, which overlies the Cambrian at a slight angle. The Middle and Upper Devonian are mesa facies carbonate rocks, and the Permian is a continental shelf carbonate with siliceous rocks [12]. There are mainly NNE- and NW-trending regional structures (Figure 1b); the former were formed in the Caledonian–Indosinian, and the latter were formed in the Indosinian–Himalayan [12]. The Nanwenhe and Laojunshan granites are the main igneous rocks, both of which are closely related to tin and zinc polymetallic mineralization in the area [10,13,14]. The Nanwenhe granites are known as the Song Chay granites in the Vietnamese part, and are also known as the Song Chay metamorphic dome [7,8]. They intruded during the late Silurian (420–440 Ma) [7,15,16], and then underwent deformation and metamorphism during the Indosinian, forming gneissic, banded, and eyeball-shaped structures [12]. The Laojunshan granites intruded during the Cretaceous (83–117 Ma) [13,14,17–21].

2.2. Ore Deposit Geology

2.2.1. Strata

The main strata exposed in the mining area are Devonian (Figure 2). The Lower Devonian Pojiao Formation (D_1p) comprises shallow continental shelf clastic rocks; the Lower Devonian Gumu Formation (D_1g) is a carbonate mesa marginal facies deposit; the Middle Devonian Donggangling Formation (D_2d) is a sub-tidal sedimentary of the mesa; the Upper Devonian Gedang Formation (D_3g) is a shallow facies carbonate mesa deposit. The Batang Wedge (*bw*) is an informal stratigraphic unit that belongs to the late Early Devonian–early Middle Devonian carbonate mesa slope facies [22]. The Pojiao Formation is the main ore-hosting layer in the mining area, which is in extensive contact with the overlying Batang Wedge, and has a transitional relationship with the Lower Posongchong Formation. It is mudstone, marl locally interspersed with quartz sandstone, carbonaceous mudstone, marl limestone lens, and has been metamorphosed into mica schist, quartz schist, siliceous dolomite, and locally siliceous rocks.

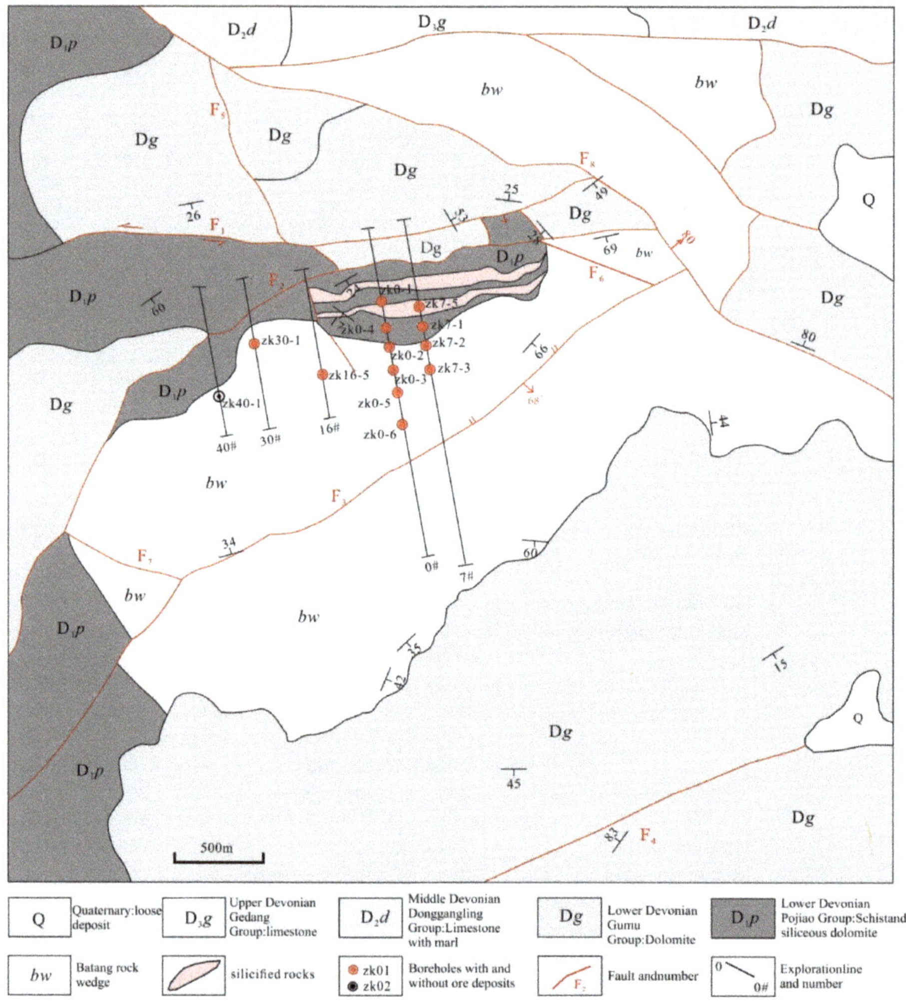

Q	Quaternary:loose deposit	D_3g	Upper Devonian Gedang Group:limestone	D_2d	Middle Devonian Donggangling Group:Limestone with marl	Dg	Lower Devonian Gumu Group:Dolomite	D_1p	Lower Devonian Pojiao Group:Schistand siliceous dolomite
bw	Batang rock wedge		silicified rocks	zk01 / zk02	Boreholes with and without ore deposits	$F_?$	Fault andnumber	0 / 0#	Explorationline and number

Figure 2. Geological map of mining area.

The Pojiao Formation can be subdivided into two sections [23]. The lower section (D_1p^1) can be divided into three beds from bottom to top, as follows: the first bed is quartz schist and siliceous dolomite; the second is quartz schist mixed with siliceous rocks; the third is carbonaceous mica schist and quartz mica schist. The upper section (D_1p^2) can be divided into four beds from bottom to top, as follows: the first bed is thick bedded siliceous rocks sandwiched between banded quartz schist; the second is quartz schist and siliceous dolomite interspersed with banded siliceous rocks; the third is siliceous dolomite interspersed with siliceous rock; the fourth is quartzite mica schist with quartz schist, locally sandwiched with thin siliceous bands, and locally contains striped pyrite.

2.2.2. Tectonic

The overall structural form of the mining area is a monocline that strikes EW and dips south (Figure 2). A secondary steep slope of compressive tensional NNW, near EW and NE faults developed, off the Xingjie fault (Fs). The strata of the mine area were strongly compressed, resulting in a series of soft wrinkles and folds, which were caused by the above-mentioned fault activity.

2.2.3. Ore Body

A total of eight conformable Pb–Zn strata-bound ore bodies have been found in the mining area, in the siliceous dolomite rocks layer of the Lower Devonian Pojiao Formation (D_1p) (Figure 3). Ores have banded and laminated structures (Figure 4). The main ore body is spread along the banded siliceous dolomite on top of the siliceous rocks [23], which can be divided into lower and upper ore-bearing sections.

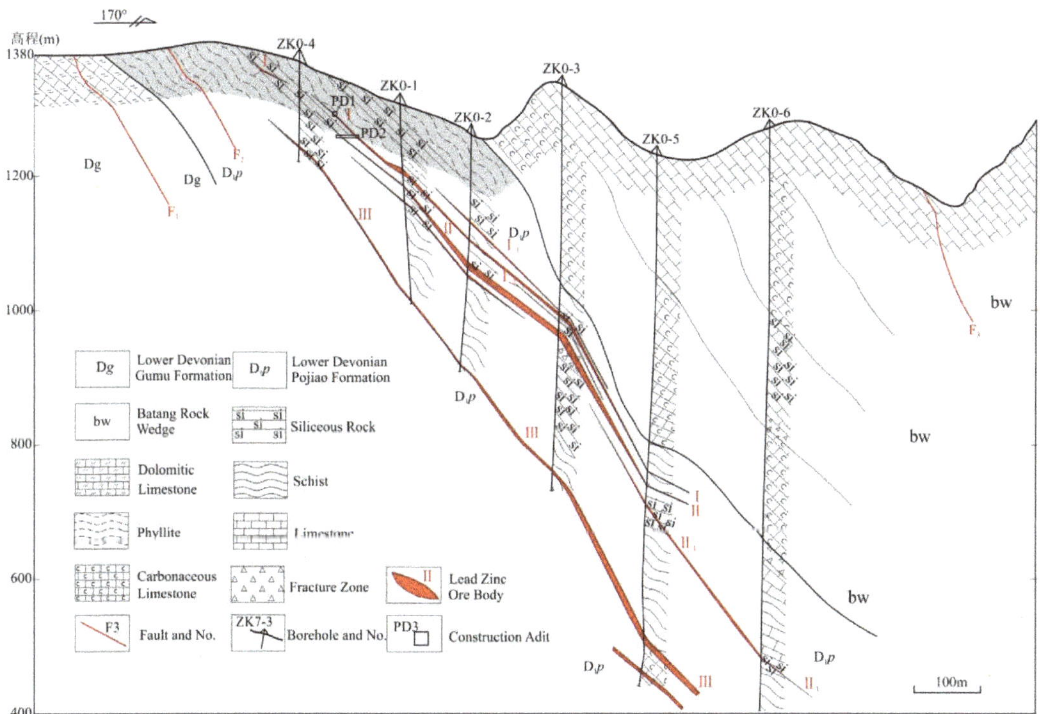

Figure 3. Profile of exploration line 0# in the Caiyuanzi Pb–Zn mining area.

Figure 4. Strata-bound ore body (**a**) and ores with laminated structure (**b**). Abbreviations are as follows: Py, pyrite; Gn, galena; Sph, sphalerite.

2.2.4. Texture and Structure

The ore minerals are mainly pyrite, chalcopyrite, sphalerite, and galena, with a small amount of hematite, pyrrhotite, and magnetite, and the gangue minerals are mainly calcite, quartz, and epidote. The ore minerals have euhedral granular, allomorphic granular and metasomatic residual textures. The sulfide ores have massive, disseminated, veined disseminated, banded, and laminated structures (Figure 5).

Figure 5. Hand specimens and microscopic photos of the ores from the Caiyuanzi deposit. (**a**) Disseminated-banded Pb–Zn ore sample (CYZ-1); (**b**) disseminated-net vein Pb–Zn ore (CYZ-2); (**c**) pyrite replaced and enclosed by galena and sphalerite; (**d**) sphalerite, galena, and chalcopyrite, forming network veins occurring along the wall rock fractures, where galena and chalcopyrite alternate with sphalerite. Abbreviations are as follows: Py, pyrite; Gn, galena; Sph, sphalerite; Qz, quartz; Ccp, chalcopyrite; Ep, epidote; Cal, calcite.

The galena, sphalerite, and chalcopyrite are mostly disseminated, veined, banded, and laminated. Pyrite is subhedral to euhedral granular, forming locally fine-grained aggregates; sphalerite and chalcopyrite are anhedral granular and aggregate; galena is allomorphic granular and aggregate. Pyrite is encapsulated and cemented by other sulfides or alternatively metasomatized, suggesting that pyrite formed early. Sphalerite, galena, and chalcopyrite often occur together where galena and chalcopyrite are replaced with sphalerite, indicating that the formation of sphalerite is later than galena and chalcopyrite, but the formation sequence of galena and chalcopyrite is difficult to determine (Figure 5).

2.2.5. Altered Wall Rocks

The wall rock alteration mainly includes silicification, skarnization, pyritization, and calcitization. The wall rock alteration has an enrichment effect on the Pb–Zn–Cu polymetallic mineralization in the mining area. Skarns include actinolite epidote skarn, and chlorite epidote skarn, which are limited to the siliceous limestone in the Pojiao Formation.

3. Sampling and Analytical Methods

3.1. Samples

All the samples were collected from the PD2 tunnel of the Caiyuanzi ore deposit. The detailed information about those samples is listed in Table 1.

Table 1. Information of the samples.

No.	Locations	Features	Purposes
CYZ—1	PD2	Disseminated banded ores	Pb isotope analyses
CYZ—1(1)	PD2	Banded ores	Pb isotope analyses
CYZ—2	PD2	Disseminated ores	S isotope analyses
CYZ—3	PD2	Skarn ores	Pb isotope analyses
CYZ—4	PD2	Sulfide-bearing limestone	
CYZ—5	PD2	Sulfide-bearing calcium siliceous rocks	S and Pb isotope analyses
CYZ—6	PD2	Skarn ores	
CYZ—7	PD2	Sulfide-bearing schistose marble	S isotope analyses

3.2. Analysis Methods

The micro area in situ sulfur isotope test of sulfide was completed in Nanjing Polyspectrum Testing Technology Co., Ltd., and the galena, sphalerite, pyrite, and chalcopyrite of samples CYZ-2, CYZ-5, and CYZ-7 were selected for sulfur isotope analyses. The mass spectrometer model is the Nu Plasma II MC-ICPMS, and the laser model is Analytical Excite. The deep ultraviolet beam generated by the laser generator is focused on the sulfide surface through the homogenizing optical path. First, the gas background is collected for 40 s, and then the appropriate beam spot (pyrite 33 μm; sphalerite 40 μm; chalcopyrite 50 μm) at a 5 Hz frequency for 35 s, before the aerosol is sent out of the denudation pool by helium, mixed with argon, and then enters the MC-ICPMS (single integration time is 0.3 s, and there are about 110 groups of data within the denudation time of 35 s). We used a GBW07267 pyrite cake pressed by National Geological Experimental Testing Center of Chinese Academy of Geological Sciences ($\delta^{34}S$ = 3.6‰) and GBW07268 chalcopyrite cake pressing ($\delta^{34}S$ = −0.3‰), and NIST SRM 123 crushed zinc blender particles ($\delta^{34}S$ = 17.1‰) as the data quality control, and the long-term external reproducibility is about ±0.6‰ (1 SD).

Micro area in situ lead isotope testing of sulfide was completed in two testing units, respectively. The lead isotope composition analysis of samples CYZ-2 and CYZ-5 was completed in Wuhan Shangpu Analysis Technology Co., Ltd. The instrument model is the MC-ICPMS (Neptune Plus) with multi-receiver mass spectrometry, GeoLas HD with a 193 nm exciter laser ablation system, and a beam spot of 90–120 μm. Energy intensity is 6 mJ/cm^2, the frequency is 8 Hz, the carrier gas (He) is 500 mL/min, collected data (pulses) are 500, and the recommended values of standard samples (Sph HYLM) are $^{208}Pb/^{204}Pb$

(38.519), $^{207}\text{Pb}/^{204}\text{Pb}$ (15.764), and $^{206}\text{Pb}/^{204}\text{Pb}$ (18.217). The lead isotope analyses of samples CYZ-1 and CYZ-3 were completed in the National Key Laboratory of Continental Dynamics, Northwest University. The mass spectrometer model is Nu Plasma II MC-ICPMS, the laser model is Quantronix Integra HE Ti 266 nm NWR UP Femto (ESI, Hartland, WI, USA), and the erosion radius is 15–65 μm. The laser frequency is 5–50 Hz, the erosion mode is 3 μm/s lines scanning, and the He airflow is 0.7 L/min. The sample standard sample cross method is adopted. The standard sample is NIST610, and the analysis error is better than 0.003 (1 σ).

4. Results

4.1. In Situ S Isotopic Compositions

The results of in situ S isotopic compositions of pyrite, chalcopyrite, and sphalerite are shown in Table 2 and Figure 6. Pyrite, chalcopyrite, and sphalerite have $\delta^{34}\text{S}$ values between 0.1‰ and 6‰, with an average of 4.7‰ (Figure 7). Pyrite has $\delta^{34}\text{S}$ values ranging from 4.3‰ to 6‰ (except 0.1‰ for one point), with an average of 5.40‰; sphalerite has $\delta^{34}\text{S}$ values are between 4.7‰ and 5.3‰, with an average of 5.08‰; chalcopyrite has $\delta^{34}\text{S}$ values are between 4.3‰ and 4.9‰, with an average of 4.67‰.

Table 2. In situ sulfur isotopic compositions of ore sulfides.

No.	Point No.	Mineral	$\delta^{34}\text{S}$ (‰)
cyz-2	cpy-sp-1	Sphalerite	4.70
	cpy-sp-3	Chalcopyrite	4.50
	cpy-sp-4	Chalcopyrite	4.90
	cpy-sp-5	Chalcopyrite	4.90
	py-sp-1	Pyrite	0.10
	py-sp-2	Pyrite	5.00
	py-sp-3	Sphalerite	4.70
	py-sp-4	Pyrite	6.00
cyz-5	cpy-Gn-1	Chalcopyrite	4.70
	cpy-Gn-2	Chalcopyrite	5.10
	pyd-pyx-sp-cpy-1	Pyrite	5.30
	pyd-pyx-sp-cpy-2	Pyrite	5.90
	pyd-pyx-sp-cpy-3	Pyrite	5.50
	pyd-pyx-sp-cpy-4	Sphalerite	4.90
	pyd-pyx-sp-cpy-5	Sphalerite	5.20
	sp-cpy-1	Sphalerite	5.30
	sp-cpy-2	Chalcopyrite	4.90
cyz-7	cpy-1	Chalcopyrite	4.40
	cpy1-1	Chalcopyrite	4.80
	cpy-2	Chalcopyrite	4.30
	cpy-5	Sphalerite	4.30

4.2. In Situ Pb Isotopic Ratios

The results of LA-MC-ICPMS in situ Pb isotopes of galena are listed in Table 3. The Pb isotopic ratios of galena are relatively uniform, with $^{206}\text{Pb}/^{204}\text{Pb}$, $^{207}\text{Pb}/^{204}\text{Pb}$, and $^{208}\text{Pb}/^{204}\text{Pb}$ ratios of 18.134–18.202 (mean 18.158), 15.698–15.735 (mean 15.715), and 38.430–38.542 (mean 38.46), respectively.

Figure 6. In situ S isotope analyses of the Caiyuanzi deposit. (**a,b,d**) Microscopic photos of metal sulfide and corresponding in situ sulfur isotope values; (**c**) in situ S isotope test specimen of sulfide ores.

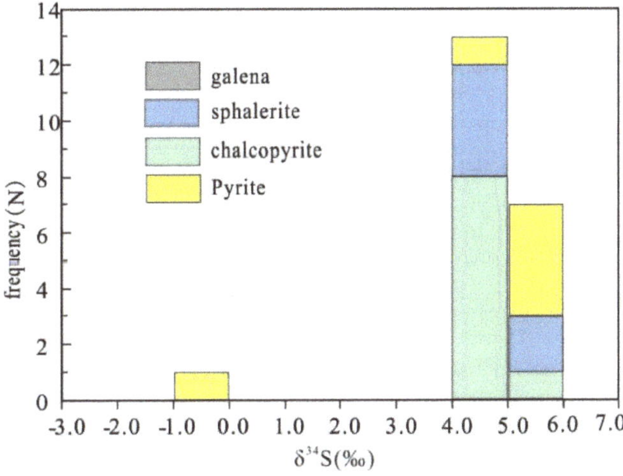

Figure 7. The histogram of S isotopes.

Table 3. In situ Pb isotopes of sphalerite, pyrite, and chalcopyrite.

No.	Deposits	Point Nos.	Mineral	^{206}Pb/^{204}Pb	1s	^{207}Pb/^{204}Pb	1s	^{208}Pb/^{204}Pb	1s
CYZ-1		GN-1	Galena	18.148	0.001	15.712	0.001	38.449	0.002
		1-GN-2	Galena	18.146	0.001	15.711	0.001	38.447	0.002
		1-GN-7	Galena	18.142	0.004	15.707	0.004	38.432	0.010
		1-GN-8	Galena	18.134	0.001	15.698	0.001	38.410	0.003
CYZ-1	Caiyuanzi	GN-1	Galena	18.153	0.001	15.719	0.001	38.464	0.002
		GN-10	Galena	18.147	0.001	15.710	0.001	38.445	0.002
		GN-11	Galena	18.148	0.001	15.711	0.001	38.444	0.002
		GN-2	Galena	18.151	0.001	15.717	0.001	38.461	0.002
		GN-3	Galena	18.153	0.001	15.718	0.001	38.463	0.002
		GN-4	Galena	18.147	0.001	15.712	0.001	38.449	0.003
		GN-5	Galena	18.146	0.001	15.711	0.001	38.446	0.002
		GN-6	Galena	18.145	0.001	15.709	0.001	38.437	0.002
		GN-7	Galena	18.143	0.001	15.707	0.001	38.435	0.002
		GN-8	Galena	18.144	0.001	15.708	0.001	38.438	0.002
		GN-9	Galena	18.142	0.001	15.707	0.001	38.434	0.002
		SP-1	Sphalerite	18.162	0.022	15.720	0.020	38.473	0.052
CYZ-2	Caiyuanzi	01	Pyrite	18.170	0.003	15.728	0.003	38.510	0.006
		04	Pyrite	18.167	0.002	15.721	0.002	38.492	0.006
CYZ-5	Caiyuanzi	02	Pyrite	18.147	0.006	15.720	0.004	38.467	0.011
CYZ-3	Caiyuanzi	GN-1	Galena	18.147	0.001	15.703	0.001	38.430	0.003
		GN-10	Galena	18.154	0.001	15.709	0.001	38.444	0.003
		GN-11	Galena	18.151	0.001	15.706	0.001	38.438	0.002
		GN-12	Galena	18.152	0.001	15.708	0.001	38.445	0.002
		GN-2	Galena	18.149	0.001	15.705	0.001	38.436	0.003
		GN-3	Galena	18.158	0.001	15.716	0.001	38.468	0.002
		GN-4	Galena	18.158	0.001	15.717	0.001	38.470	0.002
		GN-5	Galena	18.157	0.001	15.716	0.001	38.469	0.003
		GN-6	Galena	18.153	0.001	15.711	0.001	38.455	0.003
		GN-7	Galena	18.151	0.001	15.708	0.001	38.445	0.003
		GN-8	Galena	18.149	0.001	15.706	0.001	38.443	0.002
		GN-9	Galena	18.156	0.001	15.711	0.001	38.453	0.002

5. Discussion

5.1. Source and Formation Mechanism of Reduced Sulfur

Sulfur isotopes are one of the most important bases for determining the source of sulfur and the formation process of sulfide deposits [24–61]. Three sources of sulfur have been proposed, as follows: (1) mantle-derived sulfur, $\delta^{34}S$ = −3‰ to 3‰ (average 0‰); (2) sedimentary sulfur (marine sulfate), which could form reduced sulfur by thermochemical sulfate reduction (TSR) or bacterial sulfate reduction (BSR) [32,37]; and (3) mixed sulfur of the above two types [31].

The ore mineral assemblages of the Caiyuanzi Pb–Zn deposit are simple, with mainly pyrite, sphalerite, and galena and other sulfides. The $\delta^{34}S$ values of the Caiyuanzi deposit are relatively homogeneous (0.1‰ to 6‰, with a mean value of 4.7‰) and positive, which may represent the $\delta^{34}S_{\Sigma S}$ of the ore-forming hydrothermal fluids.

The $\delta^{34}S$ values differ significantly from the values found in typical Mississippi Valley-type (MVT) Pb–Zn deposits, whose reduced sulfur was mainly formed by TSR and/or BSR; for example, the $\delta^{34}S$ values of sulfides in the Daliangzi (MVT) Pb–Zn deposit are mainly 10‰ to 20‰ [39,41], while the in situ $\delta^{34}S$ values of sulfides in the Maoping (MVT) Pb–Zn deposit ore are −20.4‰ to 25.6‰ [33,35,40].

In addition, the sulfur isotopic compositions of the Caiyuanzi Pb–Zn deposit are similar to those of the adjacent Gejiu Sn and Dulong Sn–Zn polymetallic deposits (Figure 8), whose sulfur was mainly derived from the magmatic rocks, with less marine sulfate [13,14]. For example, the $\delta^{34}S$ values of the Gejiu deposit are mainly −3.1‰ to 8.4‰ [51], and the $\delta^{34}S$

values of the Dulong deposit are mainly 4.2‰ to 12.4‰, most being 5.2‰–9.4‰ [10,42]. Hence, we propose that the sulfur for the Caiyuanzi deposit is mainly derived from the magmatic rocks, although some contribution from the wall rock cannot be excluded.

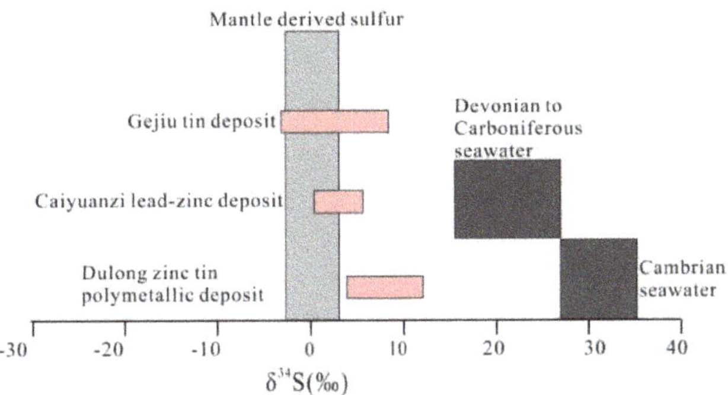

Figure 8. Comparison of sulfur isotopic composition between Pb–Zn deposits in strata of different ages in the Laojunshan region and seawater- and mantle-derived sulfur in the same period (the isotopic composition range of mantle derived sulfur is according to [52]; the sulfur isotopic composition range of seawater in the same period is according to [54]; the sulfur isotopic composition range of ore deposits is according to [10,42,51]).

5.2. Source of Metals

Due to the low contents of U and Th, the proportion of radiogenic Pb in sulfide minerals is negligible. Therefore, Pb isotopes of galena could represent the Pb isotopes of the ore-forming fluids without age correction [49,50]. The in situ Pb isotopic ratios of galena from the Caiyuanzi Pb–Zn deposit obtained in this study has a narrow range (Table 3), suggesting either a single source or a high degree of homogenization in the ore-forming metals in this deposit [39]. In this paper, we collected Pb isotopic data from the Laojunshan granites, marble, schist, and ores in Dulong (Table 3). The samples of the Caiyuanzi deposit fall on the average upper crustal growth curve and mantle curve in the corresponding Figure 9a,b, respectively. The whole rock Pb isotope ratios of marble and schist are significantly different from those of the Caiyuanzi deposit (Figure 9), so the wall rocks (marble and schist) may not have provide lead to the deposit. The data of Yanshanian granites are concentrated between the orogenic belt and the upper crust, close to the upper crust, and its ^{208}Pb/^{204}Pb and ^{207}Pb/^{204}Pb ratios are consistent with the data of Caiyuanzi sample points. The Pb isotope ratios of the Caiyuanzi and Dulong deposits and the Yanshanian granites have the same distribution range and trend and are projected between the orogenic belt and the upper crustal evolution curve, indicating that the Laojunshan granites might have provided metals for the Caiyuanzi deposit. Another end member should be the underlying Proterozoic rocks, with relatively unradiogenic crustal Pb.

In addition, the μ values (^{238}U/^{204}Pb) of the Caiyuanzi deposit range from 9.71 to 9.76, which are between the mantle or lower crust Pb (μ = 7.86–7.94) and upper crust Pb (μ = 9.81), and so could be a mixture between them. The average value of ω is 39.14, which is closer to the upper crust Pb between normal lead (ω = 35.55 ± 0.59) and the upper crust Pb (41.860) [38]. The Th/U average value is 3.89, which is close to normal Pb (Th/U = 3.92 ± 0.9), slightly higher than the upper crust of the Chinese mainland (Th/U = 3.76). In the corresponding Pb isotope △β–△γ genetic classification diagram [55], the data point of the Caiyuanzi deposit falls in the upper crust Pb source area (Figure 10).

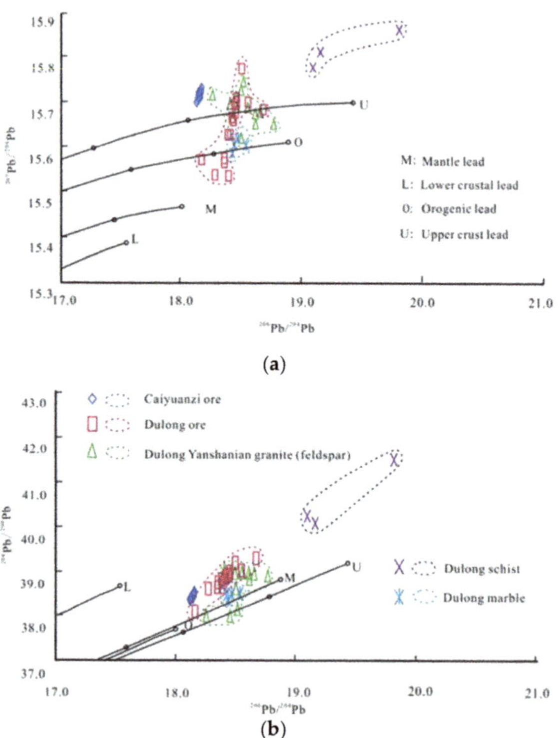

Figure 9. The Pb model diagram for the Caiyuanzi deposit (base map is modeled after [38,62], and lead isotope data of samples in the Dulong ore area are quoted from [43]). (**a**) The samples of the Caiyuanzi deposit fall on the average upper crustal growth curve. (**b**) The mantle curve in the corresponding.

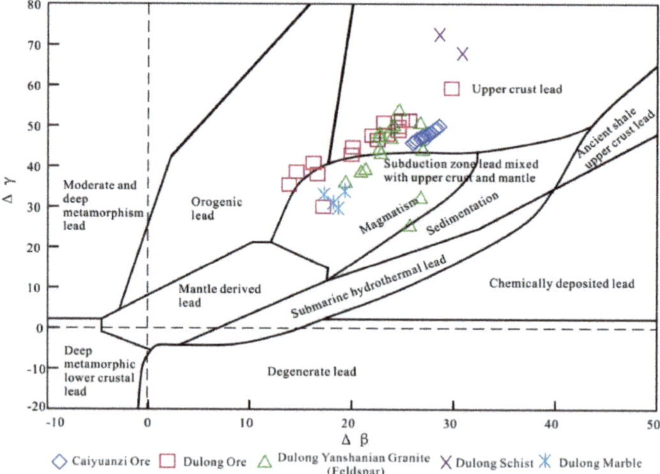

Figure 10. Lead isotope Δβ-Δγ genetic classification diagram (base map is modeled after [55], and lead isotope data of samples in the Dulong ore area are quoted from [43]).

The Pb isotopes of the Dulong deposit and the Yanshanian granites span two source areas of upper crustal Pb and magmatic Pb, and generally show a trend from magmatic Pb to crustal Pb, suggesting that the intermediate-acidic magma rich in deep-source low μ-value Pb has been contaminated by shallow-source high μ-value Pb during ascent. The μ values of the Dulong deposit (9.56) and the Dulong Yanshanian granites ($\mu = 9.62$) are similar to that of the Caiyuanzi deposit ($\mu = 9.72$).

The degree of crustal contamination of magmatic hydrothermal fluids is positively correlated with the μ values [44], suggesting that the intermediate-acid magmatic hydrothermal fluids related to the Caiyuanzi Pb–Zn deposit are greatly contaminated by crustal materials. In addition, the Pb isotope of sulfides in the Caiyuanzi deposit is significantly higher in U/Pb ($^{208}Pb/^{204}Pb > 18.000$, $^{207}Pb/^{204}Pb > 15.300$), and slightly lower in Th/Pb ($^{208}Pb/^{204}Pb < 39.000$), suggesting that the ore-forming material is dominated by upper crust lead, with a small amount of deep crust-derived magmatic Pb, showing the characteristics of orogenic belt Pb.

5.3. Ore Genesis

At present, the ore genesis of the Pb–Zn deposits in the Laojunshan area is still controversial. The focus is whether it belongs to a SEDEX deposit or a magmatic hydrothermal deposit. In this paper, the in situ S and Pb isotopes of the Caiyuanzi Pb–Zn deposit show that the sulfur was mainly derived from the mixed sources of magmatic rocks and marine sulfate, and the source of metal Pb is the upper crust. The ore bodies are strata-bound and stratiform, which resembles the SEDEX deposits (Table 4). However, the Caiyuanzi Pb–Zn deposit is characterized by epigenetic mineralization with extensive pyrrhotite and skarnization, which can be compared to the general metallogenic characteristics of magmatic hydrothermal deposits. Most of ore bodies and ores underwent some fractures and deformation. Therefore, all of these observations suggest that the Caiyuanzi Pb–Zn deposit was a product of syn-sedimentary hydrothermal exhalative and superimposed magmatic–hydrothermal ore-forming processes.

Table 4. Summary and comparison of principal characteristics of SEDEX, MVT, magmatic hydrothermal vein-type, and the Caiyuanzi Pb–Zn deposit.

Features	Sedimentary Exhalative (SEDEX)	Mississippi Valley-Type (MVT)	Magmatic Hydrothermal Vein-Type	Caiyuanzi Deposit
Ore-forming age	Syngenetic—early diagenetic	Epigenetic	Epigenetic	Syngenetic, epigenetic
Geological setting	Extensional first and second-order basins	Carbonate platform sequences and thrust belts, rare occurrences in extensional basins	Varied	Thrust belt
Host rocks	Varied. Mainly sandstones, siltstones, limestones, dolomites, cherts, and turbidites	Limestones, dolostones, and rare micrites	Varied. Sandstone, siltstone, and carbonates	Siliceous dolomite, quartz schist
Structural controls	Syn-sedimentary faults controlling sub-basins and associated fractures and breccias	Normal, trans-tensional, and wrench faults and associated fractures and breccias	Fault zone/strata	Lithologic interface
Associated igneous activity	No direct association with igneous activity, but tuffs related to synchronous distal volcanism may be present	Not associated with igneous activity	Associated with igneous activity	Associated with igneous activity
Ore-body morphology	Single or multiple wedge- or lens-shaped, or sheeted/stratiform morphology	Commonly discordant on a deposit scale but strata-bound on a regional scale	Veins, stratiform-like morphology	Stratiform-like morphology

Table 4. *Cont.*

Features	Sedimentary Exhalative (SEDEX)	Mississippi Valley-Type (MVT)	Magmatic Hydrothermal Vein-Type	Caiyuanzi Deposit
Mineralogy	Sp, Gn, and Py (±Pyr) and common Brt, Ap, and very rare Fl	Sp, Gn, Py, Mar, minor Dol, Cal, Fl (rare), Cpy, and Brt (minor to absent)	Sp, Gn, Py, Cpy, and minor Brt	Sp, Gn, Py, Ccp, minor Hem, Po, Mag, Grt, and Chl
Host rock alteration	Silicification, chloritization, epidotization, albitization	Carbonatization, silicification	Silicification, pyritic and carbonate alteration	Silicification, skarnization
References	[56]	[57–59]	[60,61]	This paper

Abbreviations are as follows: Sp, sphalerite; Gn, galena; Py, pyrite; Brt, barite; Ap, apatite; Cpy, chalcopyrite; Fl, fluorite; hem, hematite; Po, pyrrhotite; Mag, magmatite, Grt, garnet, Chl, chlorite.

In South China, the Devonian is widely exposed and consists of carbonate and clastic deposition of large transgressive-regressive cycles [63]. The Devonian sedimentary rocks in South China host numerous SEDEX pyrite deposits, such as Dajiangping [64], and sedimentary reworking deposits, such as Huodehong [65]. The reducing environment within the early Devonian sea floor led to the rapid burial of organic matter. The reduction of marine sulfate by organic matter formed S^{2-} and then mixed with deeply-derived Pb and Zn, etc., and eventually formed the syn-sedimentary sulfide ore bodies. During the Cretaceous, extensional tectonics developed in South China. The southeastern Yunnan–northern Guangxi and the large-scale mineralization in Late Mesozoic in western South China were controlled by a similar continental dynamic background [48]. After 135 Ma, the movement direction of the Izanagi plate in eastern China changed, from the subduction of the Eurasian continent to rapid strike-slip along a NE direction [46], and the South China region underwent lithospheric extension. As a result of the extension of the lithosphere, the lithospheric mantle has undergone underplating and upwelling, resulting in a large amount of ferromagnesian magmatism. Upwelling of this magma and underplating of the lower crust, as well as the partial melting of the lower crust, may have produced granitic melt that invaded the upper crust [45]. During this period, a large number of granite bodies were formed in southeastern Yunnan, such as the Gejiu, Laojunshan, and Bozhushan granites. At the same time, a number of world-class W–Sn polymetallic deposits related to granites were formed, such as the Gejiu, Dulong, Dachang, and Bainiuchang deposits. The newly obtained S and Pb isotopic data from Caiyuanzi suggest that the mineralization is related to the Yanshanian granites in Laojunshan. The regional magmatic hydrothermal events contributed to the Dulong Sn (diopside, garnet, and tremolite skarn) and Zn mineralization (epidote skarn, although some parts show a lack of skarn minerals), and Caiyuanzi skarnization (epidote, garnet skarn).

6. Conclusions

(1) The sulfur and lead of ore minerals come from the upper crust and mantle.
(2) The Caiyuanzi Pb–Zn deposit is a hydrothermal deposit formed by the superimposed magma of sedimentary exhalative.

Author Contributions: Investigation, Y.J. and Y.Z.; writing–original draft preparation, Y.J.; writing–review and editing, H.N., C.Y., M.L., H.X., J.C. and H.L.; supervision, Y.C. and Y.Z.; funding acquisition, Y.Z. All authors have read and agreed to the published version of the manuscript.

Funding: This research was financially supported by the National Natural Science Foundation of China (No. 41802087) and the Basic Research Program of Yunnan Provincial Department of Science and Technology, China (No. 2019FB144).

Data Availability Statement: The data used to support this study are included within the article.

Acknowledgments: The field work was supported and assisted by workers of the No. 317 Geological Teams of Yunnan Nonferrous Geological Bureau. The experimental process was guided and assisted by the staff of the two laboratories. We would like to express our heartfelt thanks to them! We sincerely thank the Editorial Board members and anonymous reviewers for their constructive comments.

Conflicts of Interest: The authors declare that they have no known competing financial interest or personal relationships that could have appeared to influence the work reported in this paper.

References

1. Lin, Q.S. Geological characteristics and Ore-hunting Significance of the Hongshiyan Jets sedimentary Type lead-zinc-copper deposits in Xichou County, Yunnan Province. *Fujian Geol.* **2013**, *32*, 185–192.
2. Li, Z.Q.; Huan, J.G.; Han, R.S.; Ren, T.; Wang, L.; Qiu, W.L. Geochemical Characteristics and Tectonic Setting of Epidote Rocks from in the Hongshiyan Pb-Zn-Cu Polymetallic Ore Deposit, Southeastern Yunnan Province. *Geol. Explor.* **2013**, *49*, 289–299.
3. Yang, C.B.; Pu, C.; Shi, G.F.; Wang, Y.Z.; Yang, F.X.; Zheng, T.P.; Zhang, X.H. The feature and prospecting criteria of Gaji Pb-Zn Deposit in Xichou, Yunnan. *Yunnan Geol.* **2020**, *39*, 1–7.
4. Yang, C.B.; Pu, C.; Wang, Y.Z.; Shi, G.F.; Yang, F.X.; Zheng, T.P.; Zhang, X.H. Mineralization characteristics of the Gaji Cu-Pb-Zn polymetallic deposit, Xichou County, Yunnan Province, China. *Acta Mineral. Sin.* **2020**, *40*, 255–266.
5. Ma, Y.H.; Wang, J.S.; Yu, W.X.; Yang, C.B.; Zheng, X.J.; Han, H.H.; He, H. Zircon U-Pb geochronology and Hf isotopic characteristics of metamorphic rocks in the Gaji polymetallic metallogenic area in the southeastern Yunnan and their geological implications. *Acta Mineral. Sin.* **2021**, *41*, 129–140.
6. Cai, J.J.; Wang, J.D.; Li, L.X.; Yang, Z. The feature and genesis of Caiyuanzi Pb-Zn-Cu Deposit in Xichou, Yunnan. *Yunnan Geol.* **2021**, *40*, 193–198.
7. Roger, F.; Leloup, P.H.; Jolivet, M.; Lacassin, R.; Trinh, P.T.; Brunel, M.; Seward, D. Long and complex thermal history of the Song Chay metamorphic dome (Northern Vietnam) by multi-system geochronology. *Tectonophysics* **2000**, *321*, 449–466. [CrossRef]
8. Maluskia, H.; Lepvrier, C.; Jolivet, L.; Carter, A.; Roques, D.; Beyssac, O.; Trong, T.T.; Duc, T.N.; Avigad, D. Ar-Ar and fission-track ages in the Song Chay Massif: Early Triassic and Cenozoic tectonics in northern Vietnam. *J. Asian Earth Sci.* **2001**, *19*, 233–248. [CrossRef]
9. Yang, J.H.; Cawood, P.A.; Du, Y.S.; Huang, H.; Hu, L.S. Detrital record of Indosinian mountain building in SW China: Provenance of the Middle Triassic turbidites in the Youjiang Basin. *Tectonophysics* **2012**, *574–575*, 105–117. [CrossRef]
10. Niu, H.B. *The Ore-Forming Fluid and Metallogenesis of W-Sn Polymetallic Deposits in the Nanwenhe-Song Chay Dome Areas, Southestern Yunnan Province*; Guangzhou Institute of Geochemistry, Chinese Academy of Sciences: Guangzhou, China, 2021.
11. Mei, M.X.; Li, Z.Y. Sequence Stratigraphic Succession and Sedimentary Basin Evolution from Late Paleozoic to Triassic in Yunnan-Guizhou-Guangxi Region. *Geoscience* **2004**, *18*, 555–563.
12. Zhang, S.T.; Feng, M.G.; Lv, W. Analysis of the Nanwenhe Metamorphic Core Complex in Southeast Yunnan. *Reginoal Geol. China* **1998**, *17*, 390–397.
13. Liu, Y.P.; Li, Z.X.; Li, H.M.; Guo, L.G.; Xu, W.; Ye, L.; Li, C.Y.; Pi, D.H. U-Pb geochronology of cassiterite and zircon from the Dulong Sn-Zn deposit: Evidence for Cretaceous large-scale granitic magmatism and mineralization events in southeastern Yunnan province, China. *Acta Petrol. Sin.* **2007**, *23*, 967–976.
14. Xu, B.; Jiang, S.Y.; Wang, R. Late Cretaceous granites from the giant Dulong Sn-polymetallic ore district in Yunnan Province, South China: Geochronology, geochemistry, mineral chemistry and Nd-Hf isotopic compositions. *Lithos* **2015**, *218–219*, 54–72. [CrossRef]
15. Carter, A.; Roques, D.; Bristow, C.; Kinny, P. Understanding Mesozoic accretion in Southeast Asia: Significance of Triassic thermotectonism (Indosinian orogeny) in Vietnam. *Geology* **2001**, *29*, 211–214. [CrossRef]
16. Guo, L.G.; Liu, Y.P.; Li, C.Y.; Xu, W.; Ye, L. SHRIMPzircon U-Pb geochronology and lithogeochemistry of Caledonian Granites from the Laojunshanarea, south eastern Yunnan province, China: Implications for the collision between the Yangtze and Cathaysia blocks. *Geochem. J.* **2009**, *43*, 101–122. [CrossRef]
17. Feng, J.R.; Mao, J.W.; Pei, R.F.; Zhou, Z.H.; Yang, Z.X. SHRMP zircon U-Pb dating and geochemical characteristics of Laojunshan granite intrusion from the Wazhu tungsten deposit, Yunnan Province and their implications for petrogenesis. *Acta Petrol. Sin.* **2010**, *26*, 845–857.
18. Zhang, B.H.; Ding, J.; Ren, G.M.; Zhang, L.K.; Shi, H.Z. Geochronology and Geochemical Characteristics of the Laojunshan Granites in Maguan County, Yunnan Province, and Its Geological Implications. *Acta Geol. Sin.* **2012**, *86*, 587–601.
19. Feng, J.R.; Mao, J.W.; Pei, R.F. Ages and geochemistry of Laojunshan granites in southeastern Yunnan, China: Implications for W-Sn polymetallic ore deposits. *Mineral. Petrol.* **2013**, *107*, 573–589. [CrossRef]
20. Liu, Y.B.; Mo, X.X.; Zhang, D.; Que, C.Y.; Di, Y.J.; Pu, X.M.; Cheng, G.S.; Ma, H.H. Petrogenesis of the Late Cretaceous granite discovered in the Laojunshan region, southeastern Yunnan Province. *Acta Petrol. Sin.* **2014**, *30*, 3271–3286.
21. Lan, J.B.; Liu, Y.P.; Ye, L.; Zhang, Q.; Wang, D.P.; Su, H. Geochemistry and age spectrum of Late Yanshanian granites from Laojunshan Area, Southeastern Yunnan Province, China. *Acta Mineral. Sin.* **2016**, *36*, 477–491.

22. The Second Geological Brigade of Yunnan Provincial Bureau of Geological and Mineral Exploration and Development. *Regional Geological Survey Report of 1/50000 Xingjie Street*; The Second Geological Brigade of Yunnan Provincial Bureau of Geological and Mineral Exploration and Development: Wenshan, China, 1995.

23. Hunan Tiangong Mining Technology Co., Ltd. *Detailed Geological Survey Report of Ganshapo Lead Zinc (Copper) Mine in Xichou County, Yunnan Province*; Hunan Tiangong Mining Technology Co., Ltd.: Changsha, China, 2014.

24. Cheng, Y.S.; Sun, W.M. Sulfur and lead isotope geochemistry of the Dulong Sn-Zn polymetallic ore deposit, Yunnan province, China. *Nonferrous Met.* **2019**, *71*, 32–36, 65.

25. Guo, Y.W.; Zhang, Q.S.; Zhu, P.B.; Lu, D.; Wang, N.; Ai, J.B. Study on sulphur and lead isotope composition characteristics and ore material source of Wanlongshan Sn-Zn polymetallic ore deposit in Yunnan province. Contrib. *Geol. Min. Res.* **2018**, *33*, 564–572.

26. He, F.; Zhang, Q.; Wang, D.P.; Liu, Y.P.; Ye, L.; Bao, T.; Wang, X.J.; Miao, Y.L.; Zhang, S.K.; Su, H.; et al. Ore-Forming Materials Sources of the Dulong Sn-Zn Polymetallic Deposit, Yunnan, Evidences from S-C-O Stable Isotopes. *Bull. Min. Petrol. Geochem.* **2014**, *33*, 900–907.

27. Ohmoto, H. Systematics of sulfur and carbon isotopes in hydrothermal ore deposits. *Econ. Geol.* **1972**, *67*, 551–578. [CrossRef]

28. Ohmoto, H. Stable isotope geochemistry of ore deposit. *Rev. Mineral.* **1986**, *16*, 491–559.

29. Zhou, J.X.; Wang, X.C.; Wilds, S.A.; Luo, K.; Huang, Z.L.; Wu, T.; Jin, Z.G. New Insights into the Metallogeny of MVT Zn-Pb Deposits: A Case Study from the Nayongzhi in South China, Using Field Data, Fluid Compositions, and In-Situ S-Pb Isotopes. *Am. Mineral.* **2018**, *103*, 91–108. [CrossRef]

30. Machel, H.G.; Krouse, H.R.; Sassen, R. Products and distinguishing criteria of bacterial and thermochemical sulfate reduction. *Appl. Geochem.* **1995**, *10*, 373–389. [CrossRef]

31. Tan, S.C.; Zhou, J.X.; Zhou, M.F.; Ye, L. In-situ S and Pb isotope constraints in an evolving hydrothermal system, Tianbaoshan Pb-Zn-(Cu) deposit in South China. *Ore Geol. Rev.* **2019**, *115*, 103177. [CrossRef]

32. Seal, R.R. Sulfur isotope geochemistry of sulfide minerals. *Rev. Mineral. Geochem.* **2006**, *61*, 633–677. [CrossRef]

33. Tan, S.C.; Zhou, J.X.; Luo, K.; Xiang, Z.Z.; He, X.H.; Zhang, Y.H. The sources of ore-forming elements of the Maoping large-scale Pb-Zn deposit, Yunnan Province: Constrains from in-situ S and Pb isotopes. *Acta Petrol. Sin.* **2019**, *35*, 3461–3476.

34. Basuki, N.; Taylor, B.E.; Spooner, E.T.C. Sulfur isotope evidence for thermochemical reduction of dissolved sulfate in Mississippi Valley-type zinc-lead mineralization, Bongara area, northern Peru. *Econ. Geol.* **2008**, *103*, 783–799. [CrossRef]

35. Zhou, J.X.; Xiang, Z.Z.; Zhou, M.F.; Feng, Y.X.; Luo, K.; Huang, Z.L.; Wu, T. The giant upper Yangtze Pb-Zn province in SW China: Reviews, new advances and a new genetic model. *J. Asian Earth Sci.* **2018**, *154*, 280–315. [CrossRef]

36. Zheng, Y.F.; Chen, J.F. *Stable Isotope Geochemistry*; Science Press: Beijing, China, 2000.

37. Shanks, W.C. Stable isotopes in seafloor hydrothermal systems: Vent fluids, hydrothermal deposits, hydrothermal alteration, and microbial processes. *Rev. Min. Geochem.* **2001**, *43*, 469–525. [CrossRef]

38. Doe, B.R.; Zartman, R.E. *Geochemistry of Hydrothermal Ore Deposits*; John Wiley & Sons: New York, NY, USA, 1979.

39. Huang, Z.L.; Chen, J.; Han, R.S.; Li, W.B.; Liu, C.Q.; Zhang, Z.L.; Ma, D.Y.; Gao, D.R.; Yang, H.L. *Geochemistry and Genesis of the Huize Super-Large Lead-Zinc Deposit in Yunnan Province—A Discussion on the Relationship between the Emeishan Basalt and Lead-Zinc Mineralization*; Geological Publishing House: Beijing, China, 2004.

40. Ren, S.L.; Li, Y.H.; Zeng, P.S.; Qiu, W.L.; Fan, C.F.; Hu, G.Y. Effect of Sulfate Evaporate Salt Layer in Mineralization of the Huize and Maoping Lead-Zinc Deposits in Yunnan: Evidence from Sulfur Isotope. *Acta Geol. Sin.* **2018**, *92*, 1041–1055.

41. Zhou, J.X.; Huang, Z.L.; Zhou, M.F.; Zhu, X.K.; Muchez, P. Zinc, sulfur and lead isotopic variations in carbonate-hosted Pb-Zn sulfide deposits, Southwest China. *Ore Geol. Rev.* **2014**, *58*, 41–54. [CrossRef]

42. Liu, Y.P.; Li, C.Y.; Gu, T.; Wang, J.L. Isotopic constraints on the source of ore-forming materials of Dulong Sn-Zn polymetallic deposit, Yunnan. *Geol. Geochem.* **2000**, *28*, 75–81.

43. He, F.; Zhang, Q.; Liu, Y.P.; Ye, L.; Miao, Y.L.; Wang, D.P.; Su, H.; Bao, T.; Wang, X.J. Lead Isotope Compositions of Dulong Sn-Zn Polymetallic Deposit, Yunnan, China: Constraints on Ore-forming Metal Sources. *Acta Mineral. Sin.* **2015**, *35*, 309–317.

44. Yang, Y.L.; Ye, L.; Cheng, Z.T.; Bao, T.; Gao, W. A tentative discussion on the genesis of skarn Pb-Zn deposits in the Baoshan-Zhenkang terrane. *Acta Petrol. Mineral.* **2012**, *31*, 554–564.

45. Bergantz, G.W. Underplating and partial melting: Implications for melt generation and extraction. *Science* **1989**, *245*, 1093–1095. [CrossRef]

46. Mao, J.W.; Wang, Y.T.; Li, H.M.; Piraino, F.; Zhang, C.Q.; Wang, R.T. The relationship of mantle-derived fluids to gold metallogenesis in the Jiaodong Peninsula: Evidence from D-O-C-S isotope systematics. *Ore Geol. Rev.* **2008**, *33*, 361–381. [CrossRef]

47. Chen, Y.Q.; Huang, J.N.; Lu, Y.X.; Xia, Q.L.; Sun, M.X.; Li, J.R. Geochemistry of Elements, Sulphur-Lead Isotopes and Fluid Inclusions from Jinla Pb-Zn-Ag Poly-Metallic Ore Field at the Joint Area across China and Myanmar Border. *J. Earth Sci.* **2009**, *34*, 585–594.

48. Cheng, Y.B.; Mao, J.W.; Chen, X.L.; Li, W. LA-ICP-MS Zircon U-Pb Dating of the Bozhushan Granite in Southeastern Yunnan Province and Its Significance. *J. Jilin Univ.* **2010**, *40*, 869–878.

49. Carr, G.R.; Dean, J.A.; Suppel, D.W.; Heithersay, P.S. Precise lead isotope fingerprinting of hydrothermal activity associated with Ordovician to Carboniferous metallogenic events in the Lachlan fold belt of New South Wales. *Econ. Geol.* **1995**, *90*, 1467–1505. [CrossRef]

50. Zhou, J.X.; Yang, Z.M.; An, Y.L.; Luo, K.; Liu, C.; Ju, Y. An evolving MVT hydrothermal system: Insights from the Niujiaotang Cd-Zn ore field, SW China. *J. Asian Earth Sci.* **2022**, *237*, 105357. [CrossRef]

51. Cheng, Y.B.; Mao, J.W. *Study on Diagenesis and Mineralization in Gejiu Super Large Tin Polymetallic Ore Concentration Area, Yunnan*; Geological Publishing House: Beijing, China, 2014.
52. Chaussidon, M.; Albarède, F.; Sheppard, S.M.F. Sulphur isotope variations in the mantle from ion microprobe analyses of micro-sulphide inclusions. *Earth Planet. Sci. Lett.* **1989**, *92*, 144–156. [CrossRef]
53. Ohmoto, H.; Goldhaber, M.B. *Sulfur and Carbon Isotopes*, 3rd ed.; John Wiley & Sons: New York, NY, USA, 1997.
54. Claypool, G.E.; Holser, W.T.; Kaplan, I.R.; Sakai, H.; Zak, I. The age curves of sulfur and oxygen isotopes in marine sulfate and their mutual interpretation. *Chem. Geol.* **1980**, *28*, 199–260. [CrossRef]
55. Zhu, B.Q. *Theories and Application of Isotopic System in Geoscience: Crustal and Mantle Evolution in China Continent*; Science Press: Beijing, China, 1998; pp. 1–330, (In Chinese with English Abstract).
56. Leach, D.L.; Sangster, D.F.; Kelley, K.D.; Large, R.R.; Garven, G.; Allen, C.R. Sediment-hosted Pb-Zn Deposits: A global perspective. *Econ. Geol.* **2005**, *100*, 561–608.
57. Leach, D.L.; Bradley, D.C.; Huston, D.; Pisarevsky, S.A.; Taylor, R.D.; Gardoll, S.J. Sediment-hosted lead-zinc deposits in Earth history. *Econ. Geol.* **2010**, *105*, 593–625. [CrossRef]
58. Luo, K.; Zhou, J.X.; Huang, Z.L.; Caulfield, J.; Zhao, J.X.; Feng, Y.X.; Ouyang, H. New insights into the evolution of Mississippi Valley-Type hydrothermal system: A case study of the Wusihe Pb-Zn deposit, South China, using quartz in-situ trace elements and sulfides in situ S-Pb isotopes. *Am. Mineral.* **2020**, *105*, 35–51. [CrossRef]
59. Luo, K.; Zhou, J.X.; Huang, Z.L.; Wang, X.C.; Wilde, S.A.; Zhou, W.; Tian, L. New insights into the origin of early Cambrian carbonate-hosted Pb-Zn deposits in South China: A case study of the Maliping Pb-Zn deposit. *Gondwana Res.* **2019**, *70*, 88–103. [CrossRef]
60. Wang, C.; Zhang, D.; Wu, G.; Santosh, M.; Zhang, J.; Xu, Y.; Zhang, Y. Geological and isotopic evidence for magmatic-hydrothermal origin of the Ag-Pb-Zn deposits in the Lengshuikeng District, east-central China. *Miner. Depos.* **2014**, *49*, 733–749. [CrossRef]
61. Zhou, Z.; Wen, H.; Qin, C.; Liu, L. Geochemical and isotopic evidence for a magmatic-hydrothermal origin of the polymetallic vein-type Zn-Pb deposits in the northwest margin of Jiangnan Orogen, South China. *Ore Geol. Rev.* **2017**, *86*, 673–691. [CrossRef]
62. Zartman, R.E.; Doe, B.R. Plumbotectonics: The model. *Tectonophysics* **1981**, *75*, 135–162. [CrossRef]
63. Ma, X.P.; Liao, W.H.; Wang, D.M. *The Devonian System of China, with a Discussion on Sea-Level Change in South China*; Geological Society of London Special Publication: London, UK, 2009; Volume 314, pp. 241–262.
64. Qiu, W.J.; Zhou, M.F.; Li, X.; Williams-Jones, A.E.; Yuan, H. The genesis of the giant Dajiangping SEDEX-type pyrite deposit, South China. *Econ. Geol.* **2018**, *113*, 1419–1446. [CrossRef]
65. Luo, K.; Zhou, J.X.; Sun, G.T.; Nguyen, A.; Qin, Z.X. The metallogeny of the Devonian sediment-hosted sulfide deposits, South China: A case study of the Huodehong deposit. *Ore Geol. Rev.* **2022**, *143*, 104747. [CrossRef]

minerals

Article

Pore Variation Characteristics of Altered Wall Rocks in the Huize Lead–Zinc Deposit, Yunnan, China and Their Geological Significance

Yanglin Li [1], Zhigang Kong [1,*], Changqing Zhang [2], Yue Wu [3], Xue Yang [4], Yu Wang [1] and Gang Chen [2]

1. Faculty of Land Resource Engineering, Kunming University of Science and Technology, Kunming 650093, China
2. MNR Key Laboratory of Metallogeny and Mineral Assessment, Institute of Mineral Resources, Chinese Academy of Geological Sciences, Beijing 100037, China
3. College of Resources and Environment, Yangtze University, Wuhan 430100, China
4. Sichuan Huili Lead & Zinc Co., Ltd., Huili 615100, China
* Correspondence: zhigangkong@kust.edu.cn; Tel.: +86-159-2512-4696

Abstract: The porosity and permeability of the rock surrounding lead–zinc deposits are key factors for controlling the migration and precipitation of ore-forming hydrothermal fluid. In this paper, the Huize super-large lead–zinc deposit was taken as the case study, and variations in the porosity and permeability of the wall rocks and their relationship with the orebody were analyzed by using CT scanning technology. The experimental results showed that the average pore radius of dolomite with a decreasing distance to the orebody ranged from 1.60 to 1.65 μm, increasing to 1.77~2.05 μm. The CT porosity increased from 2.76%–2.81% to 3.35%–3.99%. The average pore throat length decreased from 29.57–39.95 μm to 13.57–16.83 μm. In the research, it was found that the hydrothermal fluids rich in chemical elements changed the properties of the surrounding rocks. Temperature rise will lead to dolomitization of limestone and recrystallization of dolomite. This process led to an increase in the porosity of the wall rocks. During the formation of the orebody, the metal minerals in the hydrothermal fluid entered the pores of the rock. As a result, the pore radius and pore volume of the wall rocks were reduced, along with the pore throat radius and pore throat length. Therefore, the wall rock pores near the orebody were isolated from each other, and the permeability of the surrounding rock decreased. The variation characteristics for the porosity and permeability of the dolomite at various distances from the mine can be used to discover orebodies.

Keywords: porosity; permeability; lead–zinc ore; Huize; altered wall rocks

Citation: Li, Y.; Kong, Z.; Zhang, C.; Wu, Y.; Yang, X.; Wang, Y.; Chen, G. Pore Variation Characteristics of Altered Wall Rocks in the Huize Lead–Zinc Deposit, Yunnan, China and Their Geological Significance. *Minerals* **2023**, *13*, 363. https://doi.org/10.3390/min13030363

Academic Editors: Maria Economou-Eliopoulos and Huan Li

Received: 20 January 2023
Revised: 27 February 2023
Accepted: 2 March 2023
Published: 4 March 2023

1. Introduction

The border area of Sichuan, Yunnan and Guizhou on the southwest edge of the Yangzi Platform in China is an area with abundant MVT Pb–Zn ore [1]. The lead and zinc ores in this area are mostly comprised of thick carbonate formations. The deposition ages of the ore-bearing strata range from the Sinian to the Permian. Among them, the Dengying Formation of the Upper Sinian and the Baizuo Formation of the Lower Carboniferous are the most important ore-bearing strata [2,3]. Scholars have found that MVT deposits are always formed in the dolomite. For example, Davis [4] suggested that the mineralization in the Viburnum Trend lead belt in southeastern Missouri, USA occurred mainly at the dolomitic boundary of the Cambrian Bonneterre Fm. Davies and Smith [5] and Machel and Lonnee [6] suggested that hydrothermally altered dolomite formed at higher temperatures than limestone. The wall rocks were heated during the migration of the mineralized hydrothermal fluid. This process promoted the dolomitization of the limestone. Some Chinese scholars have carried out research on the spatial relationship between the MVT-type Pb–Zn ore and dolomitization. Zhang et al. noted that the dolomite has a high porosity and permeability, which may be a primary reason why lead–zinc ore deposits preferentially occur in dolomite rock [1]. The porosity and permeability of deeply buried dolomite are usually better compared to limestone at the same depth [7]. Although some scholars realize that the porosity and permeability may be an important factor affecting mineralization, they have not studied this problem quantitatively.

Significant research has been carried out on the various deposits in the Sichuan, Yunnan and Guizhou areas [8–12]. The research has focused on the age of the mineralization, the geochemical characteristics of the ore and wall rocks, the source of the mineralized material, the regional tectonic evolution and the tectonic environment of the mineralization [11,13–17]. Han et al. [18] concluded that the Huize Pb–Zn ore is obviously controlled by tectonics and lithology through tectonic geology, mineral deposit geology and geochemistry, and they proposed the tectonic–fluid penetration mineralization model [9]. Li et al. [19] concluded from an isotopic geochemical study that the ore-forming metals in the Huize Pb–Zn field were from multiple sources, mainly derived from the enclosing carbonate strata, and the reduced sulfur was mainly derived from the reduction in the marine sulfate in the formation. Zhang [20], on the other hand, concluded that the mineralized metals were derived from basement rocks and Emeishan basalts, in addition to the enclosing carbonate strata, and were characterized by different sources of metallogenic elements. Han et al. [21] concluded that the mineral control structure of the Huize Pb–Zn deposit was "stepped". Minimal work has been carried out to correlate the physical properties, such as the porosity and permeability of the wall rocks and ore bodies. The rock pore volume determines the porosity while the pore throat characteristics determine the permeability. Together, the porosity and permeability determine the ability for porous media to move through the rocks [22].

To date, only Wang [23] has carried out tensile resistance and compression resistance tests of the dolomite and limestone in the Huize lead–zinc mine. In order to verify whether the porosity is indeed effective for ore control, this paper, based on a field survey and sampling, combines the previous research and applies rock CT scanning technology. Based on the scanning results of the rock slices, the porosity, permeability and other physical parameters of the dolomite in different alteration zonings of the Huize lead–zinc deposit were studied, and the indicative effects of these physical parameters on the mineralization center were analyzed.

2. Regional Geological Background

The Sichuan–Yunnan–Guizhou Pb–Zn ore field is located at the southwest region of the Yangzi Massif [10,24,25], the transitional zone between the Gondwana and the Laurasia. This area developed intensive structures that provide suitable conditions for mineralization [14,26]. The distribution of the lead–zinc deposits in this area is closely related to the faults. The major fault zones in the region are the Anninghe, Mile–Shizong, Xiaojiang and Weining–Shuicheng faults (Figure 1B). The Anninghe fault is more than 500 km long and nearly NS-trending. It is the boundary fault between the Kangding–Yunnan uplift zone and the Huili–Kunming subsidence zone. It is a trans-crustal fault with an early formation. It cuts through the sedimentary and basement strata and has an obvious control effect on the magmatic activity. The Mile–Shizong fault, which strikes NE, is a compression-twisted fault with a length of more than 250 km. This fault controls the intrusion and eruption of basic rocks in the Emeishan volcanic province. It was active during the Mesozoic Era. The Xiaojiang fault zone is NS-trending with a length of more than 400 km. It is still an active fault and has formed multiple secondary faults. The Weining–Shuicheng fault is a concealed fracture, which is not obvious on the surface, and strikes roughly NW. This fault was activated during the later Caledonian Era. These major faults and their secondary faults together constitute the tectonic framework of the region. Many lead–zinc deposits were formed and distributed near them [15].

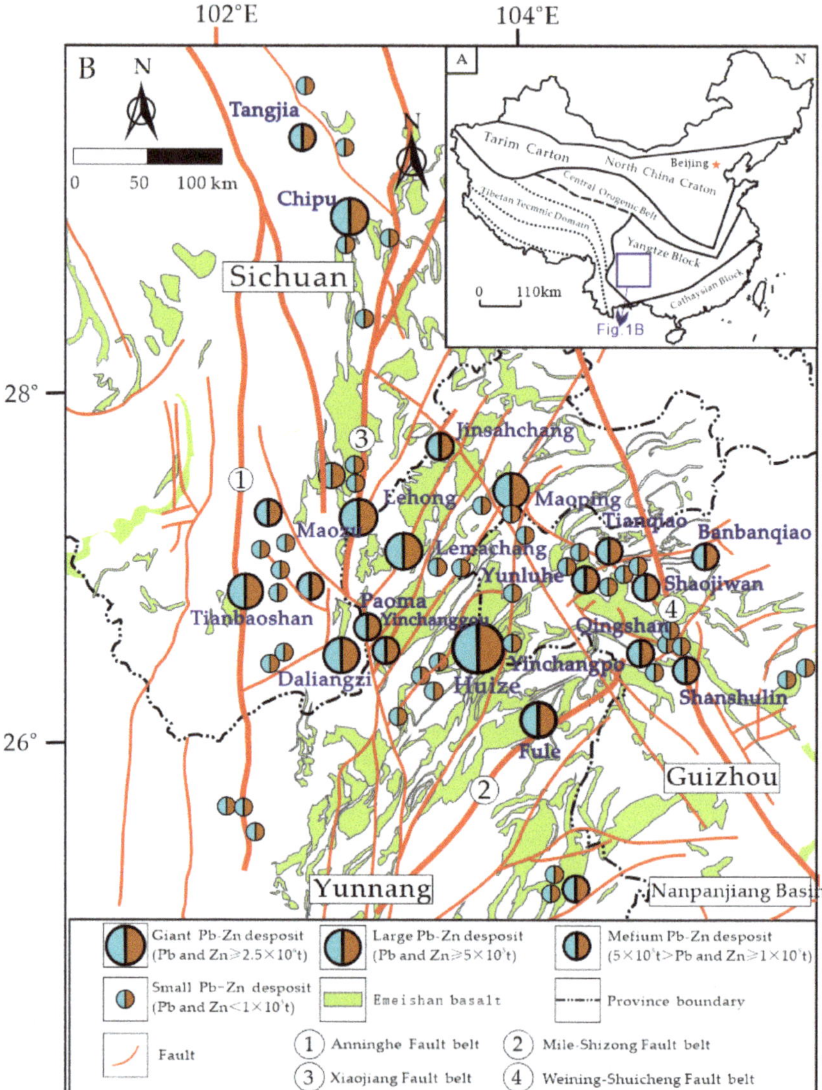

Figure 1. (**A**) Simplified tectonic map of South China (modified from [27]); (**B**) regional geological map of the Sichuan–Yunnan–Guizhou Pb–Zn triangle area (modified from [26]).

The lead–zinc deposits in this area are characterized by strata of multiple ages, but the location distribution is relatively concentrated, primarily in the northwest Guizhou, northeast Yunnan and southeast Sichuan regions [28]. The Pb–Zn deposits developed in the Sinian to Triassic strata (Table 1) and are concentrated in the carbonate strata. The ore-bearing stratigraphic lithology is mainly dolomite [2] and partly dolomitic limestone, in which large and medium-sized Pb–Zn deposits are located (Table 1). The controlling effect of lithology on the mineralization is obvious.

Table 1. Statistical table of the main host strata of the Pb–Zn deposits in the Sichuan–Yunnan–Guizhou area (modified from [29]).

Ore-Bearing Stratigraphy and Lithology of Different Ages	Large Scale	Medium Type	Smaller	Mining Sites	Mineralization Point	Representative Deposits
Triassic marlstone and limestone					3	Mine site only
Permian dolomitic limestone	1		3	9	8	Fule
Carboniferous medium-coarse crystal dolomite	2	3	13	39	12	Huize, Maoping
Upper Devonian medium-coarse crystalline dolomite	1	1	7	20	30	Zhaotong
Silurian limestone with sandy mudstone		1	5	3	5	Zhaziping
Ordovician dolomite		2	2	9	26	Butao
Upper Sinian–Lower Cambrian dolomites	3	8	18	54	96	Daliangzi, Jinsha, Maozu, Tianbaoshan

Although the Pb–Zn deposits in the area were formed during the Sinian to the Neopaleozoic and even in the Mesozoic periods, only a few are primary ore-bearing strata, such as the Upper Sinian Dengying Formation dolomite, Lower Cambrian Yuhucun Formation dolomite and Carboniferous Baizuo Formation dolomite (Figure 2). These three sets of strata host approximately 62% of the Pb–Zn deposits and 70% of the Pb–Zn resource reserves in the region (Figure 2). According to the lithology of the ore-bearing strata, the lead–zinc deposits in Sichuan, Yunnan and Guizhou can be divided into three types. The first type developed in the Upper Sinian Dengying Formation siliceous dolomite, the second type in the Lower Cambrian Yuhucun Formation phosphorus rock and phosphorus-bearing siliceous rock and the third type in the dolomite and dolomitic limestone of the Late Paleozoic Era (below the Emei Mountain basalt formation, including the Lower Carboniferous Baizuo Formation and the Lower Permian Maokou Formation) [10,24]. The Huize lead–zinc deposit belongs to the third type. The age of the ore-bearing strata in this region gradually becomes younger from west to east, and the ore-bearing rocks transition from siliceous rocks to carbonate rocks. Regionally, the carbonate rocks in the ore-bearing formation were formed in a variety of lithofacies environments, including tidal flats, closed platforms, semi-closed platforms, open platforms and lagoon facies, which do not have the uniqueness of lithofacies selection [30].

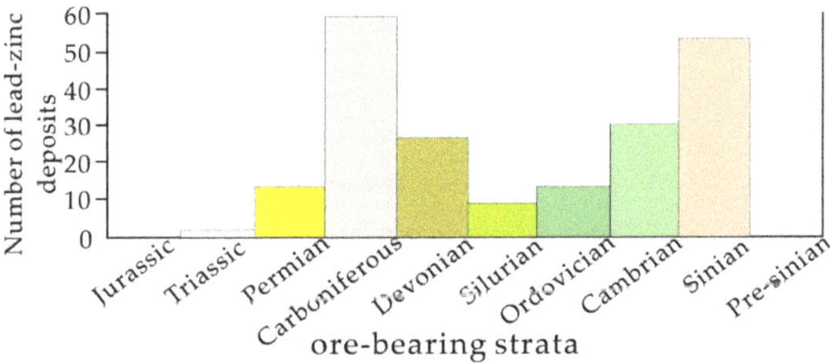

Figure 2. Statistics of the wall rocks strata of the MVT lead–zinc deposit in Sichuan, Yunnan and Guizhou (modified from [31]).

3. Geology of the Huize Lead–Zinc Deposit

3.1. Strata

The Huize Zn–Pb deposit is located in the southwest region of the Yangtze block in the eastern Kangdian oldland. The Zhaotong–Qujing fault is to the east of the Huize lead–zinc deposit and the Xiaojiang deep-seated fault zone is to the west [21]. The sedimentary strata in the mining area are dominated by carbonates and clastic rocks deposited from late Neoproterozoic Era (late Sinian) to early Permian period. In addition, the late Permian Emeishan basalts developed in the northwestern part of the mining area (Figure 1).

The ore-bearing strata of the Huize Pb–Zn deposit are mainly in the Lower Carboniferous Baizuo Formation [32]. The upper part is comprised of light gray–white crystalline dolomite. The lower part is comprised of light gray dolomite limestone [33]. The Weining Formation of the Middle Carboniferous is another ore-bearing stratum of the Huize Pb–Zn deposit. The lithology is comprised of dolomite limestone and bioclastic limestone. The dolomite in the ore-bearing part is silicified, and the wall rocks near the orebody developed large geodes due to recrystallization. Due to this phenomenon, the crystal holes become smaller the farther they are from the orebody, and there was only a weak alteration in the wall rocks located far from the orebody.

3.2. Geological Features of the Ore Bodies

There are two main ore blocks in the Huize lead–zinc mine, which are distributed in the NE direction along the Qilinchang and Kuangshanchang faults (Figure 3) [29,34]. The shape of ore body is small vein, flat columnar, cystic, tubular, network-veined and stratiform-like. The stratiform-like orebody is controlled by the NE-trending interlayer fracture zone [21]. The orebody occurs in an interbedded fracture zone, and the extension length along the dip is longer than that along the strike. The ore bodies in some areas have the characteristics of a multi-layer output [34].

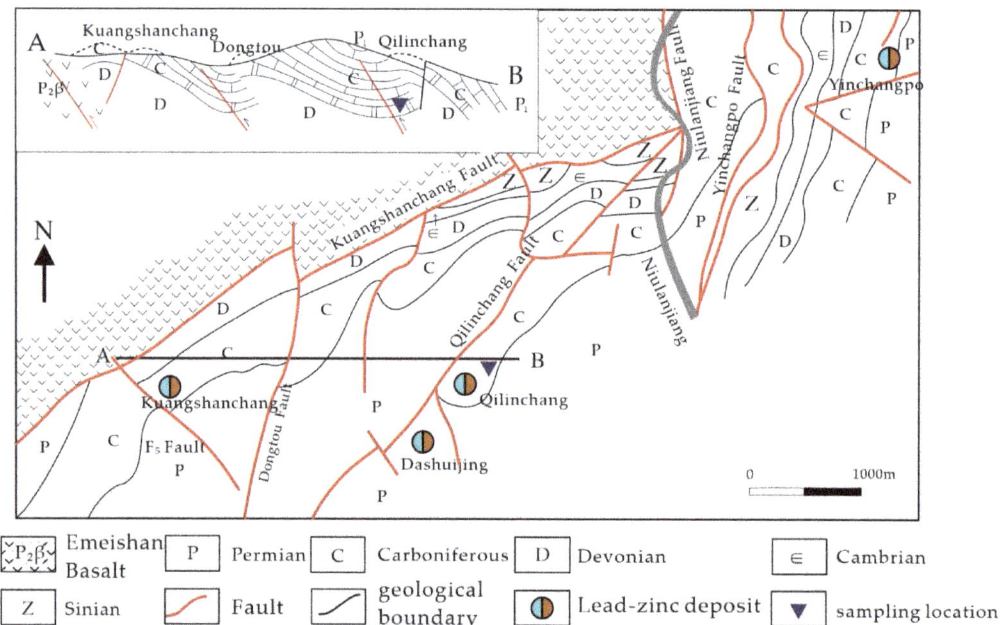

Figure 3. Geological map of the Huize lead–zinc deposit (modified from [29]). A, B: a geological section, including the mining area location, geological information and sampling location.

3.3. Mineral Assemblage

The metallic minerals in the mining area are mainly galena, sphalerite and pyrite. These metal minerals exist in the lead–zinc ore bodies, which have two forms Figure 4a shows the first form of the orebody. The contact line between the orebody and the dolomite was sharp and was formed by filling the ore-forming hydrothermal solution into the dolomite. The mineralization of another type of orebody occurred in the dolomite fracture zone, and there were fragments of dolomite in the orebody, as shown in Figure 4b. There were differences in the morphology of the two types of ore bodies, but the morphological characteristics of the two types of ore bodies combined with the characteristics under the mineral microscope indicated that the mineralization occurred after the dolomitization. As for the period of the mineral formation, the pyrite was the earliest, sphalerite was the second and galena was the third. The crystal morphology of the pyrite was mainly non-euhedral. Figure 4c–f shows that part of the pyrite was replaced by the sphalerite and galena. In Figure 4c, it can be observed that the galena replaced the sphalerite and pyrite, and black triangular pores in the galena can also be seen. It can be observed in Figure 4d that the sphalerite replaced the pyrite and that the sphalerite contained some fragments of pyrite. It can be observed in Figure 4e that the sphalerite was wrapped in pyrite, and finally that the galena replaced part of the sphalerite. Figure 4f shows that the galena filled the interior of the pyrite.

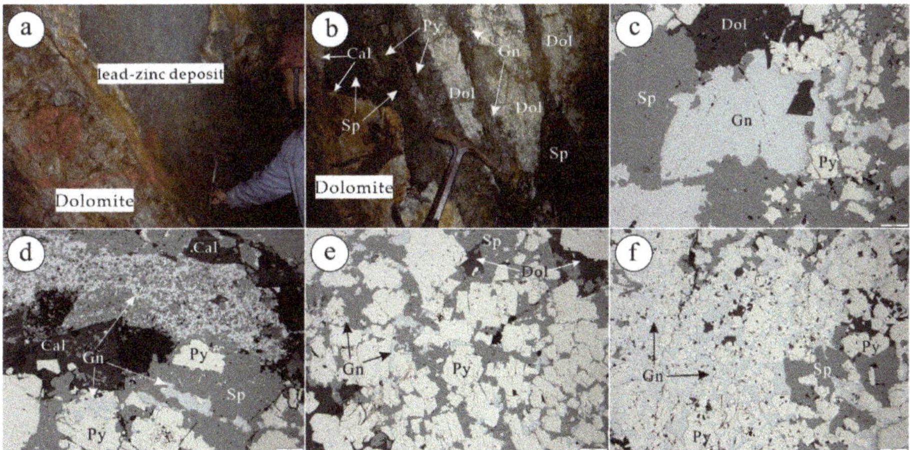

Figure 4. (**a**,**b**) Contact relationship between the wall rocks and the lead–zinc ore; (**c**–**f**) characteristics of the minerals under the microscope; Cal: calcite, Dol: dolomite, Sp: sphalerite, Gn: galena, Py: pyrite.

The gangue minerals were mainly dolomite and calcite. The crystal size of the dolomite in the mining area was 1–5 mm, and some crystals were larger than 5 mm. The crystal structure of the dolomite was granular euhedral or semi-euhedral. In Figure 4c, it can be observed that there were some fragments of pyrite and galena in the dolomite, indicating that the formation time of the dolomite was earlier than that of the metal minerals. The calcite was mainly banded and massive. Figure 4d shows that that calcite was partially replaced by the sphalerite and pyrite, and that there were some pores in the calcite.

3.4. Wall Rock Alteration Characteristics

The wall rock alteration type of the Huize lead–zinc deposit was relatively simple. The hanging wall rock and footwall rock of the orebody were both dolomitized. The footwall alteration of the orebody was more developed, and the hanging wall alteration was slightly or not developed. The dolomitization was widely developed in the middle–upper layer of the Baizuo Fm, but there was less pyritization, carbonation and silicification.

Dolomite alteration: The Lower Carboniferous Baizuo Fm was a light gray–beige, medium-coarse crystalline dolomite and light gray or gray micritic cryptocrystal limestone. Dolomite is mainly distributed in the footwall of ore body, these dolomites have many pores. The mineral composition was primarily dolomite with a content of more than 50%, followed by calcite. The main mineral of the limestone was calcite, with a content of more than 90%, however it also contained a small amount of dolomite, clay minerals and quartz. The contact boundary between the limestone and the orebody was sharp.

Pyrite replacement: Pyrite often occurs in dolomite outside the orebody or in the fracture zones near the orebody. Larger pyrite crystal has 20 mm in the bottom strata of the middle Carboniferous system, the thickness of pyrite body in some places is 5–6 m. The pyritization is strong in the areas close to the orebody and weak in the areas away from the orebody.

Carbonate alteration: The calcite was mainly developed in the Lower Carboniferous Baizuo Fm, followed by filling in the NW-trending faults. In the ore bodies and fissures, calcite occurs as lumps or veins [34].

4. Sampling and Analytical Methods

4.1. Tunnel Geological Information Record and Sample Collection

The field investigation found that there was a phenomenon of alteration zoning from the orebody to the wall rocks in the Huize lead–zinc deposit. There was silicified dolomite near the orebody (width 5–15 m) and coarse crystalline dolomite with large crystal holes and small crystal holes appeared successively in areas far from the orebody. In order to identify the relationship between the petrophysical properties of the different alteration zones and the orebody location, the tunnel passing through the orebody of stope No. 2 in the 1211 middle section in the Qilinchang lead–zinc deposit was selected as a typical profile for sampling. On the basis of the detailed field geological catalog, the mineralization and alteration characteristics of each alteration zone and their relationship with the orebody location were analyzed. The sampling was carried out at different locations, as shown in Figure 5a.

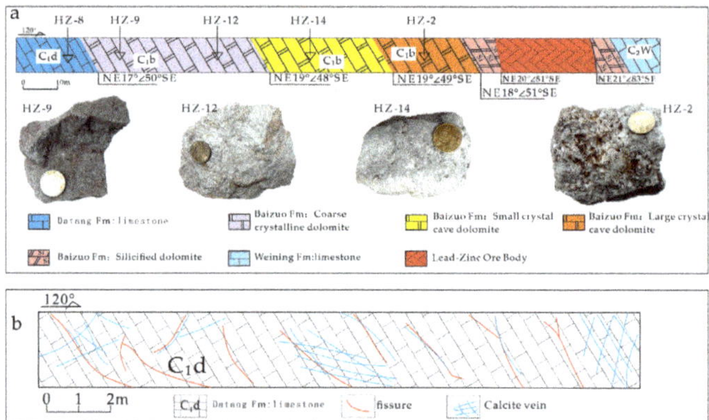

Figure 5. Geological information recording and sampling location of No. 2 stope tunnel in the 1211 middle section of the Qilinchang ore section of the Huize lead–zinc mine: (**a**) sampling location; (**b**) calcite veins and fractures in the Lower Carboniferous Datang Formation limestone.

As shown in Figure 5a, the Lower Carboniferous Datang Formation limestone was 119 m northwest of the orebody, and the color was gray. These limestones were mainly composed of calcite crystals that were less than 1 mm. In addition, in this part of the limestone, the alteration degree was very low, with nearly no dolomitization. This part of the limestone had many cracks and some vein calcite, as shown in Figure 5b. A limestone sample labelled as HZ-8 was collected 122 m to the northwest of the orebody.

The dolomite from the Lower Carboniferous Baizuo Formation, which exhibits segmentation, was located 119 m to the northwest of the position of the orebody. In the 119–70 m range, weakly altered dolomite was identified, and the color was grayish white. The composition was mainly dolomite and calcite, of which about 70% was dolomite. The size of the dolomite crystals were 0.2–2 mm, and some calcite crystals less than 0.2 mm were filled into the pores between the dolomite crystals. There was no lead–zinc mineralization in this part of the dolomite, but there were scattered pyrite and limonite. The pyrite was mainly euhedral, and the crystal form of the limonite was the same and was formed from the oxidation of the pyrite. The limonite was mainly distributed around the pyrite and dolomite crystals, which dyed the pyrite and dolomite red. There were a few cracks and geodes in this part of the dolomite, and the size of the geodes were 0.3–0.5 mm.

Sample HZ-9 was collected 108 m northwest of the orebody. The sample was gray. The main minerals were dolomite and calcite, and the content of the dolomite was more than 65%. The dolomite crystal size was 1–2 mm, and the calcite crystals were relatively small. This sample had few geodes and cracks. The radii of the geodes were approx. 0.4–0.6 mm. Sample HZ-12 was collected at a position 80 m northwest of the orebody. It was grayish white. The main minerals were dolomite and calcite, and the dolomite content was more than 70%. The size of the dolomite crystals were 1–2 mm, and the calcite crystals could only be seen using a magnifying glass. This sample had more geodes and cracks than HZ-9. The radii of the geodes was approx. 0.4–0.7 mm. At a location 70–35 m to the northwest of the orebody, strongly altered dolomite was identified with many small geodes, and the color was gray white. These dolomites were primarily composed of dolomite and calcite, and the content of the dolomite was more than 75%. The dolomite crystal size was generally larger than 2 mm, and the calcite was mainly veined, which could be seen clearly with a magnifying glass. Some galena and sphalerite were dispersed in this part of the dolomite. Sample HZ-14 was collected 53 m northwest of the orebody and was white. At this sampling location, there were some fine galena, sphalerite and pyrite distributed into the joint surface, cracks and geodes of the dolomite. The radii of the sample geodes were approx. 0.5–1 mm.

In the 35–10 m range to the northwest of the orebody, strongly altered dolomite was discovered with many large geodes. The radii of these geodes were 1.5–6 mm. The surface color of this part of dolomite was gray, and the color of the section after knocking was white. These dolomites were primarily composed of dolomite and calcite, with a dolomite content of more than 75%. The dolomite crystals were larger than 2 mm. Sample HZ-2 was collected at a position 20 m northwest of the orebody, and the dolomite content exceeded 75%. The surface of this sample was gray, and there were more geodes than the other samples. The hole radii of the geodes were also larger than the other samples, approx. 4–6 mm. Red limonite could be seen in these geodes. From 10 m to the northwest of the orebody to the location of the orebody, silicified dolomite of the Lower Carboniferous Baizuo Formation was discovered, with a crystal size of more than 4 mm. This part of the dolomite was grayish white, and there was a small number of geodes. Due to the strong silicification, these dolomites had a relatively high hardness. Near the contact between the dolomite and the orebody, there were some sphalerite, galenite and pyrite veinlets.

The next position was the orebody in this tunnel. The occurrence of this part of the lead–zinc orebody was layered, lenticular and veined, generally distributed into the cracks of the stratum. The main minerals were sphalerite, galena and pyrite. The gangue minerals were calcite, dolomite, barite and quartz. At 6 m to the southeast of the orebody, limestone of the Upper Carboniferous Weining Formation was discovered, which was gray in color. These limestones were mainly composed of calcite (more than 80%). At this location, sphalerite and galena were not seen, but some scattered pyrite was identified.

4.2. Microscopic Characteristics of Samples

The collected samples were made into light sheets, and some phenomena were observed under the microscope. Figure 6a shows the transmitted light micrograph of the HZ-8 sample. No dolomite and metal minerals were observed, and no obvious pores were found. This sample shows the characteristics of the wall rocks before mineralization. Figure 6b shows a transmitted light micrograph of the HZ-9 sample. The mineral composition was mainly pre-mineralization dolomite (HD1), and some limestone fragments can be observed. In addition, the HD1 crystal in Figure 6b was relatively small. Figure 6c shows the transmitted light micrograph of the HZ-14 sample. It can be observed that some sphalerite was filled in the cracks of HD1. Figure 6d–f shows a microscope photo of the HZ-2 sample at the same position. Among them, Figure 6d shows a reflection light photo, Figure 6e shows a transmitted light photo and Figure 6f shows a cathodoluminescence (CL) photo of this location. It can be observed that the metal minerals replaced part of the HD1 position, and then the metal minerals were replaced by hydrothermal dolomite during the mineralization (HD2) and hydrothermal calcite formed after mineralization (HC). These figures show that the mineralization time was later than that of HD1, but earlier than that of HD2. It can be observed that HD2 replaced part of HD1 in Figure 6g, indicating that the formation of the dolomite had multiple stages. In addition, some large pores can be seen in Figure 6g–I, showing that HD2 was replaced by HC after replacing HD1. Figure 6i shows the phenomenon of HC passing through HD2, indicating that HC was the latest to form. Figure 6j–l shows a photomicrograph of another position of the HZ-2 sample. Figure 6j shows a transmission light photo, Figure 6k shows a CL photo and Figure 6l shows a reflection photo of the pore position in Figure 6k. Some large pores and the dolomite that formed at different times can also be observed in Figure 6j. After amplifying the pores, the pyrite and asphalt can be observed in the pores. This phenomenon indicates that the metal minerals entered the crystal pores of the dolomite during the mineralization. Therefore, we found that the process of the dolomitization had two stages. The process of the dolomitization before the mineralization caused the number of dolomite pores to increase, which increased the porosity and permeability of the dolomite. Then, the metal elements in the hydrothermal fluid entered the pores of the dolomite and filled them, and the porosity of the dolomite was reduced to a certain extent during the mineralization period.

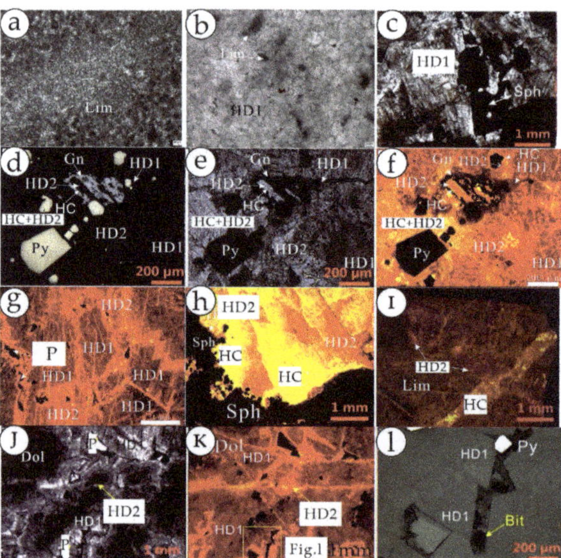

Figure 6. Dolomitization characteristics of the Huize lead–zinc deposit and its relationship to sulfide mineralization. Lim: limestone; (**a–c,e,j**) Transmission light micrograph; (**d,j**) reflection photomicrograph; (**f–i,k,l**) cathodoluminescence(CL) photo;Sph: sphalerite; Gn: galena; Py: pyrite; Bit: bitumen; HD1: pre-mineralization dolomite; HD2: hydrothermal dolomite during mineralization; HC: Hydrothermal calcite formed after mineralization; P: pore.

4.3. Analytical Methods

The porosity, permeability and connectivity of the collected samples after processing were measured using micro-scanning (CT) at the Sinopec Wuxi Institute of Petroleum

Geology. The testing instrument was a Zeiss Xradia Versa Micro-CT 520, and the test conditions were an X-ray light source of 60–140 keV and a resolution of 1.00 μm. The exposure time was 3 s, and a 3D reconstruction of the images was completed using the Avizo 9.2 software. The test conditions were as follows: sample sliced with micro-CT; light source voltage, 110 kV; exposure time, 10 s; field of view, 2 × 2 mm; size, 1024 × 1024 pixels; resolution, 1.00 μm.

In this procedure, 900 two-dimensional scanning slice images were obtained for each sample. Next, we applied the reconstruction module technology to reconstruct the 900 two-dimensional grayscale images obtained from the scanning process. Finally, the analysis algorithm in the professional image calculation and processing software was applied to calculate the images obtained from the CT scanning, and the basic pore throat structure parameter information of the sample was obtained, including the pore throat radius, pore volume, specific surface area, etc. The shape, pore distribution, pore connectivity, etc. were displayed using a three-dimensional visualization.

5. Test Results

5.1. Radius of the Pore Throat Channel

For the carbonate rocks, and the radius of the pore throat can directly affect the permeability (pore throat: the interconnected channel between pores [35]). The pore throat radius of sample HZ-9 was in the range of 0.2–2.0 μm but concentrated in the range of 0.2–0.6 μm. This showed that the pore throat radius distribution of this sample was relatively concentrated with a single peak characteristic distribution. The peak characteristics of sample HZ-9 showed that its pore throat radius tended to increase. This showed that the alteration caused some changes in the wall rocks. The pore throat radius of sample HZ-12 was in the range of 0.2–2.0 μm but concentrated in the range of 0.2–0.6 μm, showing that the pore throat radius of this sample was relatively concentrated. It showed a unimodal distribution, but a higher peak appeared in the range greater than 0.2 μm compared to the other samples. This indicates that the tendency of the weak alteration in reforming the pore throats of the wall rocks was to increase the pore throat radius. The pore throat radius of sample HZ-14 was in the range of 0.2–2.0 μm but concentrated in the range of 0.2–0.6 μm. The analysis showed that this was a relatively concentrated distribution with a single peak distribution. In addition, the number of pore throats with a radius > 1 μm in this sample was small, indicating that the throat type in this sample was relatively simple. The pore throat radius of sample HZ-2 was distributed in the range of 0.2–2.0 μm but concentrated in the range of 0.2–0.6 μm. The test data showed that this distribution was relatively concentrated with a unimodal distribution, and there were some throats with a radius greater than 1 μm. The sample had a very small number of pore throats with a radius greater than 2.0 μm, indicating that the pore throat type was relatively complex. Sample HZ-2 showed fewer throats, indicating that it was is strongly squeezed during and after diagenesis, resulting in a poor connectivity of the primary and secondary pores (Table 2 and Figure 7).

Table 2. Test results of the pore throat radius of the carbonate rock samples from the Huize lead–zinc deposit.

Sample Number	HZ-9	HZ-12	HZ-14	HZ-2
Radius (μm)	Frequency	Frequency	Frequency	Frequency
0–0.2	0	0	0	0
0.2–0.4	48	11	154	65
0.4–0.6	19	3	62	45
0.6–0.8	6	1	8	3
0.8–1.0	0	0	1	3
1.0–1.2	2	0	0	0
1.2–1.4	3	0	0	0
1.4–1.6	1	0	0	0
1.6–1.8	0	0	0	0
1.8–2.0	0	0	0	0
>2	4	3	1	7

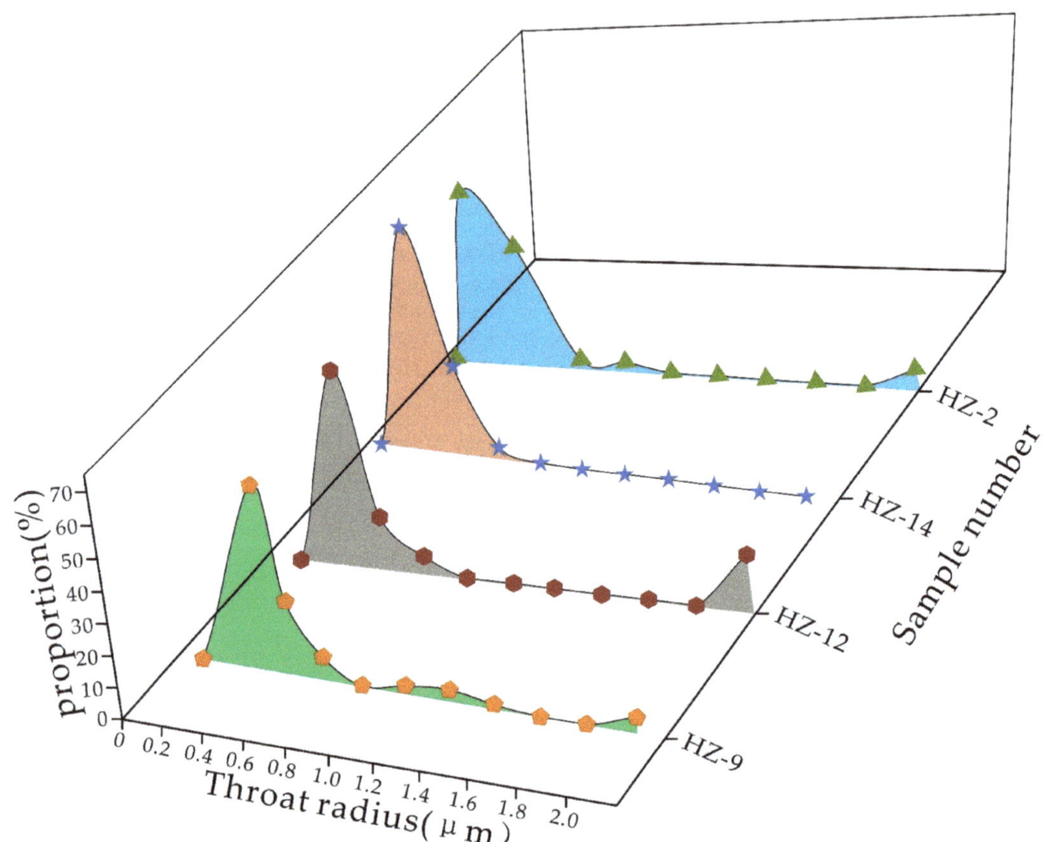

Figure 7. Ratio of the carbonate pore throat radius in the Huize lead–zinc deposit.

5.2. Pore Radius

The spaces between the mineral particles that comprise a rock are called pores. The pore radius indicates the size of the pores and has a direct effect on the porosity. The pore radius of HZ-9 was between 0.78 and 41.57 μm, among which 94.52% were 0.78–3 μm. The number of pores with a radius greater than 4 μm in this sample was very small, and the pore type was dominated by small pores, with the largest number of small pores among all the samples. The pore radius of sample HZ-12 was between 0.78 and 48.91 μm, among which 96.29% were between 0.78 and 3 μm. The proportion of the pores larger than 4 μm in HZ-12 was very low, and there were only a few pores with a large radius. This indicates that the pore types in this sample were relatively concentrated and were primarily small pores. The pore radius of sample HZ-14 was in the range of 0.89–21.88 μm, concentrated in the range of 1.0–4 μm, and 84.47% of the pores in this sample had a radius between 1 and 4 μm. The proportion of the pores with a radius greater than 4 μm in HZ-14 was higher than the other samples, and there were more large pores, indicating that the pore type in this sample was relatively complex. The pore radius of sample HZ-2 was in the range of 0.78–20 μm, concentrated in the range of 0.78–3 μm, and 94.02% of the pores had a radius of 0.78–3 μm. The proportion of the pores with radius larger than 4 μm in HZ-2 was very low. Although this sample had some large pores, the number was small. This shows that the pore types in this sample were relatively concentrated and were primarily small pores.

The number of pores with a radius larger than 20 μm in HZ-2 was the largest among all the samples (Table 3 and Figure 8).

Table 3. Test results of the pore radius of the carbonate rock samples in the Huize lead–zinc deposit.

Sample Number	HZ-9	HZ-12	HZ-14	HZ-2
Radius (μm)	Frequency	Frequency	Frequency	Frequency
0–1	1245	443	1515	1879
1–2	3412	976	8052	5977
2–3	1084	216	3692	2275
3–4	249	38	1202	418
4–5	50	5	381	86
5–6	16	6	200	36
6–7	8	5	109	18
7–8	0	2	70	13
8–9	3	0	32	18
9–10	0	0	28	8
10–11	1	0	15	7
11–12	1	0	11	6
12–13	0	0	6	6
13–14	0	3	6	4
14–15	0	0	5	2
15–16	0	0	0	5
16–17	0	0	0	3
17–18	0	0	0	3
18–19	0	0	1	3
19–20	1	0	0	0
>20	4	4	2	8

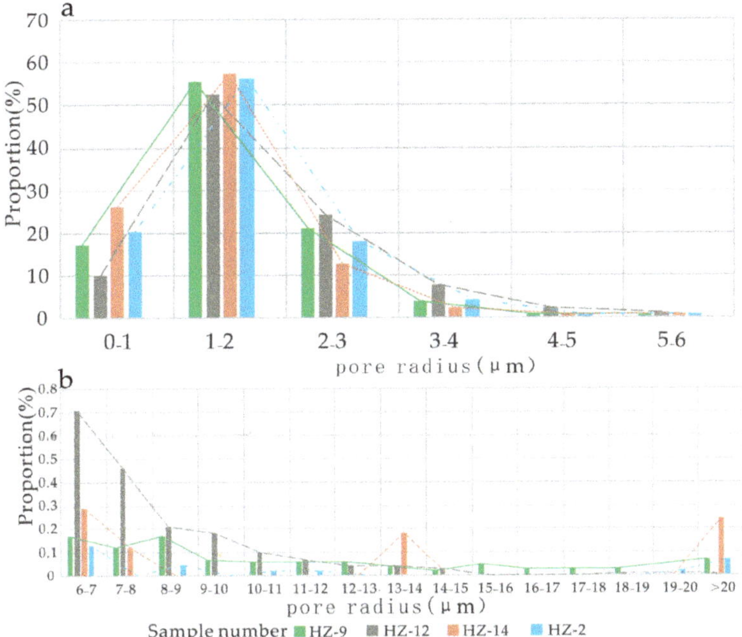

Figure 8. Pore radius ratio of the carbonate rock samples in the Huize lead–zinc deposit. (**a**) Radius 0–6 μm Histogram of the proportion of pores to the total number of pores; (**b**) Radius 6–20 μm and more than 20μm Histogram of the proportion of pores to the total number of pores.

5.3. Pore Volume

The pore volume of sample HZ-9 was in the range of 1–10^6 μm^3, concentrated in the range 10–10^2 μm^3. Similar to HZ-12, HZ-9 had some pores larger than 10^5 μm^3, but the number was small. The pore volume of HZ-12 was in the range of 1–10^6 μm^3, concentrated in the range 10–10^2 μm^3. The proportion of the pores in HZ-12 between 10^2 and 10^3 μm^3 was 3.2%, which was lower than the other samples in the same interval. The pore volume of sample HZ-14 was in the range of 1–10^5 μm^3, concentrated in the range 10–10^3 μm^3. The pore volume of HZ-14 between 10^2 and 10^3 μm^3 was 6.2%, which was higher than the other samples in the same range. The pore volume of sample HZ-2 sample was in the range of 1–10^5 μm^3 and concentrated in the range of 1–10^2 μm^3. The proportion of the pore volume greater than 10^4 μm^3 in HZ-2 was 0.26%, which was higher than the other samples (Table 4 and Figure 9).

Table 4. Test results of the pore volume of the carbonate rock samples in the Huize lead–zinc deposit.

Sample Number	HZ-9	HZ-12	HZ-14	HZ-2
Volume (μm^3)	Frequency	Frequency	Frequency	Frequency
0–10	2694	890	4560	4236
10–10^2	3012	740	8436	5772
10^2–10^3	352	55	2071	668
10^3–10^4	11	7	250	71
10^4–10^5	1	2	10	28
10^5–10^6	4	4	0	0

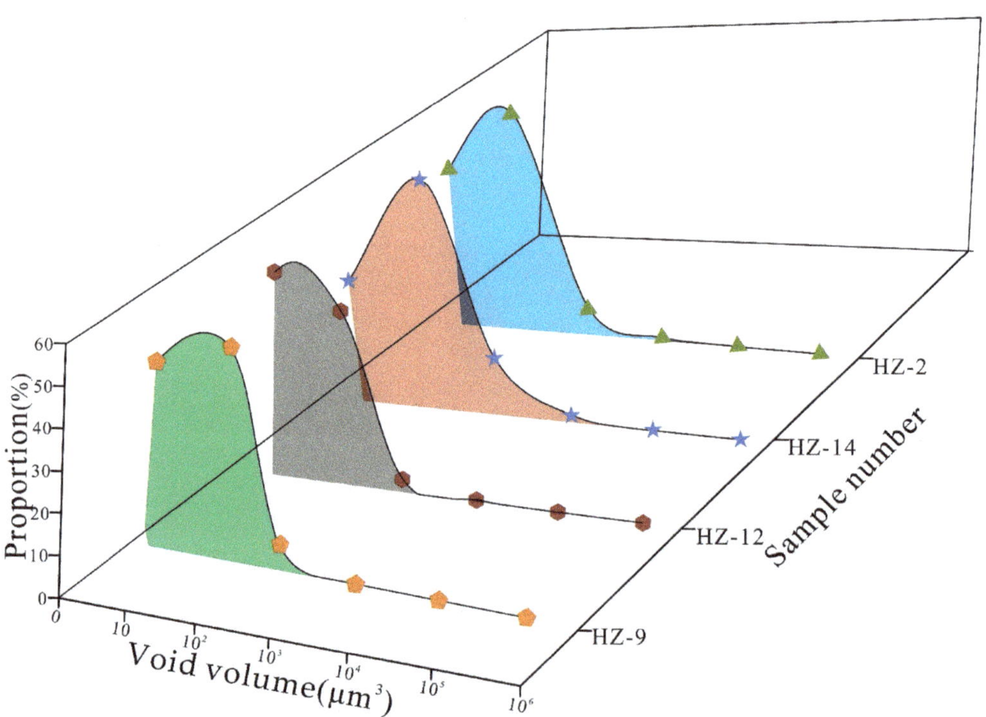

Figure 9. Pore volume ratio of the carbonate rock samples in the Huize lead–zinc deposit.

5.4. Pore Surface Area

The pore surface area of sample HZ-9 was between 1 and 10^5 μm^2, concentrated between 10 and 10^2 μm^2. The pore surface area of sample HZ-12 was between 1 and 10^5 μm^2, concentrated between 10 and 10^2 μm^2. In this sample, 4.89% of the pore surface area was between 10^2 and 10^3 μm^2, which was lower than the other samples. However, 0.29% of the pore surface area of this sample was 10^3–10^4 μm^2, higher than the other samples. Further, 94.58% of sample HZ-12 had a pore surface area less than 10 μm^2, which was the highest among all the samples. The pore surface area of sample HZ-14 was between 1 and 10^5 μm^2, concentrated between 10 and 10^3 μm^2. In this sample, 94.14% of the pore surface area was between 10 and 10^3 μm^2, which was higher than the other samples. The pore surface area of sample HZ-2 was between 1 and 10^5 μm^2, concentrated between 10 and 10^3 μm^2. In this sample, 77.25% of the pores had a surface area of 10–10^2 μm^2, slightly higher than the other samples (Table 5 and Figure 10).

Table 5. Test results of the pore surface area of the carbonate rock samples in the Huize lead–zinc deposit.

Sample Number	HZ-9	HZ-12	HZ-14	HZ-2
Area (μm^2)	Frequency	Frequency	Frequency	Frequency
0–10	857	315	687	1244
10–10^2	4633	1291	11,481	8324
10^2–10^3	566	83	2954	1115
10^3–10^4	14	5	202	78
10^4–10^5	3	4	3	14
10^5–10^6	1	0	0	0

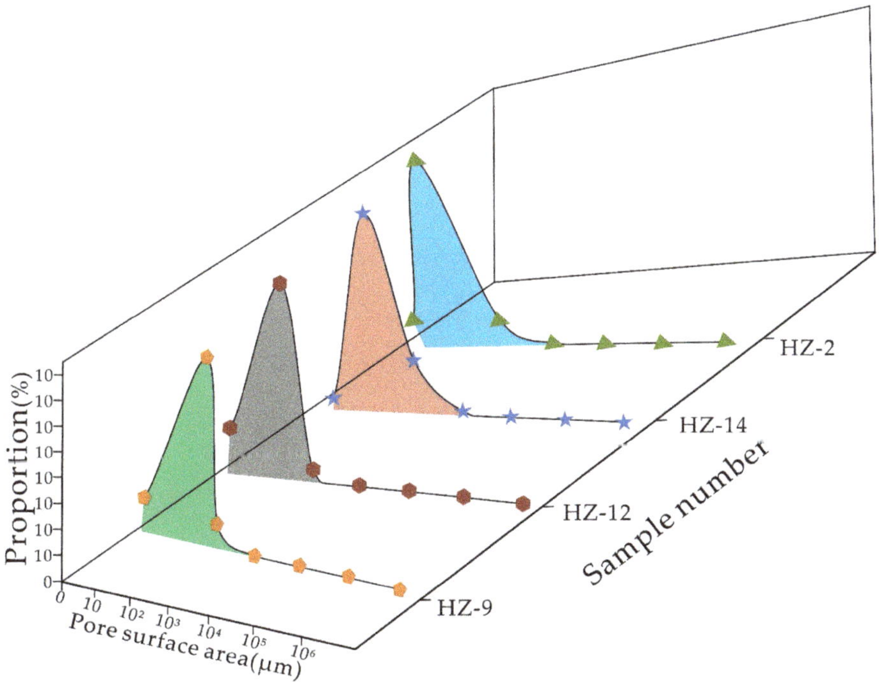

Figure 10. Pore surface area ratio of the carbonate rock samples in the Huize lead–zinc deposit.

5.5. Face Rate

The face porosity is the ratio of the pore area to the rock section area as seen under a microscope. The porosity of a rock can be calculated using the face rate of multiple slices, and this process was automatically completed by the computer. Each sample was sliced 299 times and scanned using CT to obtain the face rate for each slice. The face rates of the samples are shown in Figure 11, and the porosity of the samples was calculated from the face rate of all the sections (Table 6).

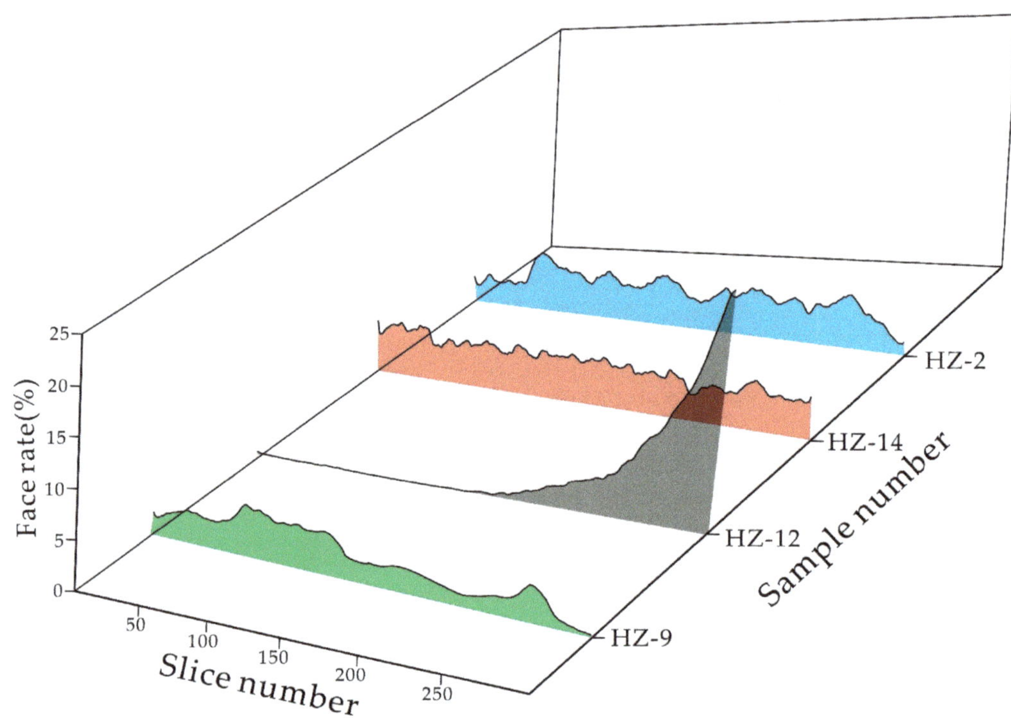

Figure 11. Face rate variation of the carbonate rock samples in the Huize lead–zinc deposit.

Table 6. Pore throat structure parameters of the carbonate rock samples in the Huize lead–zinc deposit.

Sample Number	Average Pore Radius (µm)	Average Pore Throat Radius (µm)	Average Pore Volume (µm³)	Average Specific Surface Area (µm²)	Average Pore Throat Length (µm)	CT Porosity (%)
HZ-9	1.65	0.077	179.87	91.01	29.57	2.81
HZ-12	1.60	1.02	857.55	159.76	39.95	2.76
HZ-14	2.05	0.38	110.86	107.55	13.57	3.99
HZ-2	1.77	0.55	120.46	92.11	16.83	3.35

The face rate of sample HZ-9 was between 0.25 and 5.48%. The face rate of sample HZ-12 was between 0.01 and 22.31%. The variation range of the porosity of this sample was the largest, indicating that the face rate was extremely heterogeneous, and the degree of the pore change was large. The face rate of sample HZ-14 was between 2.40 and 5.31%. The face rate of this sample changed minimally, indicating that the heterogeneity of the sample was small, and the degree of the pore change was low. The face rate of sample HZ-2 was between 0.98 and 5.50%. The face rate of HZ-9 and HZ-2 also changed greatly in the different sections, indicating that the two samples had a large heterogeneity and a large pore change.

The calculations show that the porosity of sample HZ-12 was the lowest (Table 6), but the face rate varied greatly (Figure 11). Although the porosity of sample HZ-9 was not high, the face rate was relatively uniform (Figure 11). This shows that the alteration far from the orebody had a great influence on the porosity of the wall rocks, promoting the production of larger pores in the wall rocks. However, the wall rocks close to the orebody became homogeneous again due to the recrystallization and strong mineralization. The porosity of the HZ-14 and HZ-2 samples were relatively uniform, showing the characteristics of a uniform pore distribution in the wall rocks near the orebody. The other statistical data show that the average pore radius increased with the decreasing distance to the orebody. The average pore volume decreased with the decreasing distance to the orebody. The variation characteristics of the average porosity increased with the decreasing distance to the orebody. The result of the porosity increase conforms to the law of the dolomitization process. The changed rules of the pore radius and the pore volume indicate that there should be changes in the number of pores used in this process. In terms of the test data of the pore throats, the average pore throat radius and the average pore throat length became smaller with the decreasing distance to the orebody. This result shows that the connectivity between the pores became weak.

6. Discussion

6.1. Variation in the Porosity with the Decreasing Distance from Ore Bodies

The micro-CT scan of sample HZ-9 is shown in Figure 12, in which the black part represents the pores, and the gray part represents the rock matrix.

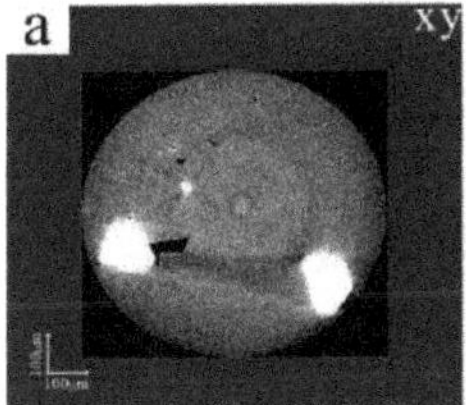
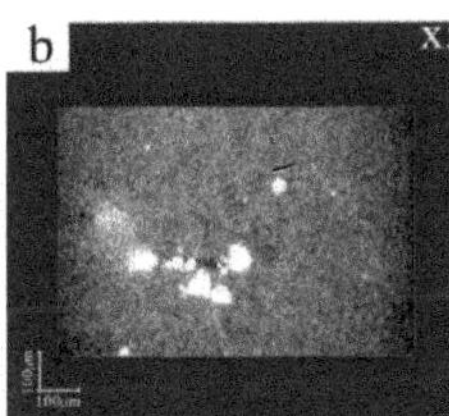
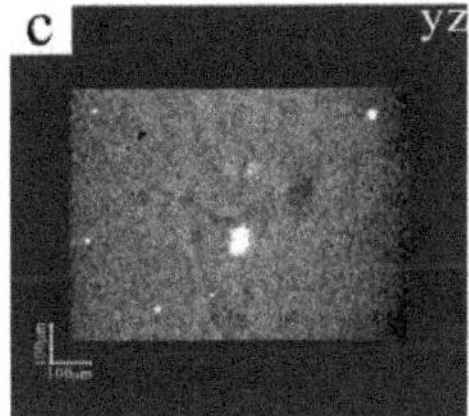
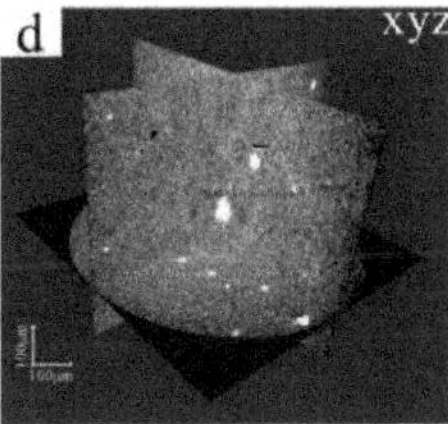

Figure 12. Micro-CT slice image of sample HZ-9 in the Huize lead–zinc deposit.(a) Sample xy section Micro-CT scan image; (**b**) Sample xz section Micro-CT scan image; (**c**) Sample yz section Micro-CT scan image; (**d**) Micro-CT scanning mosaic image of sample xyz section.

The skeleton extraction was performed on the pore throat network of the slice using the application software, and its spatial distribution model is shown in Figure 13a–c. Sample HZ-9 was compared with the other three samples and was found to have less pore development, relatively more throats and a better connectivity between the pores. The lowest gray part (pore and pore throat) in the gray image was extracted and segmented using threshold segmentation to separate the pore and the pore throat from the rock matrix. The Avizo 9.2 software was used to segment and extract the distribution positions, distribution forms and distribution of the pores and throat networks in the sample space, and a three-dimensional display was constructed. The results are shown in Figure 13d–f. By analyzing the three-dimensional diagram, we can conclude that the degree of the pore development of HZ-9 was low. The pores of this sample were mainly macropores, and there were multiple small pores. The porosity of this sample was calculated to be 2.81% (Table 6), which is in the ultra-low range. Based on the separate extraction of the pores from the matrix and particles, the separate object technique was used to analyze the spatial connectivity distribution characteristics of the pores, as shown in Figure 13g–i. Figure 13 shows that sample HZ-9 had a low degree of pore development, with large pores occupying a larger space, in addition to some small pores.

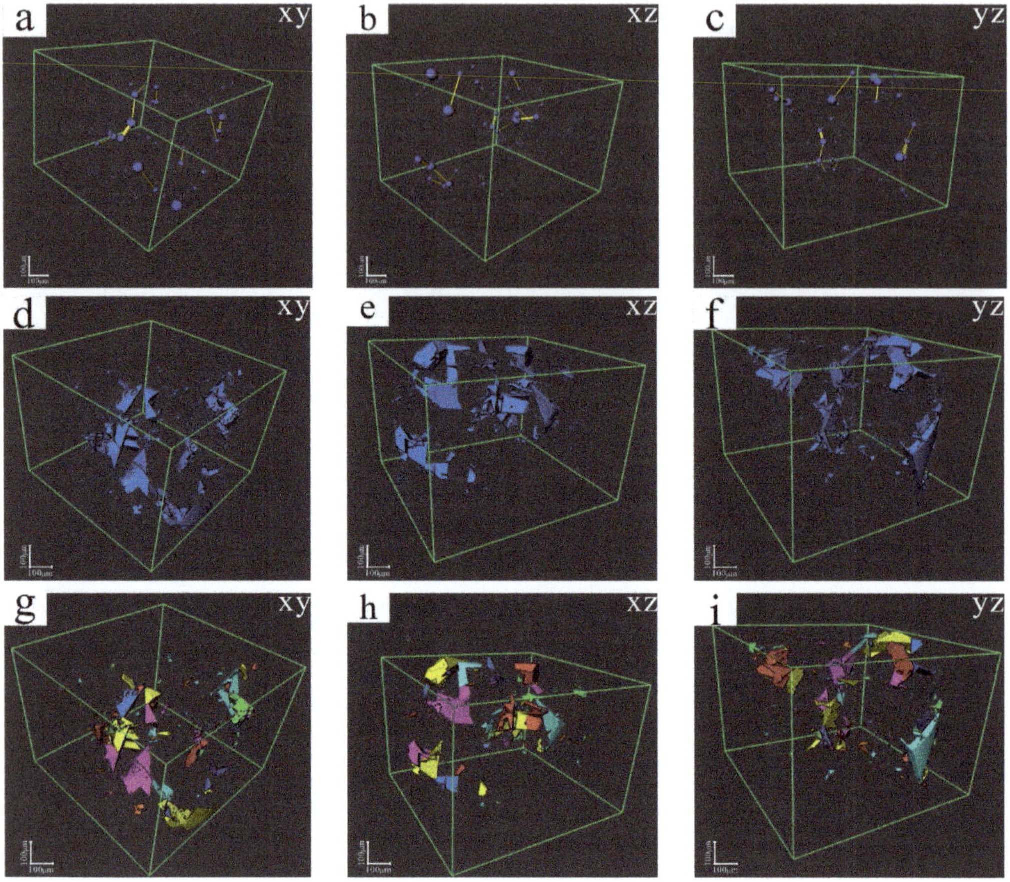

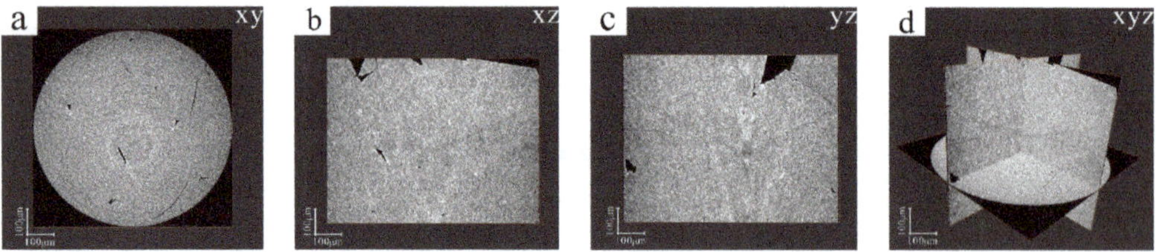

Figure 13. Pore and pore throat structure imaging of sample HZ-9: (**a–c**) the blue spheres represent the pores, and the yellow lines represent the pore throats; (**d–f**) blue represents the spatial position of the pores and pore throats; (**g–i**) the pores with similar colors in adjacent regions are connected.

The micro-CT scan of sample HZ-12 is shown in Figure 14, in which the black part represents the pores, and the gray part represents the rock matrix. Although HZ-12 had fewer pores, there were some dispersed macropores.

Figure 14. Micro-CT slice image of sample HZ-12 in the Huize lead–zinc deposit.(**a**) Sample xy section Micro-CT scan image; (**b**) Sample xz section Micro-CT scan image; (**c**) Sample yz section Micro-CT scan image; (**d**) Micro-CT scanning mosaic image of sample xyz section.

The skeleton extraction was performed on the pore throat network of the slice using the application software, and its spatial distribution model is shown in Figure 15a–c. Sample HZ-12 had many large pores with a good connectivity and generally isolated small pores. The lowest gray part (pore and pore throat) in the gray image was extracted and segmented using threshold segmentation to separate the pore and the pore throat from the rock matrix. The Avizo 9.2 software was used to segment and extract the distribution positions, distribution forms and distribution of the pores and throat networks in the sample space, and a three-dimensional display was constructed. The results are shown in Figure 15d–f. The porosity of this sample was calculated to be 2.76% (Table 6), which is in the ultra-low range. Based on the separate extraction of the pores from the matrix and particles, the separate object technique was used to analyze the spatial connectivity distribution characteristics of the pores, as shown in Figure 15g–i. Figure 15 shows that the pore connectivity of HZ-12 was poor. Some large pores were interconnected and the small pores were poorly connected.

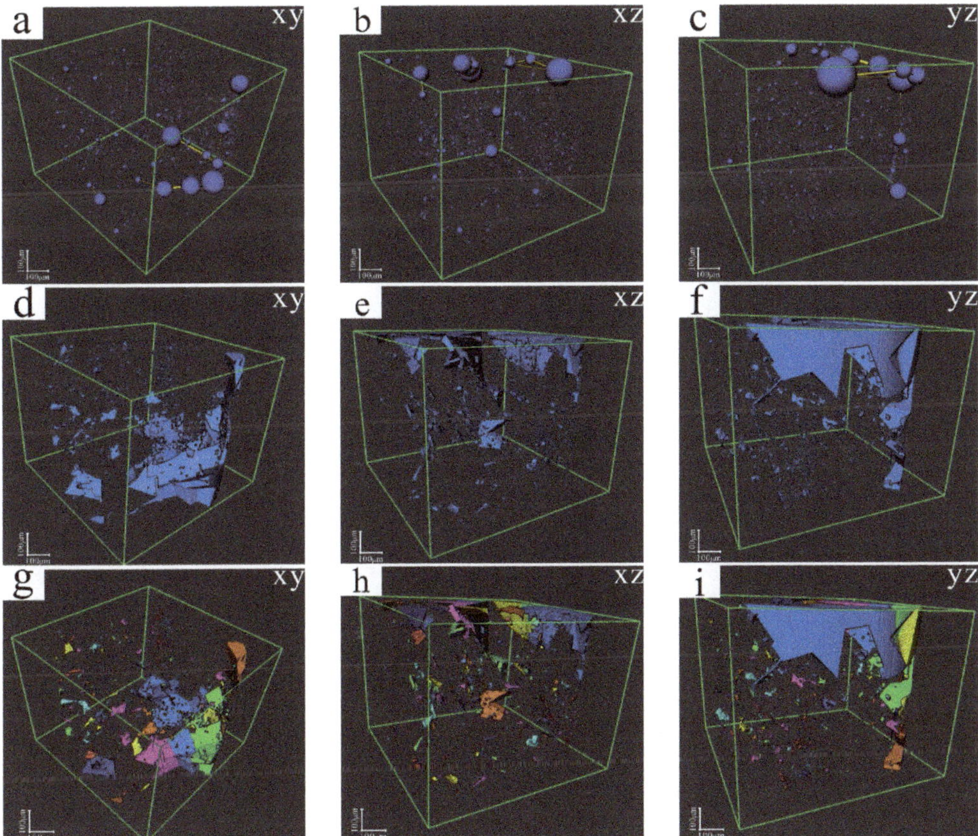

Figure 15. Pore and pore throat structure imaging of sample HZ-12: (**a–c**) the blue spheres represent the pores, and the yellow lines represent the pore throats; (**d–f**) blue represents the spatial position of the pores and pore throats; (**g–i**) the pores with similar colors in adjacent regions are connected.

The micro-CT scan of sample HZ-14 is shown in Figure 16, where the black part represents the pores, and the gray part represents the rock matrix. The pores of HZ-14 were generally small and dispersed.

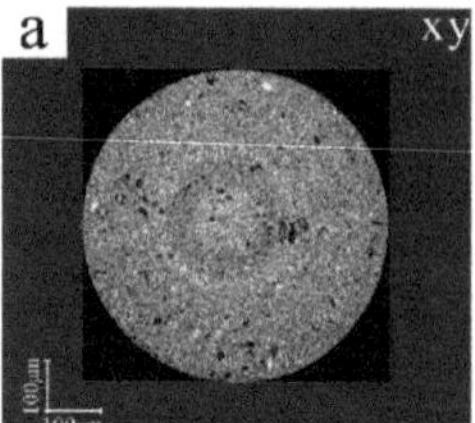 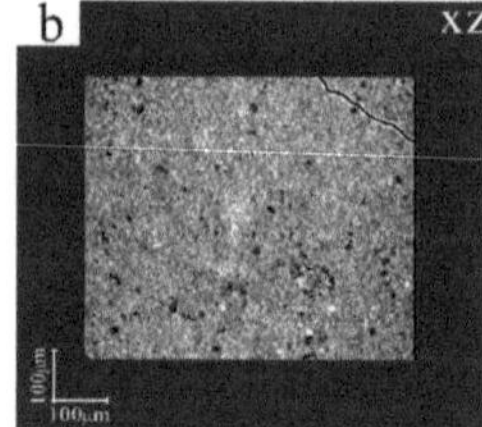 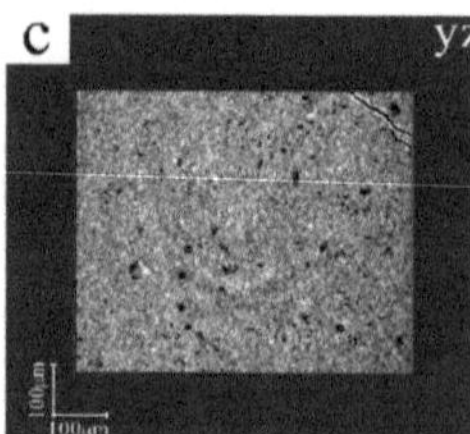 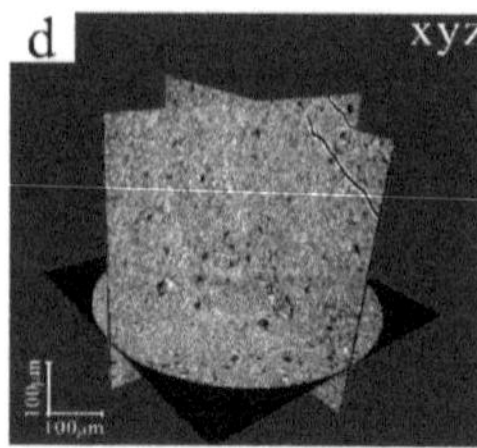

Figure 16. Micro-CT slice image of sample HZ-14 in the Huize lead–zinc deposit. (**a**) Sample xy section Micro-CT scan image; (**b**) Sample xz section Micro-CT scan image; (**c**) Sample yz section Micro-CT scan image; (**d**) Micro-CT scanning mosaic image of sample xyz section.

The skeleton extraction was performed on the pore throat network of the slice using the application software, and its spatial distribution model is shown in Figure 17a–c. The number of small pores in sample HZ-14 was particularly large, but the number of pore throats between the pores was small and the pore connectivity was poor. The lowest gray part (pore and pore throat) in the gray image was extracted and segmented using the threshold segmentation technique to separate the pore and the pore throat from the rock matrix. The Avizo 9.2 software was used to segment and extract the distribution positions, distribution forms and distribution of the pores and throat networks in the sample space, and a three-dimensional display was constructed. As shown in Figure 17d–f, sample HZ-14 had a relatively high degree of pore development and dense pores. The porosity of this sample was calculated to be 3.99% (Table 6), which is in the ultra-low range. Based on the separate extraction of the pores from the matrix and particles, the separate object technique was used to analyze the spatial connectivity distribution characteristics of the pores, as shown in Figure 17g–i. Figure 17 shows that the pore connectivity of HZ-14 was relatively poor, and most pores showed an isolated distribution.

The micro-CT scan of sample HZ-2 is shown in Figure 18, in which the black part represents the pores, and the gray part represents the rock matrix. It can be observed that HZ-2 mainly had scattered small pores.

The skeleton extraction was performed on the pore throat network of the slice using the application software, and its spatial distribution model is shown in Figure 19a–c. The lowest gray part (pore and pore throat) in the gray image was extracted and segmented using the threshold segmentation technique to separate the pore and the pore throat from the rock matrix. The Avizo 9.2 software was used to segment and extract the distribution positions, distribution forms and distribution of the pores and throat networks in the sample space, and a three-dimensional display was constructed. As shown in Figure 19d–f, the degree of the pore development of HZ-2 was relatively low. There were some superimposed large pores also in addition to a large number of small pores. The porosity of this sample was calculated to be 3.35% (Table 6), which is in the ultra-low range. Based on the separate extraction of the pores from the matrix and particles, the separate object technique was used to analyze the spatial connectivity distribution characteristics of the pores, as shown in Figure 19g–i. Comparing the results of the four samples, the number of pores in the HZ-9 and HZ-12 samples was relatively small. However, there were many pore throat connections between these pores (Figures 13a–c and 15a–c). Therefore, the pore connectivity of the HZ-9 and HZ-12 samples were relatively better (Figures 13g–i and 15g–i). There were many pores in the HZ-14 and HZ-2 samples. However, there were few pore throats between these pores (Figures 17a–c and 19a–c). So, the pore connectivity of the HZ-14 sample, and the HZ-2 sample were relatively weak (Figures 17g–i and 19g–i).

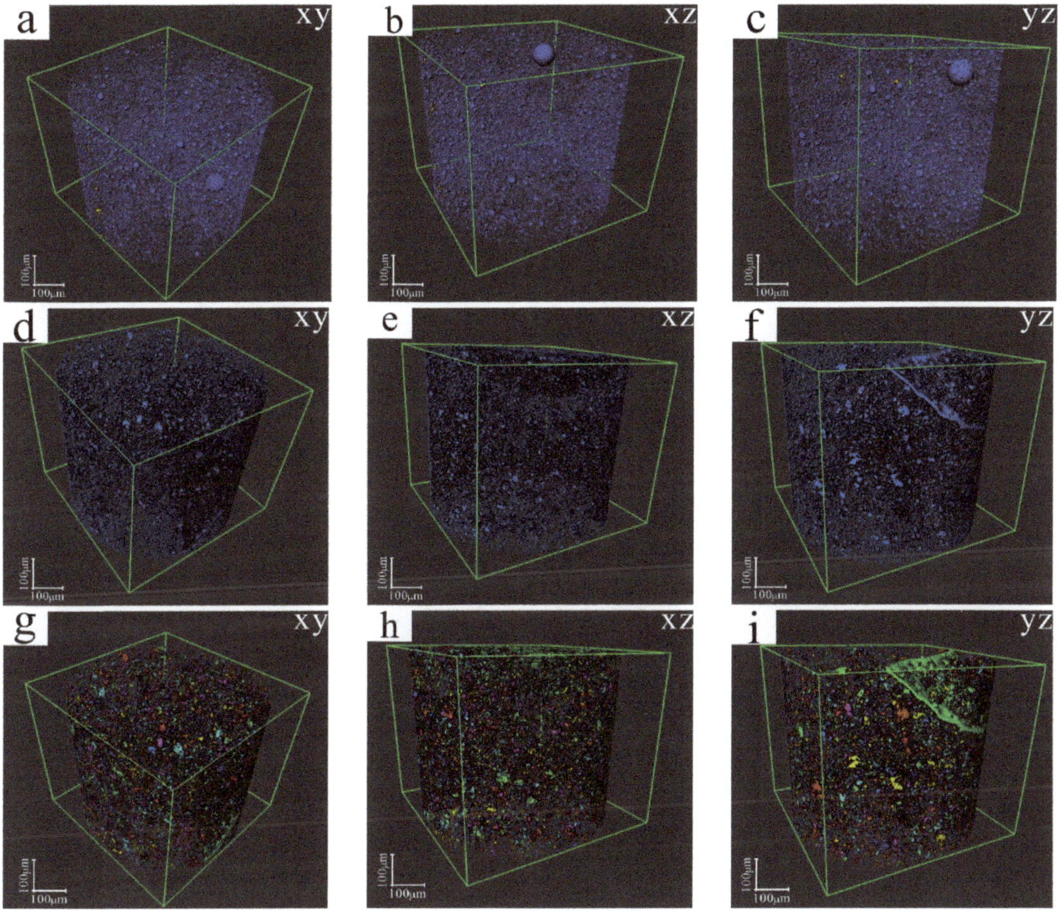

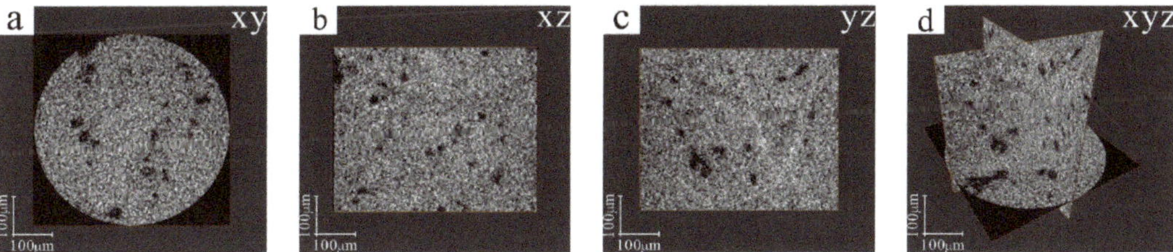

Figure 17. Pore and pore throat structure imaging of sample HZ-14: (**a–c**) the blue spheres represent the pores, and the yellow lines represent the pore throats; (**d–f**) blue represents the spatial position of the pores and pore throats; (**g–i**) the pores with similar colors in adjacent regions are connected.

Figure 18. Micro-CT slice image of sample HZ-2 in the Huize lead–zinc deposit. (**a**) Sample xy section Micro-CT scan image; (**b**) Sample xz section Micro-CT scan image; (**c**) Sample yz section Micro-CT scan image; (**d**) Micro-CT scanning mosaic image of sample xyz section.

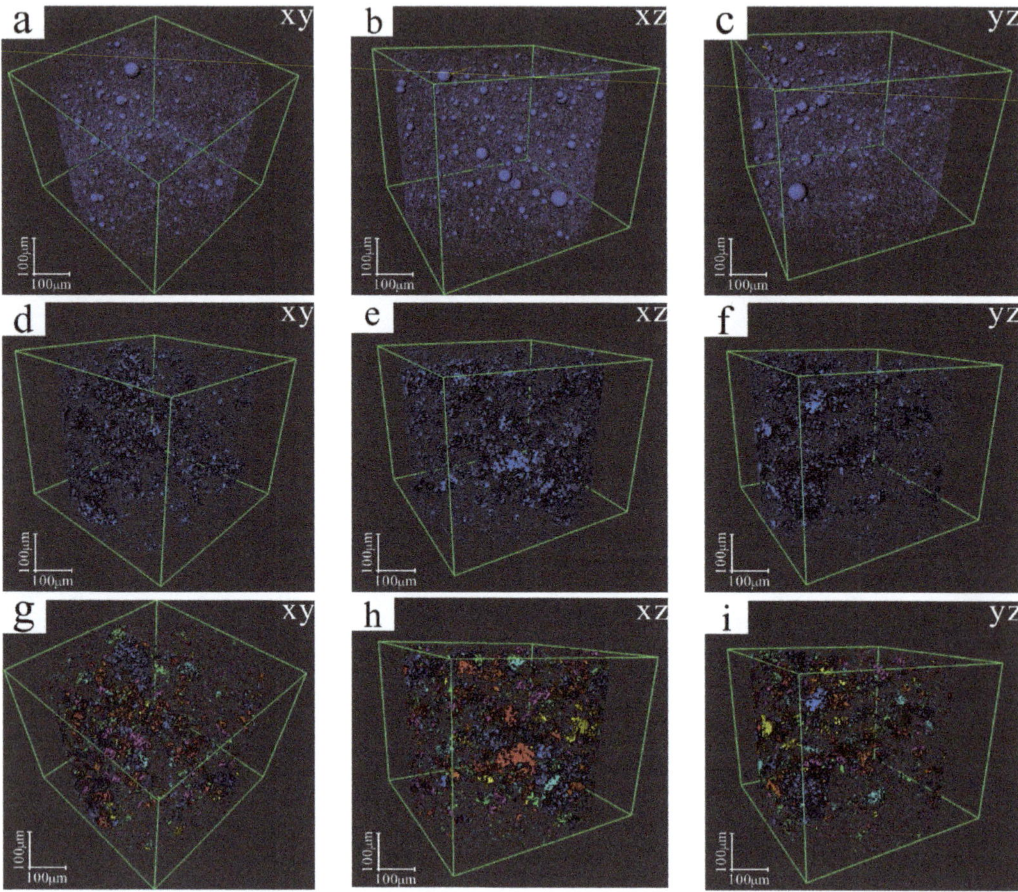

Figure 19. Pore and pore throat structure imaging of sample HZ-2: (**a–c**) the blue spheres represent the pores, and yellow lines represent the pore throats; (**d–f**) blue represents the spatial position of the pores and pore throats; (**g–i**) the pores with similar colors in adjacent regions are connected.

6.2. Relationship between the Alteration, Porosity and Ore Location

Some scholars believe that the coarse-grained dolomite of Baizuo Fm, Zaige Fm and Dengying Fm that developed throughout the Huize lead–zinc deposit was formed by the combined action of tectonic and hydrothermal alteration. The fine-grained dolomite, dolomite limestone and limestone in the deposit after alteration, became coarse-grained dolomite. Reticulated dolomite limestone is a transitional type of rock between dolomite and limestone formed through incomplete alteration [36]. From the test results, the wall rocks around the Huize lead–zinc mine were positioned far from the orebody to near the orebody. As the alteration strengthened, the porosity of the dolomite increased. The dolomite far from the orebody had fewer pores, a larger pore volume variation, a larger throat length and better connectivity between the pore throats. Toward the position close to the orebody, the dolomite samples had more pores and fewer pore throats, with only a few independent small pores, except for geodes or fissures. (Figures 16 and 18). This shows that near the orebody, after the hydrothermal fluids have penetrated the wall rocks on both sides, the fluid reacted strongly with the wall rocks. In this process, a large number of calcite, dolomite and other minerals gradually precipitated and occupied the gaps between the dolomite particles. As a result, the connectivity of the dolomite near the orebody became poor.

The connectivity of the dolomite is primarily related to the pores and throats, which determine the permeability [22]. The permeability is a key geological parameter and a major controlling factor for the fluid flow and heat transfer [37], so variations in the permeability can indicate the presence of mineralization. The permeability is largely related to the size and type of the pores, the shape and size of the pore throats and the special surface of the pores [38]. Therefore, the study of the characteristics of the carbonate pores and throats can reflect the characteristics of the permeability to some extent. Dolomite sample HZ-9, far from the orebody, was compared with samples HZ-12 and HZ-14, closer to the orebody. Its porosity was small, there were some overlapping large pores and the pore connectivity was good (Figures 13, 15 and 17). Dolomite sample HZ-2, closest to the orebody, had a large number of small pores. These small pores were isolated from each other, with less pore throat development and a poor connectivity between the pores. This difference may be related to the alteration and recrystallization of the dolomite caused by mineralization, resulting in the formation of geodes and an increased number of pores, after which the hydrothermal fluids occupied the pores and throats. The end result is that the closer to the orebody, the smaller the pore radius of the sample, the fewer the throats and the worse the pore connectivity.

7. Conclusions

(1) As the surrounding rock decreases in distance to the orebody, the porosity tends to increase. The CT porosity of the two dolomite samples far away from the orebody were 2.76%–2.81%. The 3D image shows that the number of dolomite pores at this location was relatively small. The CT porosity of the two dolomite samples near the orebody were 3.35%–3.99%. It can be seen from the 3D images that the number of pores in the two samples was large. It can be found that the porosity of the dolomite increased with a decreasing distance to the orebody. This increase was related to the increase in the number of pores. In the microscopic photos, we observed that there were some metal minerals filling the sample pores near the orebody. This indicates that the metal minerals enter and fill the dolomite pores during the mineralization.

(2) The average pore throat radii of the two samples far away from the orebody were 0.77 μm and 1.02 μm, and the average pore throat lengths were 29.57 μm and 39.95 μm. From the 3D image, it can be observed that the number of pore throats in these two samples was large, and many of the pores were interconnected. The average pore throat radii for the two samples near the orebody were 0.38 μm and 0.55 μm, and the average pore throat lengths were 13.57 μm and 16.83 μm. Compared to the samples far away from the orebody, their pore throat radii and pore throat lengths were smaller, which means that the samples near the orebody had low permeability and poor connectivity between the pores. It can also be observed from the 3D image that the number of sample pore throats near the orebody was small and the pores were isolated. This phenomenon indicates that the shorter the distance from the orebody, the worse the permeability of the dolomite. Based on the combined analysis of the microscopic photos and the 3D images, the author believes that the surrounding rock near the orebody underwent two stages of dolomitization due to the influence of hydrothermal activity. In this process, the number of pores in the surrounding rock increased, so the porosity of the dolomite increased. However, the hydrothermal minerals (Sph, Gn, Py) filled a part of the pores and pore throats, resulting in a smaller pore radius and pore throat radius and a smaller pore volume and pore throat length. Therefore, the connectivity between the pores of the samples near the orebody was worse than the samples far away from the orebody. The final result is that, with a decreasing distance to the orebody, the porosity of the dolomite increases and the permeability decreases. According to the above test results and analysis, the author believes that the gray-white coarse-porous crystalline dolomite developed in the Huize lead–zinc mine

area can be used as one of the main indicators for future prospecting. The direction of the dolomite porosity increase and the permeability decrease may represent the direction of the orebody.

Author Contributions: Conceptualization, Y.L. and Z.K.; investigation, Y.L., Z.K., C.Z. and Y.W. (Yue Wu); methodology, Y.L., Y.W. (Yu Wang); data curation, Y.L. and X.Y.; writing—original draft, Y.L.; writing—review and editing, Z.K., C.Z. and Y.W. (Yue Wu); Visualization, Y.L., Y.W. (Yu Wang) and G.C.; Funding acquisition, Z.K., C.Z.; Project administration, Y.W. (Yue Wu) and X.Y. All authors have read and agreed to the published version of the manuscript.

Funding: This work was supported by the National Key R&D Program (No. 2017YFC0602502), Kunming University of Science and Technology Introduced Talents Research Startup Fund Project (No. KKZ3202221023), the National Natural Science Foundation (No. 41672093), the Yunnan Major Scientific and Technological Projects (No. 202101BC070001-003), the China Geological Survey Project (No. DD20190182).

Data Availability Statement: The data used to support this study are included within the article.

Acknowledgments: We are very grateful for the support and assistance provided by Yunnan Chihong Zn & Ge Co., LTD. Huize Branch in our field work. The authors would like to acknowledge the editors-in-chief and the four anonymous reviewers for their constructive comments, which helped to significantly improve the manuscript.

Conflicts of Interest: The authors declare no conflict of interest.

References

1. Zhang, C.Q.; Yu, J.J.; Mao, J.; Rui, Z.Y. Advances in the study of Mississippi valley-type deposits. *Miner. Depos.* **2009**, *28*, 195–210.
2. Mao, J.; Li, X.; Li, H.; Qu, X.; Zhang, C.; Xue, C.; Wang, Z.; Yu, J.J.; Zhang, Z.; Feng, C.; et al. Types and characteristics of endogenetic metallic deposits in orogenic belts in China and their metallogenic processes. *Acta Geol. Sin.* **2005**, *79*, 342–371.
3. Zhen, S.; Zhu, X.; Li, Y. A tentative discussion on Mississippi valley-type deposits. *Miner. Depos.* **2013**, *32*, 367–379.
4. Davis, J.H. Genesis of the Southeast Missouri lead deposits. *Econ. Geol.* **1977**, *72*, 443–450. [CrossRef]
5. Davies, G.; Smith, L. Structurally controlled hydrothermal dolomite reservoir facies: An overview. *AAPG Bull.* **2006**, *90*, 1641–1690. [CrossRef]
6. Machel, H.G.; Lonnee, J. Hydrothermal dolomite—A product of poor definition and imagination. *Sediment. Geol.* **2002**, *152*, 163–171. [CrossRef]
7. Huang, S.J. *Diagenesis of Carbonate Rocks*; Geological Publishing House: Beijing, China, 2010.
8. Zhang, Z.; Li, C.; Tu, G.; Xia, B.; Wei, Z. Geotectonic evolution background and ore-forming process of Pb-Zn deposits in Chuan-Dian-Qian area of Southwest China. *Geotecton. Metallog.* **2006**, *30*, 343–354.
9. Han, R.; Wang, F.; Hu, Y.; Wang, X.; Ren, T.; Qiu, W.; Zhong, K. Metallogenic tectonic dynamics and chronology constrains on the huize-type (HZT) germanium-rich silver-zinc-lead deposits. *Geotecton. Metallog.* **2014**, *38*, 758–771.
10. Zhang, C.Q. *Distribution, Characteristics and Genesis of Mississippi Valley-Type Lead-Zinc Deposits in the Triangle Area of Sichuan-Yunnan-Guizhou Provinces*; China University of Geosciences: Beijing, China, 2005.
11. Zhang, C.Q. *The Genetic Model of Mississippi Valley-Type Deposits in the Boundary Area of Sichuan, Yunnan and Guizhou Provinces*; Chinese Academy of Geological Sciences: Beijing, China, 2008.
12. Wang, J. Localization Rules of Large Pb-Zn Deposits in the Southwestern Margin of the Upper-Middle Yangtze Block. Ph.D. Thesis, China University of Geosciences, Beijing, China, 2018.
13. Wu, J.; Li, G.; Li, Y.; Wei, J.; Wang, H.; Li, P. Chronology research progress of the MVT lead-zinc deposit in the border area of Sichuan-Yunnan-Guizhou and the background of metallogenic tectonics. *Geol. Sci. Technol. Info.* **2019**, *38*, 134–144.
14. Kong, Z.; Wu, Y.; Zhang, F.; Zhang, C.; Meng, X. Sources of hydrothermal fluids of typical Pb-Zn deposits in the Sichuan-Yunnan-Guizhou metallogenic province: Constraints from the S-Pb isotopic compositions. *Earth Sci. Front.* **2018**, *25*, 125–137.
15. Kong, Z.; Zhang, B.; Wu, Y.; Zhang, C.; Liu, Y.; Zhang, F.; Li, Y. Structural control and metallogenic mechanism of the Daliangzi Ge-rich Pb-Zn deposit in Sichuan province, China. *Earth Sci. Front.* **2022**, *29*, 143.
16. Wu, Y. The Age and Ore-Forming Process of MVT Deposits in the Boundary Area of Sichuan-Yunnan-Guizhou Provinces, Southwest China. Ph.D. Thesis, China University of Geosciences, Beijing, China, 2013; pp. 1–167.
17. Cui, Z.; Liu, X.; Zhou, J. Fractal characteristics of faults and its geological significance in Sichuan-Yunnan-Guizhou Pb-Zn metallogenic province, China. *Glob. Geol.* **2021**, *40*, 75–92.
18. Han, R.; Liu, C.; Huang, Z.; Chen, J.; Ma, D.; Li, Y. Genesis modeling of Huize lead-zinc ore deposit in Yunnan. *Acta Mineral. Sin.* **2001**, *21*, 674–680.
19. Li, W.B.; Huang, Z.L.; Zhang, G. Sources of the ore metals of the Huize ore field in Yunnan province: Constraints from Pb, S, C, H, O and Sr isotope geochemistry. *Acta Petrol. Sin.* **2006**, *22*, 2567–2580.

20. Zhang, C.; Zhang, Z.L.; Huang, Z.L.; Yan, Z.F. Study on the sources of Pb and Zn in Huize lead-zinc ore deposits. *Gansu Geol.* **2008**, *17*, 26–31.
21. Han, R.; Chen, J.; Li, Y.; Ma, D.; Zhao, D.; Ma, G. Ore-controlling tectonics and prognosis of concealed ores in Huize Pb-Zn deposit, Yunnan. *Acta Mineral. Sin.* **2001**, *21*, 265–269.
22. Erik, F. *Carbonate Micro-Characteristics*; Geological Publishing House: Beijing, China, 2006.
23. Wang, C. Experimental Study on Mechanical Properties of Dolomite Limestone in Huize Lead-Zinc Mine. Master's Thesis, Kunming University of Science and Technology, Kunming, China, 2013.
24. Zhang, C.Q.; Mao, J.; Wu, S.P.; Li, H.M.; Liu, F.; Guo, B.J.; Gao, D.R. Distribution, characteristics and genesis of Mississippi valley-type lead-zinc deposits in Sichuan-Yunnan-Guizhou area. *Miner. Depos.* **2005**, *24*, 336–348.
25. Zhou, J.-X.; Xiang, Z.-Z.; Zhou, M.-F.; Feng, Y.-X.; Luo, K.; Huang, Z.-L.; Wu, T. The giant upper yangtze Pb–Zn province in SW China: Reviews, new advances and a new genetic model. *J. Asian Earth Sci.* **2018**, *154*, 280–315. [CrossRef]
26. Kong, Z.; Wu, Y.; Liang, T.; Zhang, F.; Meng, X.; Lu, L. Sources of ore-forming material for Pb-Zn deposits in the Sichuan-Yunnan-Guizhou triangle area: Multiple constraints from C-H-O-S-Pb-Sr isotopic compositions. *Geol. J.* **2018**, *53*, 159–177. [CrossRef]
27. Zhang, C.; Wu, Y.; Hou, L.; Mao, J. Geodynamic setting of mineralization of Mississippi Valley–type deposits in world–class Sichuan–Yunnan–Guizhou Zn–Pb triangle, Southwest China: Implications from age–dating studies in the past decade and the Sm–Nd age of Jinshachang deposit. *J. Asian Earth Sci.* **2015**, *103*, 103–114. [CrossRef]
28. Tu, G.Z. Two unique mineralization areas in southwest China. *Bull. Mineral. Petrol. Geochem.* **2002**, *21*, 1–2.
29. Han, R.; Hu, Y.; Wang, X.; Huang, Z.; Chen, J.; Wang, F.; Wu, P.; Li, B.; Wang, H.; Dong, L.; et al. Mineralization model of rich Ge-Ag-bearing Zn-Pb polymetallic deposit concentrated district in northeastern Yunnan, China. *Acta Geol. Sin.* **2012**, *86*, 280–294.
30. Zhou, J.X.; Huang, Z.L.; Lv, Z.C.; Zhu, X.K.; Gao, J.G.; Mirnejad, H. Geology, isotope geochemistry and ore genesis of the shanshulin carbonate-hosted Pb–Zn deposit, southwest China. *Ore Geol. Rev.* **2014**, *63*, 209–225. [CrossRef]
31. Li, H.; Zhang, C. The genetic relationship between the H_2S-bearing gas in Sichuan basin and lead-zinc-copper deposits around the basin. *Geol. Rev.* **2012**, *58*, 495–510.
32. Chen, J.; Han, R.S.; Gao, D.R.; Zhao, D.S. Geologicai characteristics of huize Pb-Zn deposit, yunnan and model of oreprospectibg method. *Earth Environ.* **2001**, 124–129.
33. Shi, X.W.; Jia, F.J.; Ke, L.Y.; Zou, C. The geochemical characteristics of the C-O isotope of the Huize mine area of Yunnan province, China. *Acta Mineral. Sin.* **2021**, *41*, 657–667.
34. Wang, Z.Q. Enrichment Regularity of Dispersed Elements in Huize Super-Large Pb-Zn Deposit. Master's Thesis, China University of Geosciences, Beijing, China, 2017; pp. 1–65.
35. Wardlaw, N.C.; Li, Y.; Forbes, D. Pore-throat size correlation from capillary pressure curves. *Transp. Porous Media* **1987**, *2*, 597–614. [CrossRef]
36. Wen, D.X.; Han, R.S.; Wu, P.; He, J.J. Altered dolomite features and petro-geochemical prospecting indicators in the Huize lead-zinc deposit. *Geol. China* **2014**, *41*, 235–245.
37. Manning, C.E.; Ingebritsen, S.E. Permeability of the continental crust: Implications of geothermal data and metamorphic systems. *Rev. Geophys.* **1999**, *37*, 127–150. [CrossRef]
38. McCreesh, C.A.; Eris, E.L.; Brumfield, D.S.; Ehrlich, R. *Relating Thin Sections to Permeability, Mercury Porosimetry, Formation Factor, and Tortousity*; American Association of Petroleum Geologists AAPG/Datapages: Tulsa, OK, USA, 1988.

Article

The Genesis of Pyrite in the Fule Pb-Zn Deposit, Northeast Yunnan Province, China: Evidence from Mineral Chemistry and In Situ Sulfur Isotope

Meng Chen, Tao Ren * and Shenjin Guan *

Faculty of Land and Resources Engineering, Kunming University of Science and Technology, Kunming 650093, China
* Correspondence: rentao@kust.edu.cn (T.R.); guansj@kust.edu.cn (S.G.)

Abstract: The Fule deposit is a typical Cd-, Ge- and Ga-enriched Pb-Zn deposit located in the southeast of the Sichuan–Yunnan–Guizhou Pb-Zn polymetallic ore province in China. Zoned, euhedral cubic and pentagonal dodecahedral and anhedral pyrites were observed, and they are thought to comprise two generations. First generation pyrite (Py1) is homogeneous and entirely confined to a crystal core, whereas second generation pyrite (Py2) forms bright and irregular rims around the former. Second generation pyrite also occurs as a cubic and pentagonal dodecahedral crystal in/near the ore body or as an anhedral crystal generally closed to the surrounding rock. The content of S, Fe, Co, and Ni in Py1 are from 52.49 to 53.40%, 41.91 to 44.85%, 0.19 to 0.50% and 0.76 to 1.55%, respectively. The values of Co/Ni, Cu/Ni and Zn/Ni are from 0.22 to 0.42, 0.02 to 0.08 and 0.43 to 1.49, respectively, showing that the Py1 was formed in the sedimentary diagenetic stage. However, the contents of S, Fe, Co, and Ni in Py2 are in the range from 51.67 to 54.60%, 45.01 to 46.52%, 0.03 to 0.07% and 0.01 to 0.16%, respectively. The Co/Ni, Cu/Ni and Zn/Ni values of Py2 are from 0.40 to 12.33, 0.14 to 13.70 and 0.04 to 74.75, respectively, which is characterized by hydrothermal pyrite (mineralization stage). The different δ^{34}S values of the Py1 (-34.9 to $-32.3‰$) and the Py2 (9.7 to 20.5‰) indicate that there are at least two different sources of sulfur in the Fule deposit. The sulfur in Py1 was derived from the bacterial sulfate reduction (BSR), whereas the sulfur in the ore-forming fluids (Py2) was derived from the thermochemical sulfate reduction (TSR). The main reasons for the different morphologies of pyrite in the regular spatial distribution in the Fule deposit are temperature and sulfur fugacity.

Keywords: in situ S isotope; elemental geochemistry; pyrite; Fule Pb-Zn deposit; northeast Yunnan; China

Citation: Chen, M.; Ren, T.; Guan, S. The Genesis of Pyrite in the Fule Pb-Zn Deposit, Northeast Yunnan Province, China: Evidence from Mineral Chemistry and In Situ Sulfur Isotope. *Minerals* **2023**, *13*, 495. https://doi.org/10.3390/min13040495

Academic Editor: George M. Gibson

Received: 14 February 2023
Revised: 24 March 2023
Accepted: 29 March 2023
Published: 30 March 2023

1. Introduction

The Sichuan–Yunnan–Guizhou (SYG) polymetallic ore concentration area in the southwestern margin of the Yangtze block is an essential part of the low-temperature metallogenic domain in South China and one of the major production bases of Pb-Zn-Ag, and sphalerite contains significant amounts of trace elements, including Ge, Ga and Cd [1–3]. More than five hundred Pb-Zn polymetallic deposits have been produced in the region [1–3].

The Fule deposit has a mining history spanning more than 300 years. It is representative of the many large-scale lead–zinc deposits in the Sichuan–Yunnan–Guizhou Pb-Zn polymetallic metallogenic province. Since 1955, extensive research has been conducted on the deposit, including analysis of the ore field structure [1], the trace elements enrichment mechanism [4,5], the characteristics and evolution of the ore-forming fluids [5], the source of the ore-forming materials [6,7], and the metallogenic chronology [8,9]. Although much research work has been determined, there are still controversies in understanding the age of the ore formation, the mechanism of ore formation and the type of deposit. Some researchers believe that the deposit is MVT, while others believe that it is genetically related to Emeishan basalt [6–8]. Most researchers have the opinion that the deposit was formed in the late Indosinian period (191.9~222 Ma) [8,9], while some researchers consider that it was formed in the Himalayan period (20.4~34.7 Ma) [8,9].

Pyrite is one of the most abundant minerals in various deposits. More and more studies show that pyrite, with a complex internal structure and morphology, often records mineralization information [10–13]. Therefore, the study of pyrite can be used not only to reconstruct the hydrothermal evolution process [14], but also to define the genesis of the deposit [10,13–16]. Previous studies on pyrite have mainly involved the Carlin-type gold deposit [15,17,18], the epithermal deposit [19], the porphyry copper deposit [16] and the VMS-type deposit [20]. However, only the Huize deposit in the SYG polymetallic area was investigated for pyrite genesis [21,22].

This work found zoned, euhedral, and anhedral pyrites in the Fule deposit. As mentioned, the pyrites' major element contents were analyzed by electron microprobe. In addition, the in situ sulfur isotope analyses of pyrite were carried out by LA-MC-ICP-MS, and the environment, sulfur sources and genesis of pyrites were discussed.

2. Geological Background

The Fule is a large-sized Pb-Zn deposit with high Cd, Ge, and Ga contents (Figure 1a; [6,7,22]). The exploration results show that the lead and zinc metal reserves of the Fule deposit are 0.6 Mt. The deposit contains very high ore grades (up to 60 wt% Zn + Pb, average 15–20 wt%) [23]. In addition, the deposit contains metal reserves of approximately 4567 t Cd, 329 t Ge and 177 t Ga, with average grades of 0.127 wt% Cd, 0.012 wt% Ge and 0.007 wt% Ga, respectively [6].

The Permian Yangxin Formation (P_2y) is an ore-host stratum in the deposit, mainly composed of dolomite intercalated with limestone. It can be divided into the following three lithological sections. The lower section (P_2y^1) is composed of light gray limestone (Figure 1b). The middle section (P_2y^2) consists mainly of light gray limestone interbedded with dolomite, and locally contains siliceous dolomite, which is the main host rock of the deposit. The upper section (P_2y^3) is composed of gray medium-thick layered crystalline limestone, and a small amount of dolomitic limestone with chert strips. The igneous rocks in the area are Emeishan basalt, which is a series of continental rift tholeiite assemblages containing dense massive basalt and basaltic tuff. The main structures in the mining area are the Tuoniu-Duza anticline and the Mile-Shizong fault [22]. Together, they control the distribution of regional strata, secondary structures and mineralization. The Tuoniu-Duza anticline has a flat shape with a dip angle of 10° to 12° [22].

It mainly consists of three ore blocks: Laojuntai, Xinjuntai and Tonniu. The Fule deposit is buried about 150 m~200 m below the surface. Currently, 28 lead and zinc orebodies have been delineated with the NE strike, with the dip angle of 10° in the SE (Figure 1c) extending more than 3000 m [22]. The orebodies occur as stratiform to lentiform shapes or as veins along fractures within the Yangxin Formation. Metallic minerals in the deposit mainly include sphalerite, galena, pyrite, and a little chalcopyrite, tetrahedrite, tennantite, millerite, vaesite, gersdorffite and polydymite. Secondary oxides include cerussite and malachite.

According to the mineral assemblages and in combination with the previously published geological data [22], the ore-forming process of the Fule deposit can be divided into diagenetic and hydrothermal periods. The hydrothermal period can be further divided into sulfide-and-carbonate and carbonate stages.

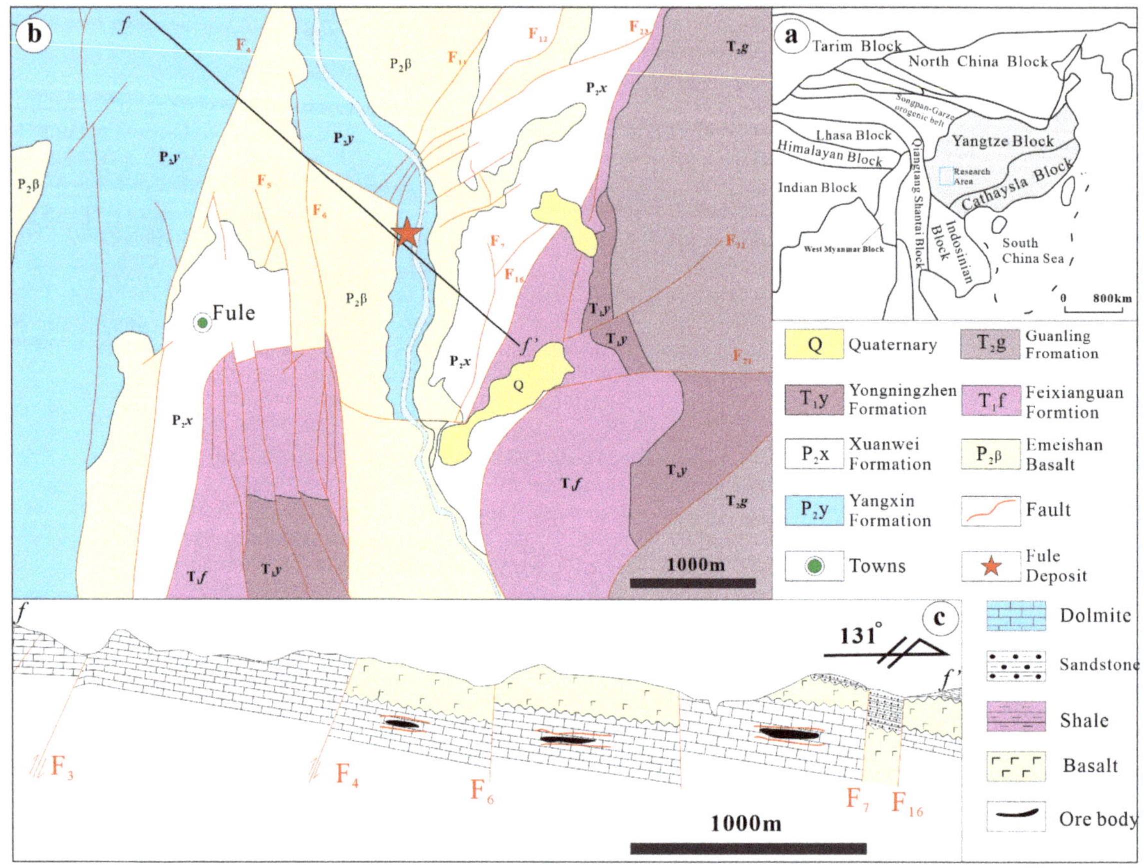

Figure 1. Regional geological setting of SW China (**a**), geological sketch map (**b**) and f-f' geological section (**c**) of the Fule Pb-Zn deposit (after Ref. [22]).

3. Sampling and Analytical Methods

All samples in this study were collected from the 1440 level of the Fule Pb-Zn deposit. Representative samples were selected for major elements and in situ sulfur isotope composition analysis. The in situ major element analysis of pyrite was carried out in the State Key Laboratory of Geochemistry, Institute of Geochemistry, Chinese Academy of Sciences. The instrument was a JXA-8230 electron probe, with an acceleration voltage of 25 kV, a current of 10 nA and a beam spot diameter of 1–5 μm. The SPI#02753-AB was used as the standard sample. The detection limits were from 100 to 200 ppm. Precisions for major elements and trace elements were approximately ±2% and ±10%, respectively. In situ sulfur isotope analysis was performed using a Nu plasma II multi receiver inductively coupled plasma mass spectrometer (MC-ICP-MS) equipped with a resolution S-155 Ar-F laser ablation system with a resolution of 193 nm in the GPMR laboratory of the China University of Geosciences (Wuhan). Working conditions included a laser energy density of 3 J/cm^2, a spot diameter of 33 μm and a single-point ablation time of 40 s. Natural pyrite WS-1 ($\delta^{34}S_{V\text{-}CDT}$ = 1.1 ± 0.2‰) was used to calibrate the sulfur isotope deviation, and V-CDT (Vienna-Cañon Diablo troilite) was used as a standard for the measured sulfur isotope data ($\delta^{34}S$).

4. Mineralogical Characteristics of Pyrite

Based on the morphology and chemical composition, pyrites are zoned and of two generations, with the first generation occupying the crystal core (Py1), whereas the other forms rims (Py2). Py2 formed in the second stage is intimately related to Pb-Zn mineralization, including cubic, pentagonal dodecahedral and anhedral crystals (Figure 2).

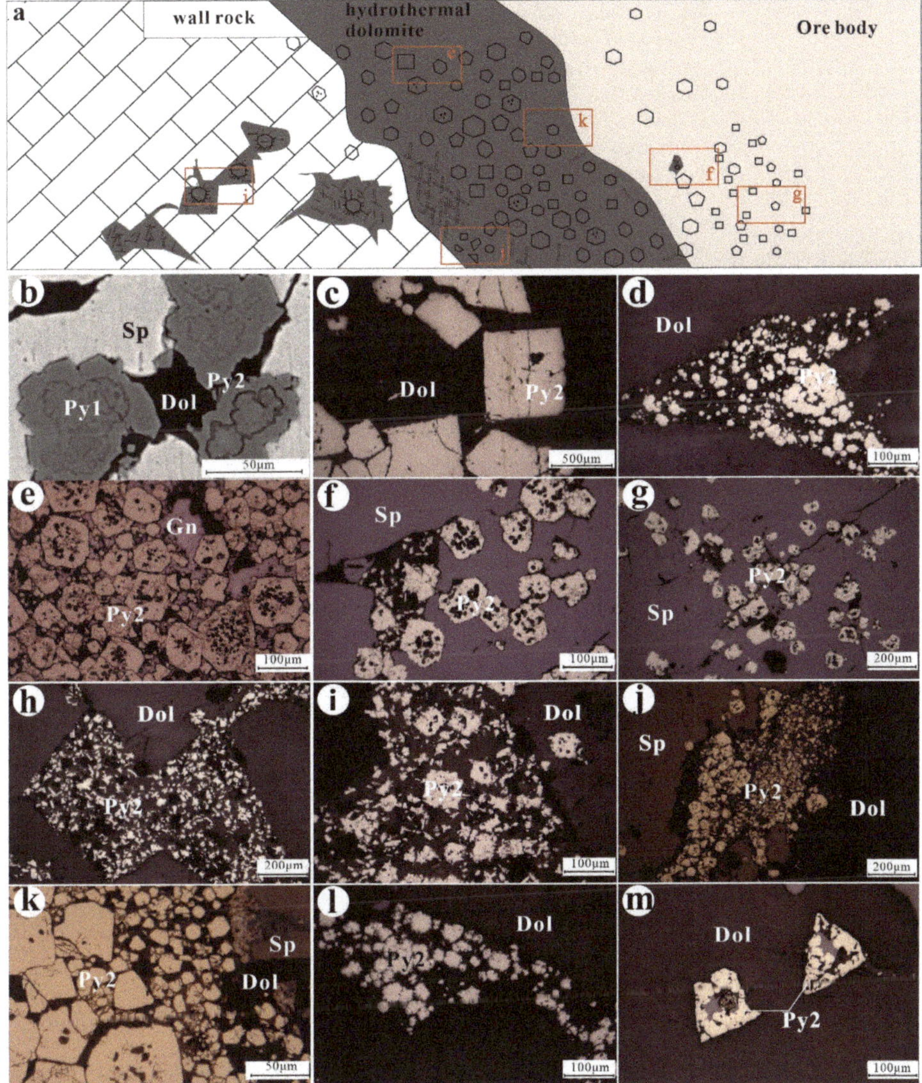

Figure 2. Mineral assemblage sketch (**a**) and microscopic images of pyrite (**b–m**) in the Fule deposit. (**b**) Zoned pyrite contains a Py1 core and a Py2 mantle; (**c**) cubic pyrite in the hydrothermal dolomite vein; (**d**) pentagonal dodecahedral pyrite filled in a fissure; (**e,f**) euhedral and subhedral pyrite containing subrounded dolomite, black voids in the pyrite are dolomite; (**g**) subhedral pyrite wrapped in sphalerite; (**h,i**) anhedral pyrite filled in the cavity; (**j,k**) anhedral pyrite aggregates occurring at the contact zone of sphalerite and dolomite; (**l**) granular pyrite aggregates; (**m**) intergrowth of pyrite and sphalerite filled in the cavity of dolomite; Abbreviations: Gn—galena; Sp—sphalerite; Py—pyrite; Dol—dolomite.

(1) Zoned pyrite (Figure 2b) consists of a homogeneous core (Py1) and a bright mantle (Py2) with a particle size of 50–80 μm.

(2) Euhedral pyrite includes pentagonal dodecahedral and cubic crystals (Figure 2c,d) with significant variations in crystal size, distributed in the center of the hydrothermal dolomite vein (Figure 2a). Pentagonal dodecahedral pyrite is also present in the fracture of recrystallized dolomite or the contact with sphalerite (Figure 2d). The large cubic crystal occurs in the fissure of recrystallized dolomite (Figure 2c). A small amount of cubic crystal occurs at the margin of the hydrothermal dolomite vein or is enveloped in sphalerite.

(3) Subhedral euhedral pyrite occurs at the margin of the pyrite-bearing hydrothermal dolomite vein or filled in the fissure of recrystallized dolomite (Figure 2e–g).

(4) Anhedral pyrite (Figure 2h–l) occurs in the hydrothermal dolomite vein near the recrystallized dolomite (Figure 2l,m).

5. Results

5.1. Major Elements

The results of the electron microprobe analysis of pyrite are shown in Table 1. The variation ranges of S and Fe in Py1 are from 52.49 to 53.40% and from 41.91 to 44.85%, respectively. Py1 is rich in Co, Ni and As, with contents of 0.19~0.50%, 0.76~1.55% and 0.44~1.37%, respectively. The S content (51.67 to 54.60%) of Py2 is slightly lower than that of Py1. Py2 has an Fe content of 45.01 to 46.52%. Py2 is relatively rich in Pb and Se, with values of 0.02~1.61% and 0.01~0.09%, respectively. The contents of Co, Ni and As in Py2 are 0.01 to 0.16%, 0.03 to 0.07% and 0.01 to 0.65%, respectively.

5.2. In Situ S Isotope Analysis

The results of the in situ S isotope analysis of pyrite are presented in Table 2 and Figure 3. Sulfur isotope values of pyrite in the Fule deposit are from -34.9 to 20.5‰ ($n = 29$). The δ^{34}S values in Py1 are from -34.9 to -32.3‰ ($n = 2$), while in Py2 they are from 9.7 to 20.5‰ ($n = 26$). The δ^{34}S values gradually increased from 10.9 to 13.6‰ from the core to the rim in a cubic pyrite crystal (Figure 4).

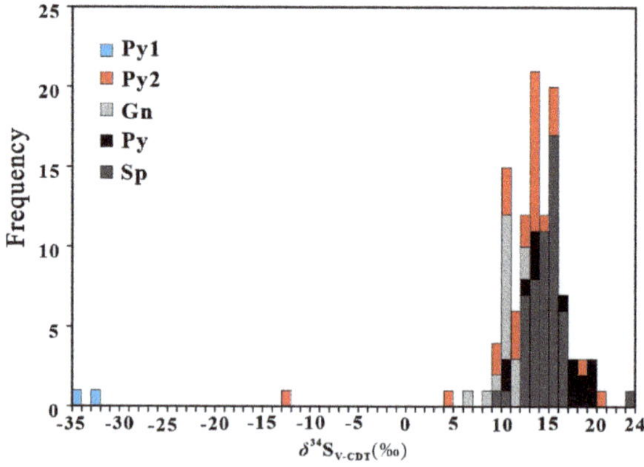

Figure 3. Histogram of the sulfur isotopic compositions of sulfide minerals from the Fule deposit. Data are taken from Refs. [7,9,22] and this paper. Abbreviations: Gn—galena; Sp—sphalerite; Py—pyrite; Dol—dolomite.

Table 1. Chemical composition of pyrite in the Fule deposit.

Sample No.	Pyrite Type		Fe	Cu	Zn	Ni	Se	As	S	Cd	Pb	Co	Total	Co/Ni	Cu/Ni	Zn/Ni	S/Fe
								%									
fl17-3-core	Py1	Zoned pyrite	44.85	0.06	1.13	0.76	0.00	0.44	53.03	0.01	0.01	0.19	100.48	0.25	0.08	1.49	2.06
fl17-4-core			44.29	0.07	1.09	1.29	0.03	0.99	52.49	0.01	0.04	0.28	100.59	0.22	0.06	0.85	2.06
fl17-5-core			41.91	0.04	0.66	1.55	0.01	1.37	52.96	-	-	0.50	99.00	0.32	0.03	0.43	1.10
fl17-6-core			42.15	0.02	0.48	1.05	0.06	1.14	53.40	-	-	0.45	98.76	0.42	0.02	0.46	1.10
S.D.			0.28	0.01	0.02	0.27	0.02	0.28	0.27	-	0.02	0.05	0.05	0.02	0.01	0.32	0.00
Median			44.57	0.07	1.11	1.03	0.02	0.72	52.76	0.01	0.03	0.24	100.54	0.24	0.07	1.17	2.06
Mean			44.57	0.07	1.11	1.03	0.02	0.72	52.76	0.01	0.03	0.24	100.54	0.24	0.07	1.17	2.06
fl2001-02	Py2	Subhedral-euhedral pyrite	46.20	0.05	0.10	0.08	0.04	0.38	53.32	0.02	0.03	0.06	100.30	0.78	0.58	1.28	2.01
fl2001-03			46.15	0.01	0.09	-	0.04	0.19	53.28	0.03	0.02	0.06	99.85	-	-	-	2.01
fl2001-09			46.41	-	0.02	0.01	0.03	0.43	53.11	0.01	-	0.04	100.05	3.89	-	1.67	1.99
fl2001-10			46.30	-	0.79	-	0.05	0.52	53.67	-	-	0.05	101.38	-	-	-	2.02
fl2001-11			45.95	-	2.01	-	-	0.18	54.03	0.01	-	0.05	102.23	-	-	-	2.05
fl2001-12			45.01	-	1.86	-	0.03	0.32	52.27	0.01	-	0.05	99.55	-	-	-	2.02
fl2013-17			46.45	-	0.14	0.02	-	0.23	53.80	-	-	0.03	100.66	1.67	-	7.72	2.02
fl2013-14			46.29	0.03	0.36	-	0.09	0.25	53.73	-	0.02	0.04	100.81	-	-	-	2.02
fl2001-04			45.64	0.07	0.05	0.13	0.05	0.65	52.83	0.01	0.20	0.05	99.68	0.40	0.52	0.35	2.02
fl2001-05			45.88	0.08	0.35	0.03	-	0.21	53.66	0.02	-	0.06	100.26	1.96	2.71	12.46	2.04
fl2001-07			46.27	0.03	0.15	-	-	0.09	53.69	-	0.13	0.07	100.41	-	-	-	2.02
fl2001-08			45.64	0.04	0.05	-	-	-	52.12	0.01	0.15	0.04	98.05	12.33	13.67	18.00	1.99
fl2013-15			45.84	0.01	1.06	-	0.01	-	53.62	-	0.09	0.03	100.65	-	-	-	2.04
fl2013-19			46.11	0.04	0.27	-	-	0.59	53.46	0.01	0.05	0.06	100.57	-	-	-	2.02
fl2013-20			46.27	0.06	0.05	0.04	-	0.11	53.50	0.01	-	0.06	100.08	1.63	1.69	1.34	2.01
fl2013-22			46.20	0.01	-	-	0.01	0.13	53.59	0.02	-	0.06	100.02	-	-	-	2.02
fl2013-24			46.13	-	0.01	-	-	0.27	53.19	-	-	0.05	99.65	-	-	-	2.01
fl2013-25			46.08	0.03	-	0.04	-	0.29	53.47	-	-	0.05	99.96	1.27	0.76	0.11	2.02
fl2013-30			46.03	0.04	-	-	0.01	-	54.60	-	0.06	0.05	100.78	-	-	-	2.07
S.D.			0.33	0.02	0.63	0.04	0.02	0.16	0.55	0.01	0.06	0.01	0.82	3.66	4.69	6.27	0.02
Median			46.13	0.04	0.15	0.04	0.04	0.26	53.50	0.01	0.06	0.05	100.26	1.65	1.23	1.51	2.02
Mean			46.04	0.04	0.46	0.05	0.04	0.30	53.42	0.01	0.08	0.05	100.26	2.99	3.32	5.37	2.02

Table 1. Cont.

Sample No.	Pyrite Type		Fe	Cu	Zn	Ni	Se	As	S	Cd	Pb	Co	Total	Co/Ni	Cu/Ni	Zn/Ni	S/Fe
								%									
fl10-01			46.52	0.03	0.05	-	0.03	-	53.59	-	-	0.05	100.27	-	-	-	2.01
fl10-02			45.92	0.14	0.08	0.01	-	0.04	53.76	-	-	0.06	100.01	6.30	13.70	8.00	2.04
fl10-03			46.18	0.09	0.72	-	0.03	0.01	53.81	-	-	0.06	100.89	-	-	-	2.03
fl2002-01			45.91	0.02	-	-	-	0.13	52.93	0.01	0.07	0.05	99.12	-	-	-	2.01
fl2002-02			46.40	-	0.01	-	-	-	53.25	-	0.05	0.06	99.75	-	-	-	2.00
fl2002-03		Anhedral	45.66	0.04	0.02	0.02	-	0.10	51.67	-	-	0.05	97.56	2.70	1.75	1.10	1.97
fl2002-04		pyrite	45.84	0.01	0.02	-	-	0.03	53.33	-	-	0.04	99.28	-	-	-	2.03
fl17-10			45.68	0.33	0.02	-	-	-	53.08	-	0.07	0.05	99.23	-	-	-	2.02
fl17-11			45.06	0.29	0.18	-	-	-	51.85	0.01	1.61	0.06	99.06	-	-	-	2.00
fl2001-01			46.24	0.03	0.02	-	-	-	54.07	-	0.07	-	100.44	-	-	-	2.04
S.D.	Py2		0.40	0.11	0.22	0.01	-	0.05	0.76	-	0.62	0.01	0.89	1.80	5.98	3.45	0.02
Median			45.92	0.04	0.02	0.02	0.03	0.04	53.29	0.01	0.07	0.05	99.52	4.50	7.73	4.55	2.02
Mean			45.94	0.11	0.12	0.02	0.03	0.06	53.13	0.01	0.37	0.05	99.56	4.50	7.73	4.55	2.02
fl2001-13			45.96	0.02	0.01	0.16	-	0.43	53.25	0.01	0.03	0.07	99.92	0.44	0.14	0.04	2.02
fl2013-18		Pentagonal	45.58	0.00	0.39	-	0.04	0.52	52.22	0.03	0.02	0.05	98.88	-	-	-	2.00
S.D.		dodecahe-	0.19	0.01	0.19	0.00	0.00	0.05	0.50	0.01	0.01	0.01	0.52	0.00	0.00	0.00	0.01
Median		dral	45.77	0.01	0.20	0.16	0.04	0.48	52.76	0.02	0.03	0.06	99.40	0.44	0.14	0.04	2.01
Mean		pyrite	45.77	0.01	0.20	0.16	0.04	0.48	52.76	0.02	0.03	0.06	99.40	0.44	0.14	0.04	2.01
fl17-1-rim			45.78	0.00	1.50	0.02	-	0.08	53.41	0.03	-	0.05	100.87	2.30	0.20	74.75	2.03
fl17-2-rim		Zoned pyrite	45.54	0.03	1.63	0.05	-	0.10	53.96	-	0.04	0.05	101.41	0.98	0.47	30.81	2.06
S.D.			0.12	0.02	0.06	0.02	-	0.01	0.28	0.00	0.00	0.00	0.27	0.66	0.14	21.97	0.02
Median			45.66	0.02	1.57	0.04	-	0.09	53.69	0.03	0.04	0.05	101.14	1.64	0.34	52.78	2.05
Mean			45.66	0.02	1.57	0.04	-	0.09	53.69	0.03	0.04	0.05	101.14	1.64	0.34	52.78	2.05

Table 2. In situ LA-MC-ICP-MS sulfur isotopic composition of pyrite from the Fule deposit (‰).

Sample No.	Description	Pyrite Type	δ^{34}S
fl17-2-core		Py1	−32.9
fl-17-3-core	Zoned pyrite		−34.3
fl17-1-rim			−12.5
fl-20-1-rim			13.6
fl-20-2-rim			12.0
fl-20-3-core	Cubic pyrite		10.9
fl-20-4-rim			13.0
fl-20-5-rim			13.6
fl-20-6-rim			13.1
fl20-13-8			13.6
fl20-13-1	Pentagonal dodecahedral pyrite		13.5
fl20-13-3			14.0
fl20-13-12		Py2	9.8
fl20-13-13			9.7
fl20-13-6			10.0
fl20-13-5			11.1
fl20-13-2			12.1
fl20-13-4	Subhedral-euhedral pyrite		10.5
fl20-13-7			11.3
fl20-13-9			13.7
fl20-13-10			13.3
fl20-13-11			15.0
fl17-8			11.4
fl10-5			13.4
Fl17-2			18.1
fl17-3	Anhedral pyrite		13.1
fl17-4			15.4
fl17-5			15.2
fl17-6			20.5

6. Discussion

6.1. Chemical Composition and Sulfur Sources of Pyrite

Pyrite contains more than 30 kinds of trace elements, including chalcophile, lithophile and siderophile elements [10–13]. The trace element contents are closely related to the type and genetic type of the deposit, as well as the temperature and pressure conditions [24]. Pyrites formed in high-temperature hydrothermal deposits are generally rich in siderophile and lithophile elements. They also have high amounts of Bi, Cu, Zn and As. Under moderate temperatures, pyrite is mainly rich in Cu, Au, Pb, Zn, Bi, Ag, etc. Pyrite in epithermal deposit has high Hg, Sb, Ag and As content. Compared to other types of deposits, the pyrite in the Fule deposit is characterized by the enrichment of Cu, As, Co, Ni and Se, and the contents of most trace elements are low, indicating that the deposit was formed under medium-low temperature conditions. It is consistent with the homogenization temperature of fluid inclusion in Fule sphalerite [5]. Cu, Ni and Co enrichment in Py2 and Cd, Ge and Ga enrichment in sphalerite indicate that the deposit may be an MVT deposit. Ore bodies host in dolomite and have simple mineral assemblages (mainly of sphalerite, galena and pyrite), which are also basically consistent with the geological characteristics of typical MVT deposits.

Many sulfur isotope analyses have been conducted on sphalerite, galena and pyrite in the Fule deposit, and the δ^{34}S values are concentrated from 10.04 to 19.30‰ [7,22]. This indicates that there is a seawater sulfate reservoir existing in the strata, and thermochemical sulfate reduction (TSR) is mainly the formation mechanism of reduced sulfur in the deposit [10]. However, some researchers still believe that there may be multiple sulfur sources [7,22]. In this study, Py1 with δ^{34}S of −34.9‰ to −32.3‰ was first found in the Fule deposit. Therefore, based on the previous research and the sulfur isotope results obtained in this work, sulfur in the Fule deposit has at least two different sources (Figure 4).

TSR and bacterial sulfate reduction (BSR) are the two main mechanisms for the conversion of sulfate to reduced sulfur [25]. H_2S of TSR origin inherits the sulfur isotopic composition of sulfate, and the resulting sulfide generally has high sulfur isotopic value [26] and a relatively high formation temperature (100 to 140 °C). The Py2 formed in the metallogenic stage has a $\delta^{34}S$ of 10 to 20‰, which is similar to that of sphalerite and galena reported in previous studies [22]. This isotopic value is close to the Permian, Carboniferous and Cambrian marine sulfate (11‰, 14‰ and 17‰, respectively, [22]) in northeastern Yunnan, indicating that the sulfate in the sedimentary strata is the main source of sulfur. TSR may be the main mechanism for reducing sulfur formation in the ore-forming stage.

BSR can occur near the surface or in shallow burial environments at low temperatures (<80 °C; [27–29]). BSR can result in significant isotopic fractionation, generally between 4 to 46‰, and 65 ‰ in extreme cases [30,31], such that H_2S produced by BSR has sulfur isotope values of −50‰ and 30‰ [31]. The $\delta^{34}S$ of the Py1 in the Fule deposit ranges from −34.9 to −32.3‰. This value is about 40‰ lower than that of the hydrothermal sulfide, indicating that the sulfur in Py1 is of BSR origin.

The $\delta^{34}S$ values of the cubic pyrite (Py2) show a gradual increase from the core to the rim. According to previous studies, when sulfide precipitated in a hydrothermal solution dominated by H_2S and containing a small amount of SO_4^{2-}, its $\delta^{34}S$ value was similar to that of the initial solution in the early stage, but higher than that of the initial solution in the late phase [32]. This may be the reason for the change in the sulfur isotope of cubic pyrite in the Fule deposit.

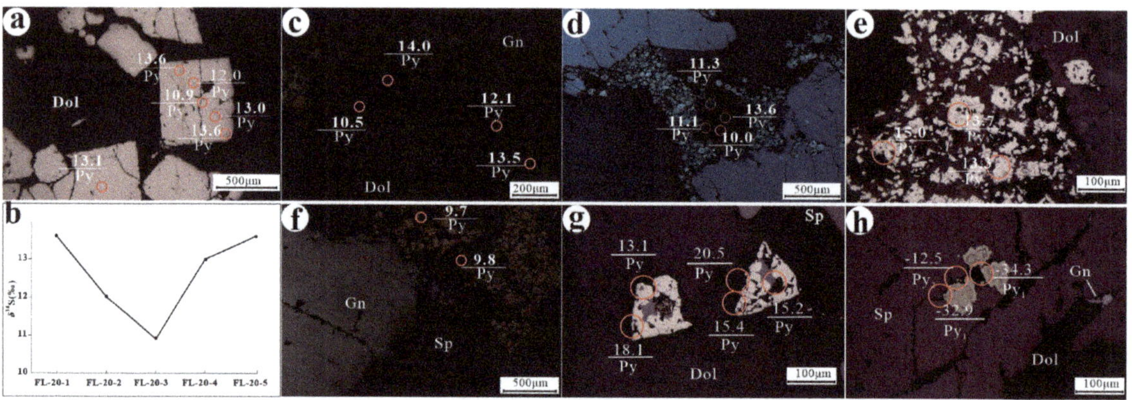

Figure 4. Photographs showing in situ LA-MC-ICP-MS sulfur isotope analysis spots and $\delta^{34}S$ values. (**a**) cubic pyrite in hydrothermal dolomite; (**b**) the sulfur isotope values in the cubic pyrite core are relatively lower than that of rim; (**c**,**d**) anhedral pyrite aggregate; (**e**) anhedral pyrite filled in the cavity; (**f**) subhedral–anhedral pyrite filled in the contact zone of galena and sphalerite; (**g**) intimate growth of pyrite and sphalerite; (**h**) zoned pyrite has a core and a mantle filled in the fissure of sphalerite. Abbreviations: Gn—galena; Sp—sphalerite; Py—pyrite; Dol—dolomite.

6.2. Genesis of Pyrite

The Co/Ni ratio in pyrite could be used to effectively indicate the genesis of pyrite [33,34]. The contents of Co and Ni in sedimentary pyrite are generally low and the Co/Ni values are less than 1. The Co and Ni contents and Co/Ni values of hydrothermal pyrite vary greatly, with 1 < Co/Ni < 5. Pyrite formed in volcanic exhalative massive sulfide deposits has high Co and low Ni contents and high Co/Ni values (5~50) [34–36].

Py1 in the Fule deposit had relatively higher Co and Ni contents than Py2. In addition, the content of Ni was much higher than that of Co, and the Co/Ni values were between 0.22

and 0.42, which indicate that Py1 was formed during the depositional period (deposited at the same time as the ore-hosted rock). However, the Co/Ni ratio in Py2 was between 0.40 and 12.33 (mostly are from 1 to 5), which is consistent with the value of hydrothermal pyrite. The Cu/Ni and Zn/Ni values from Py1 in the Fule deposit were from 0.02 to 0.08 and from 0.43 to 1.49, respectively. These values plot in the range of sedimentary pyrite (0.01 to 2 and 0.01 to 10) [37,38]. In contrast, the ratios from Py2 ranged from 0.14 to 13.70 and 0.04 to 74.75, also indicating a hydrothermal origin.

A systematic study of pyrite formed under natural and synthetic conditions showed that temperature and (or) sulfur saturation really influenced the pyrite morphology. The crystal appearance experienced a change trend from columnar → cube → pentagonal dodecahedron or octahedron → irregular granular [39]. As mentioned above, hydrothermal pyrite in the Fule deposit had a significant morphological zone in space (Figure 2a). Columnar pyrite distributed in the margin of hydrothermal dolomite near the wall rock shows that the crystal selects the {100} and extends growth rapidly along the <001> direction. It can be formed in a large temperature gradient and low material conditions. Meanwhile, cubic and pentagonal dodecahedral pyrites appear inside the ore body or in the center of hydrothermal dolomite due to suitable temperature, sufficient material supply and high sulfur fugacity.

7. Conclusions

(1) Pyrite in the Fule deposit has various morphologies, including zoned, euhedral (cubic and pentagonal dodecahedral), subhedral-euhedral and anhedral crystals. The pyrite core was formed during the sedimentary stage, and the rim, euhedral and anhedral pyrite was formed in the metallogenic stage.

(2) The high S/Fe ratio, As, Cu, and Zn contents reflect that Py2 was formed in a medium-low temperature environment. The sulfur isotopic composition of the pyrite in the deposit shows that there are at least two different sulfur sources. The sulfur in Py1 was derived from BSR, while the sulfur in Py2 was derived from the TSR.

(3) Pyrite morphology in the Fule deposit goes through the changing trends of columnar, cube, pentagonal dodecahedron, and irregular granular morphology. Temperature and (or) sulfur saturation dominate the morphological change.

Author Contributions: Data curation, investigation, writing—original draft, M.C.; conceptualization, methodology, writing—review and editing, T.R. and S.G. All authors have read and agreed to the published version of the manuscript.

Funding: This study was supported by the National Natural Science Foundation of China (NSFC) project (42163005).

Data Availability Statement: Data are contained within the article.

Acknowledgments: We would like to thank Zhaojun Lv for the assistance during fieldwork.

References

1. Liu, T.T.; Zhu, C.W.; Yang, G.S.; Zhang, G.S.; Fan, H.F.; Zhang, Y.X.; Wen, H.J. Primary study of germanium isotope composition in sphalerite from the Fule Pb-Zn deposit, Yunnan province. *Ore Geol. Rev.* **2020**, *120*, 103466. [CrossRef]
2. Zhou, J.X.; Huang, Z.L.; Zhou, M.F.; Li, X.B.; Jin, Z.G. Constraints of C-O-S-Pb isotope compositions and Rb-Sr isotopic age on the origin of the Tianqiao carbonate-hosted Pb-Zn deposit, SW China. *Ore Geol. Rev.* **2013**, *53*, 77–92. [CrossRef]
3. Wang, C.M.; Deng, J.; Carranza, E.J.M.; Lei, X.R. Nature, diversity and temporal-spatial distributions of sediment-hosted Pb-Zn deposit in China. *Ore Geol. Rev.* **2014**, *56*, 327–351. [CrossRef]
4. Hu, R.Z.; Fu, S.L.; Huang, Y.; Zhou, M.F.; Fu, S.H.; Zhao, C.H.; Wang, Y.J.; Bi, X.W.; Xiao, J.F. The giant South China Mesozoic low-temperature metallogenic domain: Reviews and a new geodynamic model. *J. Asian Earth Sci.* **2017**, *137*, 9–34. [CrossRef]
5. Nian, H.L.; Cui, Y.L.; Li, Z.L.; Jia, F.J.; Chen, W.; Yang, S.X.; Yang, Z. Features of sphalerite-hosted fluid inclusions of Fule lead-zinc mining area and outskirts in Luoping area, eastern Yunnan Province, China. *Acta Mineral. Sin.* **2017**, *37*, 469–474. (In Chinese)
6. Zhu, C.W.; Wen, H.J.; Zhang, Y.X.; Fu, S.H.; Fan, H.F.; Cloquet, C. Cadmium isotope fractionation in the Fule Mississippi Valley-type deposit, Southwest China. *Miner. Deposita* **2017**, *52*, 675–686. [CrossRef]

7. Zhou, J.X.; Luo, K.; Wang, X.C.; Simon, A.W.; Wu, T.; Huang, Z.L.; Cui, Y.L.; Zhao, J.X. Ore genesis of the Fule Pb-Zn deposit and its relationship with the Emeishan large igneous province: Evidence from mineralogy, bulk C-O-S and in situ S-Pb isotopes. *Gondwana Res.* **2018**, *54*, 161–179. [CrossRef]

8. Liu, Y.; Qi, L.; Gao, J.; Ye, L.; Huang, Z.; Zhou, J. Re-Os dating of galena and sphalerite from lead-zinc sulfide deposits in Yunnan Province, SW China. *J. Earth Sci.* **2015**, *26*, 343–351. [CrossRef]

9. Lyu, C.; Gao, J.F.; Qi, L.; Huang, X.W. Re-Os isotope system of sulfide from the Fule carbonate-hosted Pb-Zn deposit, SW China: Implications for Re-Os dating of Pb-Zn mineralization. *Ore Geol. Rev.* **2020**, *121*, 103558. [CrossRef]

10. Craig, J.R.; Vokes, F.M.; Solberg, T.N. Pyrite: Physical and chemical textures. *Miner. Deposita* **1998**, *34*, 82–101. [CrossRef]

11. Deditius, A.P.; Reich, M.; Kesler, S.E.; Utsunomiya, S.; Chryssoulis, S.L.; Walshe, J.; Ewing, R.C. The coupled geochemistry of Au and As in pyrite from hydrothermal ore deposits. *Geochim. Et Cosmochim. Acta* **2014**, *140*, 644–670. [CrossRef]

12. Large, R.R.; Halpin, J.A.; Danyushevsky, L.V.; Maslennikov, V.V.; Bull, S.W.; Long, J.A.; Gregory, D.D.; Lounejeva, E.; Lyons, T.W.; Sack, P.J. Trace element content of sedimentary pyrite as a new proxy for deep-time ocean-atmosphere evolution. *Earth Planet. Sci. Lett.* **2014**, *389*, 209–220. [CrossRef]

13. Franchini, M.; Mcfarlane, C.; Maydagán, L.; Reich, M.; Lentz, D.R.; Meinert, L.; Bouhier, V. Trace metals in pyrite and marcasite from the Agua Rica porphyry-high sulfidation epithermal deposit, Catamarca, Argentina: Textural features and metal zoning at the porphyry to epithermal transition. *Ore Geol. Rev.* **2015**, *66*, 366–387. [CrossRef]

14. Genna, D.; Gaboury, D. Deciphering the hydrothermal evolution of a VMS system by LA-ICP-MS using trace elements in pyrite: An example from the Bracemac-McLeod deposits, Abitib, Canada, and implications for exploration. *Econ. Geol.* **2015**, *110*, 2087–2108. [CrossRef]

15. Large, R.R.; Danyushevsky, L.; Hollit, C.; Maslennikov, V.V.; Meffre, S.; Gilbert, S.; Bull, S.; Scott, R.; Emsbo, P.; Thomas, H.; et al. Gold and trace element zonation in pyrite using a laser imaging technique: Implications for the timing of gold in orogenic and Carlin-style sediment-hosted deposits. *Econ. Geol.* **2009**, *104*, 635–668. [CrossRef]

16. Reich, M.; Deditius, A.; Chryssoulis, S.; Li, J.W.; Ma, C.Q.; Parada, M.A.; Barra, F.; Mittermayr, F. Pyrite as a record of hydrothermal fluid evolution in a porphyry copper system: A SIMS/EMPA trace element study. *Geochim. Et Cosmochim. Acta* **2013**, *104*, 42–62. [CrossRef]

17. Sung, Y.H.; Brugger, J.; Ciobanu, C.L.; Pring, A.; Skinner, W.; Nugus, M. Invisible gold in arsenian pyrite and arsenopyrite from a multistage Archaean gold deposit: Sunrise Dam, Eastern Goldfields Province, Western Australia. *Miner. Deposita* **2009**, *44*, 765–791. [CrossRef]

18. Zhang, K.; Li, H.Y. Migration of trace elements in pyrite from orogenic gold deposits: Evidence from LA-ICP-MS analyses. *Acta Geol. Sin.* **2014**, *88* (Suppl. 2), 841–842. [CrossRef]

19. Winderbaum, L.; Ciobanu, C.L.; Cook, N.J.; Paul, M.; Metcalfe, A.; Gilbert, S. Multivariate analysis of an LA-ICP-MS trace element dataset for pyrite. *Math. Geosci.* **2012**, *44*, 823–842. [CrossRef]

20. Basori, M.B.I.; Gilbert, S.; Large, R.R.; Zaw, K. Textures and trace element composition of pyrite from the Bukit Botol volcanic-hosted massive sulphide deposit, Peninsular Malaysia. *J. Asian Earth Sci.* **2018**, *158*, 173–185. [CrossRef]

21. Meng, Y.M.; Hu, R.Z.; Huang, X.W.; Gao, J.F.; Christian, S. The origin of the carbonate-hosted Huize Zn-Pb-Ag deposit, Yunnan province, SW China: Constraints from the trace element and sulfur isotopic compositions of pyrite. *Mineral. Petrol.* **2019**, *113*, 369–391. [CrossRef]

22. Ren, T.; Zhou, J.X.; Wang, D.; Yang, G.S.; Lv, C.L. Trace elemental and S-Pb isotopic geochemistry of the Fule Pb-Zn deposit, NE Yunnan Province. *Acta Petrol. Sin.* **2019**, *35*, 3493–3505. (In Chinese)

23. Li, L.T. The geological feature of Fulechang Pb-Zn deposit and inference of deep prospecting in Luoping, Yunnan. *Geol. Yunnan* **2014**, *33*, 240–244. (In Chinese)

24. Keith, M.; Häckel, F.; Haase, K.M.; Schwarz-Schampera, U.; Klemd, R. Trace element systematics of pyrite from submarine hydrothermal vents. *Ore Geol. Rev.* **2016**, *72*, 728–745. [CrossRef]

25. Ohmoto, H. Stable isotope geochemistry of ore deposits. *Rev. Mineral. Geochem.* **1986**, *16*, 491–559.

26. Gomes, M.L.; Fike, D.A.; Bergmann, K.D.; Jones, C.; Knoll, A.H. Environmental insights from high-resolution (SIMS) sulfur isotope analyses of sulfides in Proterozoic microbialites with diverse mat textures. *Geobiology* **2018**, *16*, 17–34. [CrossRef] [PubMed]

27. Ohmoto, H. Biogeochemistry of sulfur and the mechanisms of sulfide-sulfate mineralization in Archean oceans. In *Early Organic Evolution: Implications for Mineral and Energy Resources*; Schidlowski, M., Golubic, S., Kimberley, M.M., McKirdy, D.M., Trudinger, P.A., Eds.; Springer: Berlin/Heidelberg, Germany, 1992; pp. 378–397.

28. Wilhelms, A.; Larter, S.R.; Head, I.; Farrimond, P.; Diprimio, R.; Zwach, C. Biodegradation of oil in uplifted basins prevented by deep-burial sterilization. *Nature* **2001**, *411*, 1034–1037. [CrossRef]

29. Head, I.M.; Jones, D.M.; Larter, S.R. Biological activity in the deep subsurface and the origin of heavy oil. *Nature* **2003**, *426*, 344–352. [CrossRef] [PubMed]

30. Canfield, D.E.; Teske, A. Late Proterozoic rise in atmospheric oxygen concentration inferred from phylogenetic and sulphur-isotope studies. *Nature* **1996**, *382*, 127–132. [CrossRef] [PubMed]

31. Habicht, K.S.; Canfield, D.E. Sulfur isotope fractionation during bacterial sulfate reduction in organic-rich sediments. *Geochim. Et Cosmochim. Acta* **1997**, *61*, 5351–5361. [CrossRef] [PubMed]

32. Loftus-Hills, G.; Solomon, M. Cobalt, nickel and selenium in sulphides as indicators of ore genesis. *Miner. Deposita* **1967**, *2*, 228–242. [CrossRef]

33. Fleischer, M. Minor elements in some sulphide minerals. *Econ. Geol.* **1955**, *50*, 970–1024.
34. Wilkin, R.T.; Barnes, H.L. Pyrite formation by reactions of iron monosulfides with dissolved inorganic and organic sulfur species. *Geochim. Et Cosmochim. Acta* **1996**, *60*, 4167–4179. [CrossRef]
35. Bajwah, Z.U.; Seccombe, P.K.; Offler, R. Trace element distribution, Co/Ni ratios and genesis of the Big Cadia iron-copper deposit, New South Wales, Australia. *Miner. Deposita* **1987**, *22*, 292–303. [CrossRef]
36. Bralia, A.; Sabatini, G.; Troja, F. A revaluation of the Co/Ni ratio in pyrite as geochemical tool in ore genesis problems. *Miner. Deposita* **1979**, *14*, 353–374. [CrossRef]
37. Gregory, D.D.; Large, R.R.; Halpin, J.A.; Steadman, J.A.; Hickman, A.H.; Ireland, T.R.; Holden, P. The chemical conditions of the late Archean Hamersley basin inferred from whole rock and pyrite geochemistry with $\Delta^{33}S$ and $\delta^{34}S$ isotope analyses. *Geochim. Et Cosmochim. Acta* **2015**, *149*, 223–250. [CrossRef]
38. Gregory, D.D.; Lyons, T.W.; Large, R.R.; Jiang, G.; Stepanov, A.S.; Diamond, C.W.; Figueroa, M.C.; Olin, P. Whole rock and discrete pyrite geochemistry as complementary tracers of ancient ocean chemistry: An example from the Neoproterozoic Doushantuo Formation, China. *Geochim. Et Cosmochim. Acta* **2017**, *216*, 201–220. [CrossRef]
39. Alonso-Azcárate, J.; Rodas, M.; Fernández-Díaz, L.; Bottrell, S.H.; Mas, J.R.; López-Andrés, S. Causes of variation in crystal morphology in metamorphogenic pyrite deposits of the Cameros Basin (N Spain). *Geol. J.* **2001**, *36*, 159–170. [CrossRef]

Article

Trace Elements of Gangue Minerals from the Banbianjie Ge-Zn Deposit in Guizhou Province, SW China

Yun-Lin An [1,2], Jia-Xi Zhou [1,3,4,*], Qing-Tian Meng [5], Guo-Tao Sun [6] and Zhi-Mou Yang [1,2]

1   Key Laboratory of Critical Minerals Metallogeny in Universities of Yunnan Province, School of Earth Sciences, Yunnan University, Kunming 650500, China
2   Zijin Mining Group Co., Ltd., Longyan 364200, China
3   Key Laboratory of Sanjiang Metallogeny and Resources Exploration & Utilization, MNR, School of Earth Sciences, Yunnan University, Kunming 650500, China
4   Yunnan Key Laboratory of Sanjiang Metallogeny and Resources Exploration & Utilization, School of Earth Sciences, Yunnan University, Kunming 650500, China
5   No. 104 Geological Team, Guizhou Bureau of Geology and Mineral Exploration and Development, Duyun 558000, China
6   College of Resources and Environmental Engineering, Guizhou University, Guiyang 550025, China
*   Correspondence: zhoujiaxi@ynu.edu.cn

Abstract: There are many dispersed element-rich Pb-Zn deposits hosted by Paleozoic carbonate rocks in the Middle-Upper Yangtze Block, China. The origin and nature of the ore-forming fluids that formed them are still much debated (syngenetic vs. epigenetic). The Banbianjie Ge-Zn deposit is located in the southeastern margin of the Yangtze Block, SW China. It is a newly discovered medium-sized Zn (Zn metal reserves > 0.39 Mt, @1.78%–9.5% Zn) and large-scale Ge deposit (Ge metal resources > 900 t, @100 $\times$ 10^{-6}–110 $\times$ 10^{-6} Ge) in the Western Hunan–Eastern Guizhou Pb-Zn metallogenic belt, SW China. Gangue minerals in the Banbianjie deposit are very developed, including calcite, dolomite and barite, which are closely associated with sulfides. Hence, the trace elements of gangue minerals could be used to trace the nature, source and evolution of ore-forming fluids, and the ore genesis of this deposit can be discussed. These gangue minerals are nearly horizontally distributed in the plot of La/Ho-Y/Ho, suggesting that they are the products of the same hydrothermal fluids. The total rare earth element ($\sum$REE) contents from calcite and dolomite to barite show an increasing trend, indicating that the REEs in the ore-forming fluids were mainly enriched in barite. Hence, the $\sum$REE of barite can approximately represent the ΣREE of the hydrothermal fluids, which are quite similar to those of the underlying strata, indicating that the ore-forming fluids were likely originated from and/or flowed through them. The Eu anomalies from dolomite (Eu/Eu* = 0.33–0.66) to calcite (Eu/Eu* = 0.29–1.13) and then to barite (Eu/Eu* = 1.64–7.71) show an increasing trend, suggesting that the ore-forming fluids experienced a shift in the ore-forming environment from reduced to oxidized. Hence, the source of the Banbianjie Ge-Zn deposit is the underlying strata, and the ore-forming physical–chemical condition has experienced a transition from reduction to oxidation during the Ge-Zn mineralization. The ore genesis of the Banbianjie Ge-Zn deposit is most likely a Mississippi Valley-type (MVT) deposit.

Keywords: Banbianjie Ge-Zn deposit; gangue minerals; nature; source and evolution of ore-forming fluids; trace elements; Western Hunan–Eastern Guizhou Pb-Zn metallogenic belt; SW China

Citation: An, Y.-L.; Zhou, J.-X.; Meng, Q.-T.; Sun, G.-T.; Yang, Z.-M. Trace Elements of Gangue Minerals from the Banbianjie Ge-Zn Deposit in Guizhou Province, SW China. *Minerals* **2023**, *13*, 638. https://doi.org/10.3390/min13050638

Academic Editor: Daniel Marshall

Received: 11 April 2023
Revised: 30 April 2023
Accepted: 2 May 2023
Published: 4 May 2023

1. Introduction

Dispersed metals (Ga, Ge, Se, Cd, In, Te, Re and Tl) are widely used in emerging industries and are critical metals [1–3]. Germanium, a typical dispersed metal, only has a CLARKE value of about 1.5×10^{-6} [1]. Industrial germanium mainly comes from lead–zinc mines and coal mines [2,3]. There are many Ge-rich Pb-Zn deposits hosted by Paleozoic carbonate rocks in the Middle-Upper Yangtze Block, China [3–5]. The genesis of these Ge-rich Pb-Zn deposits has attracted widespread attention.

The study area is located in the southeastern margin of the Yangtze Block, SW China, which is an important part of the Western Hunan–Eastern Guizhou (WHEG) Pb-Zn metallogenic belt and contains many dispersed element-rich Pb-Zn deposits (such as Cd-rich Niujiaotang and Ge-rich Zhulingou and Banbianjie) [4–8]. The distribution of these deposits is jointly controlled by the WE-trending Huangsi and NE-trending Mandong faults [9]. Previous studies have shown that the source of sulfur in these deposits is the marine sulfates within the ore-bearing strata, and the ore-forming metals (Zn, Ge, Cd, etc.) were mainly sourced from the basement rocks [10–14]. However, the nature and source of ore-forming fluids are still controversial: (i) ore-forming fluids are characterized by "multi-source" [11,15–19]; (ii) ore-forming fluids are associated with ancient oil and gas reservoirs [4,10]; (iii) stratigraphic fluids [20]; and (iv) deep circulation basin brine [6]. In addition, studies on the evolution of ore-forming fluids are rarely reported.

In our previous study, isotope geochemistry was used to trace the source of ore-forming elements (such as S and Pb) [14]. As the trace elements characteristics of hydrothermal minerals depend on the nature, composition and physical–chemical condition of ore-forming fluids [21–25], they are widely used to explore the nature, source and evolution of hydrothermal fluids. Based on the previously reported whole-rock rare earth elements (REE) of partial dolomite [14], this paper analyzed the in situ trace elements (including REE) of gangue minerals (dolomite, calcite and barite) from the Banbianjie Ge-Zn deposit to reveal this issue, which provided more abundant geochemical information for comprehending the ore genesis of the deposit.

2. Regional Geology Setting

The study area is located in the southeastern margin of the Yangtze Block and the western margin of the Jiangnan Orogenic Belt (Figure 1a). It is mainly confined by three regional deep faults: the Tongren–Sandu fault to the east, the Ziyun–Luodian fault to the south and the Guiyang–Zhenyuan fault to the north [26–28]. The regional strata include metamorphic basements and overlying sedimentary covers. Among them, the oldest stratigraphic unit is the Upper Proterozoic Banxi Group, which consists of metamorphic sandstone, slate and a small amount of carbonate rocks [29,30]. The Paleozoic–Cenozoic sedimentary strata are mainly composed of carbonate rocks, argillaceous sandstone and shales, and the spatial distribution has the characteristic of old to new from east to west (Figure 1b). The Cambrian–Devonian carbonate rocks are the main ore-bearing rocks for these dispersed element-rich Pb-Zn deposits [6,31,32].

Since the formation period of the Wuling–Jinning basement, the study area has successively experienced the tectonic evolution of the Xuefeng–Caledonian ocean–land transition, the Indosinian–Yanshanian orogeny and the Himalayan differential uplift. The main structures are folds, faults and thrust nappe structures [33–35]. The evolution of the fractured ocean basin is closely related to the formation of many endogenous hydrothermal metallic minerals (e.g., Au, Sb, Pb, Zn, Mn, Mo and Ni deposits) and sedimentary minerals resources (e.g., phosphate ores and shale gas) in this area [30]. The hydrothermal deposits are mainly structurally controlled by the NE-trending Mandong and near EW-trending Huangsi faults (Figure 1b).

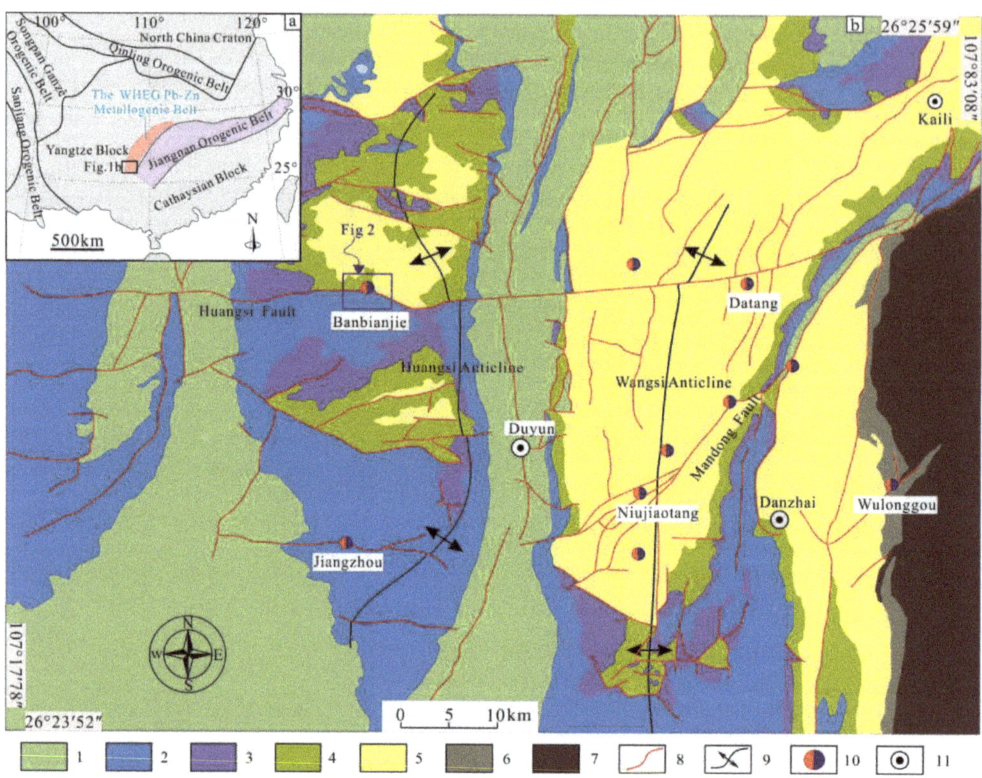

Figure 1. (**a**) Sketch map of the Western Hunan–Eastern Guizhou Pb-Zn metallogenic belt in the Yangtze Block (modified after [36]); (**b**) structural sketch map for the deposits in southern Guizhou district (modified after [37]). 1—Carboniferous–Permian sandstone and mudstone; 2—Devonian dolostone; 3—Silurian sandstone; 4—Ordovician limestone; 5—Cambrian dolostone; 6—Sinian dolomite and sandstone; 7—Banxi Group metamorphic rocks; 8—Fault; 9—Anticline; 10—Pb-Zn deposit; 11—City.

3. Geology of Ore Deposit

3.1. Strata

The strata are Silurian, Devonian, Carboniferous, Permian and Quaternary from ancient to present (Figure 2). The Silurian to Middle Devonian consists of coastal clastic sedimentary rocks, the Upper Devonian is a set of confined marine mesa facies post-reef lagoon carbonate rocks, the Carboniferous to Permian is carbonate mesa facies sedimentary rocks and the Quaternary consists of residual and slope wash (Figure 2). The Upper Devonian can be further divided into three sections (Figure 3). The Upper Devonian Gaopochang Formation second section (D_3g^2) is the main ore-bearing position of the Banbianjie Ge-Zn deposit, and its overall lithologies can be divided into three layers. The upper layer (D_3g^{2a}) is dark-gray flint-bearing clumpy fine-grained dolostone, with black argillaceous bands and geodes interspersed, and fissure and dolomite veins are developed. The middle layer (D_3g^{2b}) is dark-gray biodetritus fine-grained dolostone, with gray-yellow argillaceous bands interspersed, and fractures are relatively developed, as well as clumpy and veined dolomites, which are messily distributed. The lower layer (D_3g^{2c}) is light-gray medium-thickness siliceous fine-grained dolostone, with gray-yellow argillaceous bands interspersed (Figure 2). The ore-bearing rocks are the argillaceous bands dolostone.

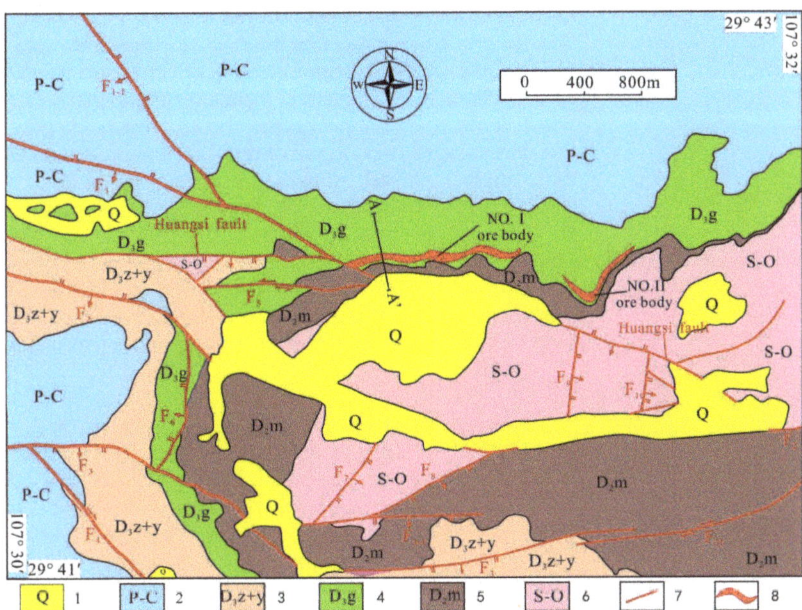

Figure 2. Sketch geological map of the Banbianjie mining area, Guizhou province (modified after [13]). 1—Quaternary; 2—Permian-Carboniferous; 3—Upper Devonian Zhewang and Yaosuo Formations; 4—Upper Devonian Gaopochang Formation; 5—Middle Devonian Mangshan Formation; 6—Silurian-Ordovician; 7—Fault; 8—ore body.

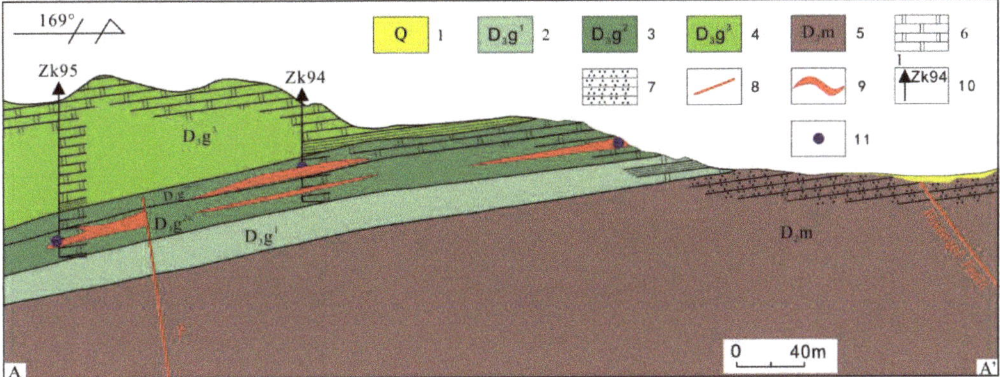

Figure 3. The cross-section (A–A′) through the Banbianjie Ge-Zn deposit (modified after [13]). 1—Quaternary; 2—first section of the Upper Devonian Gaopochang Formation; 3—second section of the Upper Devonian Gaopochang Formation; 4—third section of the Upper Devonian Gaopochang Formation; 5—Middle Devonian Mangshan Formation; 6—Dolostone; 7—sandstone; 8—fault; 9—ore body; 10—drilling; 11—sample location.

3.2. Structure

The mining area is located in the western wing of the wide Huangsi anticline in the regional septal fold assemblage. This assemblage is composed of a series of close synclines and wide anticlines in the near SN-trending and NNE-trending arranged in parallel, and the secondary folds are not developed (Figures 1 and 2). Regional faults are most developed in the near EW trending, and these faults are mostly thickness mutation zones of the Silurian and Devonian strata and also control the sedimentary evolution of the Carboniferous and Triassic strata, indicating that the faults formed in the late Caledonian. In the subsequent geological process, different degrees of activation occurred, with obvious multi-phase activity characteristics. Among them, the main representative tectonic is the Huangsi fault, which controls sedimentary facies and mineralization [9].

In addition, a new group of the NE-trending structures, with a trend of about 50° and a dip angle of 75°–80°, shows as strike-slip faults (Figure 3). The relationship between the NE-trending structures and mineralization is very obvious. Generally, the closer to this group of structures, the thicker the ore body and the higher the grade. In previous studies, the ore bodies in this deposit are stratiform and veined [38,39]. On the other side of the structures, there are veined ore bodies cutting the bedded ones, which have not been reported before. This is a new understanding of the Banbianjie deposit. The alteration in wall rocks is also developed near the NE-trending structures.

3.3. Ore Body

In the Banbianjie Ge-Zn deposit, the Zn metal reserves are more than 0.39 Mt, and the Ge metal resources are more than 900 t [37]. It mainly develops two ore bodies (I and II, Figures 2 and 3); each of them has a total thickness of 3–10 m. The No. I ore body is hosted in D_3g^{2b} and D_3g^{2c}, and the ore body floor is the ore-bearing layer floor. The No. II ore body is located in the bottom of D_3g^{2a}, and the floor is about 15 m away from the sandstone roof of the Mangshan Formation (D_2m). The No. II ore body is the main one, and its Zn metal reserve accounts for about 80% of the total Zn metal reserve. The shape of the ore bodies is strata-bound. At the near surface, the occurrence of ore bodies is basically consistent with that of the wall rocks. In the deep part, the irregularly veined ore bodies crosscut the wall rocks (Figure 4). The ore bodies extend over 600 m along the strike, and the ore grades of Zn range from 1.78% to 9.50% (average 5.1%), while the ore grades of Ge range from 100×10^{-6} to 110×10^{-6}. In general, the thicknesses and grades of the ore bodies in this deposit change a little in space.

Figure 4. Macroscopic characteristics of ore bodies in the Banbianjie Ge-Zn deposit. (**a**) Veined ore body; (**b**) stock-work ore body.

3.4. Mineralogy

The mineral compositions of the Banbianjie Ge-Zn deposit are relatively simple. Sulfide ore minerals are dominated by sphalerite and pyrite, with minor amounts of marcasite and galena. The gangue minerals are mainly dolomite, calcite and barite, with minor amounts of quartz. There are three main ore types: (i) massive ores, sphalerite coexists with dolomite and calcite (Figure 5a,b); (ii) veined ores, the veins are between 5 cm and 60 cm wide, sphalerite coexists with barite (Figure 5c); and (iii) colloidal ores, sphalerite coexists with dolomite and pyrite (Figure 5d).

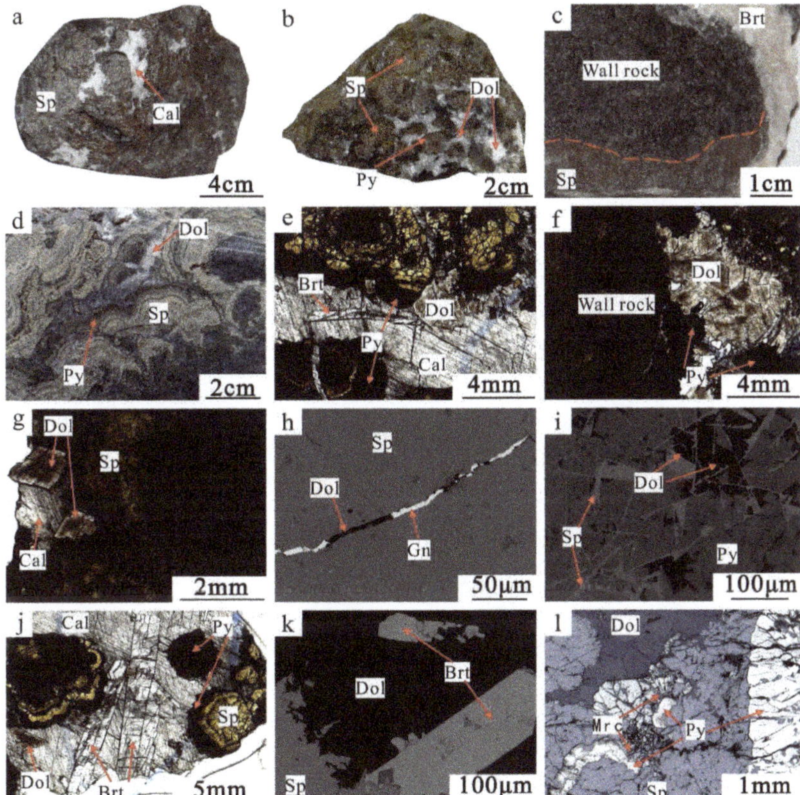

Figure 5. Texture and structure characteristics of the Banbianjie Ge-Zn Deposit. (**a**) Massive ores, sphalerite coexists calcite; (**b**) massive ores, sphalerite coexists dolomite and pyrite; (**c**) veined ores, sphalerite coexists barite; (**d**) colloidal ores, sphalerite coexists with pyrite; (**e**) dolomite/calcite cement ring-banded sulfides, columnar barite embedded in calcite; (**f**) granular dolomite cement pyrite in contact with the scarf; (**g**) granular dolomite is embedded in the sulfides; (**h**) veined galena and dolomite fill into the fracture of sphalerite; (**i**) sphalerite replaces pyrite and is cemented by dolomite; (**j**) sphalerite, pyrite, barite and dolomite are enclosed by calcite; (**k**) barite is enclosed by dolomite and coexists with sphalerite; (**l**) marcasite coexists pyrite, sphalerite and dolomite. (Sp = sphalerite; Py = pyrite; Gn = galena; Mrc = marcasite; Dol = dolomite; Cal = calcite; Brt = barite).

Sphalerite has mostly ring ribbon, colloform and metasomatic relict textures (Figure 5d–l), and the veined galena fills into the fracture of sphalerite (Figure 5h). The germanium contents in sphalerite range from 173×10^{-6} to 1553×10^{-6}. Dolomite mainly grows in the outer void of colloidal sphalerite, and the contact of dolomite and sulfides shows fine grain (Figure 5f). Dolomite and calcite exhibit symbiosis (Figure 5g). Hence, dolomite and calcite were formed in the ore-forming stage. Columnar barite is encased in dolomite and grown along the edge of sphalerite (Figure 5j,k), indicating that barite is also a product of the ore-forming stage. The veined calcite fills and cements sulfides. Irregularly, the acicular marcasite coexists with sphalerite or is metasomatized by pyrite. Dolomite with a fine-granular structure (size: 0.5–1.5 mm) is intergrown in the clearance between calcite and sulfides. Columnar barite crosscuts dolomite and calcite (Figure 5k).

3.5. Alteration in Wall Rocks

The alteration is relatively simple, and it is dominated by carbonatization and pyritization, with barization and siliconization. The pre-ore dolomite is the main mineral formed by carbonatization, which exists near or within the ore body, fault fracture zone and inter-bedded fracture zone and is closely related to Ge-Zn mineralization. It is an important prospecting indicator. Pyritization is closely related to Pb-Zn mineralization and has a wide distribution. Barization formed barite (Figure 5c), and silicification formed quartz and silicified wall rocks.

4. Analytical Methods

The whole-rock trace element analysis of barite was undertaken at the State Key Laboratory of Ore Deposit Geochemistry, Institute of Geochemistry, Chinese Academy Sciences (SKLODG-IGCAS). First, we crushed and screened the fresh samples, washed and dried them with grain sizes between 40 and 60 meshes and selected barite single mineral under the binocular microscope, with a purity greater than 99%. Then, we cleaned them with 75% alcohol and dried at low temperature. Following this, the dried single mineral samples were ground into powder less than 200 meshes in an agate mortar and were acid-dissolved. Finally, the quadrupole-inductively coupled plasma-mass spectrometry (Q-ICPMS) analysis was used to approach the contents. The precisions of REE and other trace elements are better than 5% and 10%, respectively. Please refer to [40] for the detailed analysis process. The repeated analysis results of samples are basically consistent within the error range, and the analysis results of standard samples (OU-6, AGV-2, GBPG-1) are also basically consistent with the recommended values [41–43]. The detection limits are as follows: Tb, Ho, Lu and Tm (0.01×10^{-6}); Er, Eu, Sm, Pr and Yb (0.03×10^{-6}); Ce, Gd and Dy (0.05×10^{-6}); Nd (0.1×10^{-6}); Y and La (0.5×10^{-6}) [41].

The in situ trace elements analyses of dolomite and calcite were undertaken at the SKLODG-IGCAS using laser ablation-inductively coupled plasma-mass spectrometry (LA-ICPMS) and employing a standard-sample-standard analysis approach. The instruments used were the ArF excimer laser 193 nm laser system and Agilent 7700X plasm-mass spectrometry, corrected through analysis of SRM610, SRM612 (trace elements in glass), MACS-3 (synthetic calcium carbonate microanalysis reference material), multiple external standard and single internal standard (Ca = 21.7%). The content gap is a huge difference between the Ca and REE, so we used the theoretical values of Ca in the calculation of Yb/Ca ratios. All analyses used 5 Hz laser frequency, 44 μm diameter laser beam, 50 s single measurement time, about 12 s background measurement time and the standard sample was tested once every 15-point analysis. The analysis precision was better than 10%, and the data were processed with ICP-MS DataCal software. For detailed analysis methods, please refer to [44]. The analysis results of standard samples (SRM610, SRM612 and MACS-3) were basically consistent with the recommended values. Three silicate glass reference materials, SRM610, SRM612 and MACS-3, were shown to be homogeneous [45,46]. LA-ICPMS analyses yielded results that agreed with the reference values within relative uncertainties of ca. 5%–10% at a 95% confidence level [45,46].

5. Results

The analysis results are shown in Tables 1 and 2, of which the whole-rock REE data of wall rocks (altered dolostone) were taken from [14]. The in situ contents of Mn, Fe and Sr in the dolomite and calcite from the Banbianjie deposit are more than 69×10^{-6}, 1×10^{-6} and 22×10^{-6}, respectively, and the in situ contents of Pb, Th and U are less than 8×10^{-6}, 0.7×10^{-6} and 0.7×10^{-6}, respectively. The in situ Ba contents of dolomite and calcite are between 0.155×10^{-6} and 12.4×10^{-6} (Table 1). In barite, the whole-rock contents of Zn and Sr are generally higher than their crustal abundances, while the whole-rock contents of other trace elements are lower than their crustal abundances (Table 2).

Table 1. The contents and ratios of in situ trace elements (10^{-6}) for dolomite and calcite from the Banbianjie Ge-Zn deposit.

No.	Objection	Mn	Fe	Sr	Ba	Pb	Th	U	Fe/Mn	Sr/Ba	Th/U
BBJ13-2-8	Altered dolostone	123	192	73.0	12.4	0.440	0.330	0.377	1.56	6	0.88
BBJ13-2-9	Altered dolostone	110	253	76.3	9.22	0.996	0.327	0.609	2.30	8	0.54
BBJ13-2-10	Altered dolostone	81.4	227	99.7	5.82	0.375	0.141	0.483	2.79	17	0.29
BBJ9-3-2	Calcite	93.1	14.4	92.5	1.70	0.037	0.096	0.007	0.15	54	14
BBJ9-3-3	Calcite	69.4	17.5	130	0.608	0.114	0.047	0.094	0.25	214	0.50
BBJ9-3-8	Calcite	107	75.6	98.3	0.155	0.058	0.077	0.131	0.71	634	0.59
BBJ9-3-10	Calcite	95.2	-	113	0.644	4.26	0.049	0.011		175	4.45
BBJ9-3-11	Calcite	93.8	77.4	131	0.704	0.168	0.073	0.276	0.83	186	0.26
BBJ9-3-1	Dolomite	107	-	28.8	0.583	-	0.118	-		49	
BBJ9-3-4	Dolomite	115	15.3	25.6	0.721	0.005	0.085	-	0.13	36	
BBJ9-3-5	Dolomite	107	49.8	26.3	0.898	0.744	0.101	0.006	0.47	29	17
BBJ9-3-12	Dolomite	121	-	28.0	1.53	0.070	0.079	-		18	
BBJ9-3-14	Dolomite	121	-	22.7	1.12	0.035	0.118	0.001		20	118
BBJ9-3-15	Dolomite	127	-	25.5	-	0.055	0.143	0.002			72
BBJ2-3-1	Dolomite	134	-	28.5	1.05	0.195	0.098	0.006		27	16
BBJ2-3-2	Dolomite	110	6.15	29.1	-	0.015	0.109	0.005	0.06		22
BBJ2-3-3	Dolomite	122	1.67	29.8	1.34	0.131	0.100	0.002	0.01	22	50
BBJ2-3-4	Dolomite	109	1.75	36.9	4.54	0.088	0.171	0.003	0.02	8	57
BBJ2-3-5	Dolomite	115	11.5	23.6	1.61	0.024	0.141	-	0.10	15	
BBJ2-3-6	Dolomite	104	10.0	46.5	-	1.08	0.055	0.004	0.10		14
BBJ2-3-7	Dolomite	143	9.44	23.9	1.19	0.051	0.128	0.002	0.07	20	64
BBJ2-3-8	Dolomite	148	35.1	32.8	0.303	0.039	0.124	0.002	0.24	108	62
BBJ2-3-9	Dolomite	146	-	29.3	2.01	0.028	0.187	-		15	
BBJ2-3-10	Dolomite	151	22.8	25.0	1.66	0.016	0.256	0.005	0.15	15	51
BBJ-15-1	Dolomite	86.6	-	118	1.75	1.88	0.335	0.003		67	112
BBJ-15-2	Dolomite	95.8	76.4	114	4.60	7.70	0.602	0.024	0.80	25	25
BBJ-15-3	Dolomite	89.2	34.7	103	2.26	0.806	0.329	0.008	0.39	46	41
BBJ-15-4	Dolomite	87.9	11.0	104	3.51	0.150	0.241	0.001	0.13	30	241
BBJ-15-5	Dolomite	140	77.7	35.8	0.235	10.9	0.340	0.013	0.56	152	26
BBJ-15-2-1	Dolomite	119	39.8	78.9	3.74	0.042	0.204	0.007	0.33	21	29
BBJ-15-2-2	Dolomite	159	18.8	37.7	1.62	0.005	0.230	-	0.12	23	
BBJ-15-2-3	Dolomite	112	-	26.1	0.668	0.048	0.139	-		39	
BBJ-15-2-4	Dolomite	135	22.6	24.4	0.166	0.005	0.187	-	0.17	147	
BBJ-15-2-5	Dolomite	112	-	24.3	1.37	0.003	0.101	0.003		18	34
BBJ-15-2-6	Dolomite	116	-	26.9	1.17	0.019	0.122	-		23	
BBJ-15-2-7	Dolomite	134	29.8	62.3	1.30	0.019	0.136	0.001	0.22	48	136
BBJ-15-2-8	Dolomite	136	90.2	26.2	1.44	2.76	0.254	0.009	0.66	18	28
BBJ-15-2-9	Dolomite	122	11.3	26.6	1.72	0.010	0.104	0.001	0.09	15	104

(Note: "-" indicates that the element contents of the tested sample are lower than those of the detection limitation).

Table 2. The whole-rock contents and ratios of trace elements (10^{-6}) for barite from the Banbianjie Ge-Zn deposit.

No.	Li	Be	Sc	V	Cr	Co	Ni	Cu	Zn	Ga	Ge	As
BBJ-16	0.613	0.014	0.820	0.451	1.22	6.71	38.8	5.27	869	0.497	1.12	0.632
DY29-1	0.827	0.019	1.53	1.40	0.947	4.07	11.6	8.31	193	0.541	0.495	0.505
BBJ-13	0.693	0.022	0.934	1.10	1.22	5.71	22.1	5.68	66.7	0.546	0.235	0.594
BBJ-14	0.672	0.023	0.906	1.54	1.81	5.87	23.6	5.53	1048	0.397	1.58	0.539
BBJ-1	0.628	0.015	1.07	1.72	1.33	4.82	10.4	7.99	615	0.346	0.714	0.442
BBJ 1	0.551	0.011	1.45	2.13	1.32	3.94	12.4	7.57	348	0.272	0.507	0.443
DY26-3	2.19	0.033	6.51	4.76	2.72	5.85	20.8	14.1	1607	0.901	3.21	1.09
DY2-1	1.01	0.025	1.02	1.82	2.25	4.77	11.9	8.87	115	0.379	0.372	0.690
DY13	0.450	0.015	0.674	1.10	1.11	4.05	18.0	7.28	34.7	0.444	0.177	0.545
Crustal abundances	20	3.0	11	60	35	10	20	25	71	17	1.6	1.5

No.	Ag	Cd	In	Sn	Sb	Cs	Ba	Hf	Ta	W	Tl	Pb
BBJ-16	0.039	1.15	0.005	0.230	0.173	0.010	22426	0.011	0.048	0.215	0.074	110
DY29-1	0.016	0.202	0.009	0.124	0.105	0.020	67975	0.021	0.013	0.110	0.019	3.93
BBJ-13	0.062	0.089	0.006	0.144	0.140	0.015	19052	0.059	0.036	0.197	0.036	8.16
BBJ-14	0.022	1.21	0.007	0.335	0.175	0.009	20889	0.015	0.029	0.191	0.13	28.2
BBJ-1	0.019	0.982	0.006	0.130	0.158	0.009	59948	0.020	0.013	0.117	0.037	4.81

Table 2. *Cont.*

BBJ-4	0.020	0.412	0.005	0.119	0.245	0.010	54232	0.018	0.011	0.105	0.068	4.95
DY26-3	0.142	3.53	0.040	0.312	0.378	0.016	128938	0.040	0.025	0.149	0.126	29.6
DY2-1	0.049	0.209	0.015	0.154	0.137	0.032	87013	0.044	0.016	0.106	0.015	32.6
DY13	0.021	0.033	0.005	0.107	0.077	0.014	63624	0.091	0.028	0.163	0.002	0.544
Crustal abundances	50	98	50	5.5	0.2	3.7	550	5.8	2.2	2.0	750	20

No.	Rb	Sr	Zr	Nb	Mo	Bi	Th	U	Sr/Ba	Th/U
BBJ-16	0.153	356	0.105	0.071	0.272	0.009	0.074	0.036	0.02	2
DY29-1	0.139	774	0.078	0.024	0.150	0.005	0.054	0.008	0.01	7
BBJ-13	0.138	378	0.094	0.046	0.198	0.004	0.113	0.014	0.02	8
BBJ-14	0.135	180	0.108	0.040	0.215	0.004	0.107	0.014	0.01	8
BBJ-1	0.145	1073	0.117	0.018	0.190	0.002	0.104	0.016	0.02	7
BBJ-4	0.138	896	0.094	0.028	0.187	0.004	0.071	0.011	0.02	6
DY26-3	0.210	577	0.153	0.042	0.206	0.011	0.155	0.012	0.00	13
DY2-1	0.217	1047	0.476	0.072	0.203	0.009	0.094	0.025	0.01	4
DY13	0.115	919	0.111	0.031	0.173	0.005	0.084	0.035	0.01	2
Crustal abundances	112	350	190	25	1.5	127	10.7	2.8		

(Note: crustal abundances after [1]).

The whole-rock total rare earth element ($\sum$REE) contents of altered dolostone range from 10.1×10^{-6} to 16.8×10^{-6} [14], which are slightly lower than these of hydrothermal calcite, dolomite and barite (Table 3). Altered dolostone is rich in light rare earth elements (LREE) ($\sum$LREE/$\sum$HREE = 7.4–9.4, La/Yb$_N$ = 12.5–18.6) (Figure 6a) and has negative Eu (Eu/Eu* = 0.43–0.74) and Ce anomalies (Ce/Ce* = 0.72–0.80) [14].

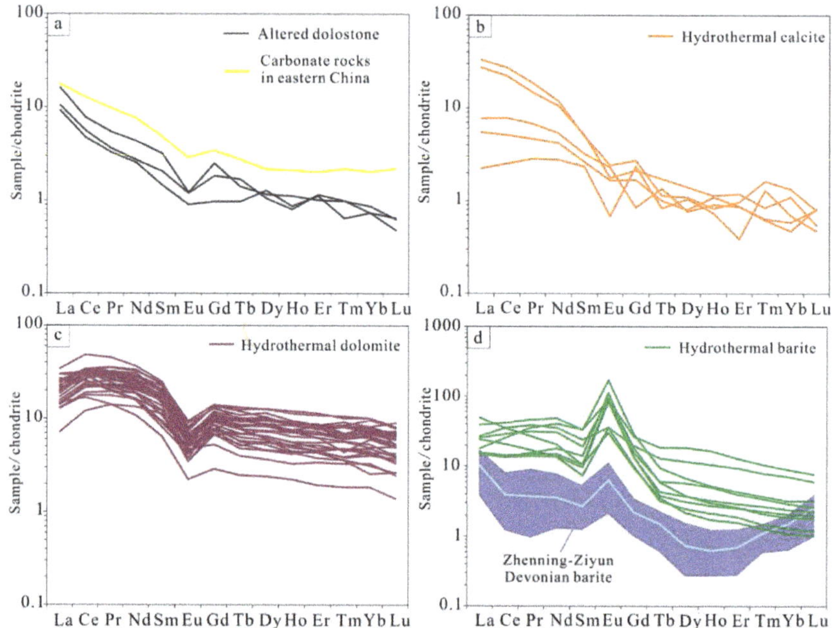

Figure 6. Chondrite-normalized REE patterns of whole-rock altered dolostone (**a**), in situ hydrothermal calcite (**b**) and dolomite (**c**), and whole-rock hydrothermal barite (**d**) from the Banbianjie Ge-Zn deposit (Chondrite values from [47]; whole-rock altered dolostone data are from [14]; carbonate rock data in Eastern China are from [48]; Zhenning–Ziyun Devonian barite data are from [49]).

Table 3. REE contents (10^{-6}) and statistical parameters of altered dolostone, and hydrothermal calcite, dolomite and barite from the Banbianjie Ge-Zn deposit.

Sample	Altered Dolostone			Average	Hydrothermal Calcite					Average	Hydrothermal Dolomite		
	BBJ13-2-8	BBJ13-2-9	BBJ13-2-10		BBJ9-3-2	BBJ9-3-3	BBJ9-3-8	BBJ9-3-10	BBJ9-3-11		BBJ9-3-1	BBJ9-3-4	BBJ9-3-5
La	2.86	3.25	5.00	3.70	0.692	1.71	10.2	2.41	8.50	4.71	4.46	4.07	4.90
Ce	3.81	4.50	6.21	4.84	2.02	4.12	21.9	6.30	17.8	10.4	17.5	15.2	19.4
Pr	0.398	0.441	0.666	0.502	0.347	0.571	2.29	0.833	1.81	1.17	2.81	2.39	3.13
Nd	1.53	1.63	2.61	1.92	1.67	2.54	7.18	3.25	6.39	4.21	13.7	11.3	15.2
Sm	0.284	0.399	0.619	0.434	0.460	0.515	0.985	0.629	1.01	0.721	3.11	2.69	3.50
Eu	0.066	0.088	0.089	0.081	0.050	0.122	0.178	0.169	0.131	0.130	0.463	0.275	0.467
Gd	0.251	0.474	0.644	0.457	0.609	0.438	0.705	0.221	0.551	0.505	2.60	2.38	2.91
Tb	0.046	0.080	0.067	0.064	0.040	0.048	0.054	0.065	0.082	0.058	0.408	0.363	0.458
Dy	0.410	0.339	0.374	0.374	0.339	0.259	0.353	0.248	0.454	0.331	2.74	2.38	3.38
Ho	0.062	0.058	0.080	0.067	0.053	0.079	0.060	0.065	0.082	0.068	0.535	0.461	0.619
Er	0.234	0.242	0.212	0.229	0.082	0.183	0.204	0.182	0.249	0.180	1.52	1.32	1.67
Tm	0.021	0.032	0.032	0.028	0.042	0.020	0.053	0.021	0.027	0.033	0.195	0.142	0.276
Yb	0.154	0.153	0.181	0.163	0.146	0.099	0.281	0.124	0.231	0.176	1.42	1.25	1.63
Lu	0.021	0.016	0.020	0.019	0.015	0.027	0.026	0.026	0.018	0.022	0.192	0.158	0.227
Y	2.54	2.92	3.16	2.87	1.30	2.08	1.77	1.97	2.56	1.94	14.4	12.4	17.4
ΣREE	10.1	11.7	16.8	12.9	6.56	10.7	44.5	14.6	37.3	22.7	51.7	44.4	57.8
LREE	8.94	10.3	15.2	11.5	5.24	9.58	42.7	13.6	35.6	21.4	42.1	36.0	46.6
HREE	1.20	1.39	1.61	1.40	1.33	1.15	1.73	0.95	1.70	1.37	9.62	8.45	11.2
LREE/HREE	7.45	7.40	9.44	8.10	3.95	8.30	24.63	14.28	21.02	14.43	4.38	4.25	4.17
La/Yb$_N$	12.54	14.33	18.61	15.16	3.20	11.57	24.60	13.15	24.76	15.46	2.12	2.19	2.02
La/Sm$_N$	6.34	5.12	5.08	5.51	0.95	2.08	6.55	2.41	5.27	3.45	0.90	0.95	0.88
Eu/Eu*	0.74	0.61	0.43	0.59	0.29	0.77	0.62	1.13	0.48	0.66	0.48	0.33	0.44
δCe	0.76	0.80	0.72	0.76	0.99	1.01	1.06	1.08	1.05	1.04	1.17	1.16	1.17

Sample	Hydrothermal dolomite												
	BBJ9-3-12	BBJ9-3-14	BBJ9-3-15	BBJ2-3-1	BBJ2-3-2	BBJ2-3-3	BBJ2-3-4	BBJ2-3-5	BBJ2-3-6	BBJ2-3-7	BBJ2-3-8	BBJ2-3-9	BBJ2-3-10
La	4.02	6.38	6.87	4.52	4.40	4.59	5.31	6.14	2.23	5.81	4.51	7.06	7.93
Ce	14.3	24.5	27.3	17.7	17.3	17.9	21.7	23.0	9.70	23.6	18.1	24.0	27.2
Pr	2.16	3.87	4.34	2.68	2.78	2.67	3.49	3.59	1.73	3.69	2.85	3.29	3.74
Nd	9.76	17.4	20.1	13.1	12.6	12.3	16.5	15.7	8.08	16.8	13.4	13.3	15.6
Sm	2.36	4.02	4.56	3.04	3.08	3.06	4.41	3.82	2.20	4.00	3.36	3.10	2.86
Eu	0.254	0.475	0.562	0.480	0.307	0.371	0.656	0.431	0.255	0.507	0.411	0.387	0.495
Gd	1.93	2.41	3.56	2.55	2.43	1.98	3.34	2.97	1.74	2.82	2.90	2.21	2.00
Tb	0.326	0.482	0.570	0.385	0.367	0.481	0.518	0.448	0.324	0.457	0.456	0.267	0.300

Table 3. *Cont.*

Dy	2.09	3.12	3.35	2.82	2.32	3.30	3.71	2.80	2.39	3.10	3.05	1.69	1.90
Ho	0.400	0.669	0.611	0.541	0.525	0.620	0.685	0.604	0.509	0.631	0.628	0.310	0.321
Er	1.16	1.70	1.66	1.71	1.44	1.84	1.78	1.64	1.44	1.77	1.61	0.791	0.906
Tm	0.183	0.273	0.272	0.216	0.222	0.225	0.295	0.203	0.195	0.261	0.262	0.117	0.115
Yb	1.02	2.00	1.59	1.59	1.72	1.43	1.71	1.51	1.35	1.52	1.61	0.651	0.535
Lu	0.161	0.260	0.222	0.204	0.170	0.237	0.289	0.220	0.172	0.211	0.217	0.084	0.085
Y	11.4	17.9	17.8	14.5	13.4	16.4	19.2	16.6	13.3	16.8	18.2	10.2	11.0
ΣREE	40.2	67.6	75.5	51.5	49.6	51.0	64.4	63.1	32.3	65.2	53.4	57.3	64.0
LREE	32.9	56.7	63.7	41.5	40.4	40.9	52.1	52.7	24.2	54.4	42.7	51.2	57.9
HREE	7.28	10.9	11.8	10.0	9.19	10.1	12.3	10.4	8.12	10.8	10.7	6.11	6.16
LREE/HREE	4.52	5.20	5.39	4.14	4.39	4.04	4.22	5.07	2.98	5.06	3.98	8.37	9.39
La/Yb$_N$	2.66	2.15	2.92	1.91	1.73	2.16	2.09	2.73	1.11	2.58	1.89	7.31	10.0
La/Sm$_N$	1.07	1.00	0.95	0.94	0.90	0.94	0.76	1.01	0.64	0.91	0.84	1.43	1.74
Eu/Eu*	0.35	0.43	0.41	0.51	0.33	0.43	0.50	0.38	0.39	0.44	0.39	0.43	0.60
δCe	1.17	1.17	1.18	1.21	1.17	1.22	1.18	1.17	1.14	1.20	1.20	1.21	1.21

Sample	Hydrothermal dolomite												
	BBJ15-1	BBJ15-2	BBJ15-3	BBJ15-4	BBJ15-5	BBJ15-2-1	BBJ15-2-2	BBJ15-2-3	BBJ15-2-4	BBJ15-2-5	BBJ15-2-6	BBJ15-2-7	BBJ15-2-8
La	9.29	8.36	7.87	9.35	8.02	6.70	7.51	7.62	5.99	6.11	4.91	4.64	10.7
Ce	23.2	20.8	20.3	25.6	26.0	20.6	25.8	27.0	23.8	24.6	19.0	13.5	38.7
Pr	2.63	2.43	2.44	3.00	3.71	2.51	3.56	3.93	3.41	3.91	3.17	1.75	5.54
Nd	11.1	9.84	9.46	11.5	15.7	9.52	15.0	18.6	14.9	17.5	14.2	6.52	22.0
Sm	2.14	1.85	1.59	2.40	3.20	1.76	3.13	3.66	2.89	4.36	3.05	1.25	4.89
Eu	0.326	0.366	0.306	0.424	0.467	0.345	0.455	0.462	0.441	0.543	0.392	0.164	0.600
Gd	1.94	1.74	1.78	2.03	2.23	1.37	2.48	3.10	2.76	3.57	2.76	0.748	3.47
Tb	0.243	0.250	0.226	0.273	0.328	0.189	0.368	0.483	0.397	0.597	0.438	0.117	0.463
Dy	1.81	1.48	1.47	1.78	2.07	1.18	1.96	3.17	2.66	4.13	3.05	0.774	2.78
Ho	0.350	0.315	0.300	0.372	0.403	0.235	0.369	0.608	0.512	0.868	0.638	0.160	0.515
Er	0.953	0.936	0.939	1.09	1.02	0.712	0.95	1.63	1.42	2.41	1.80	0.407	1.31
Tm	0.153	0.124	0.125	0.148	0.154	0.106	0.153	0.224	0.210	0.329	0.266	0.060	0.185
Yb	0.992	0.836	0.908	0.898	0.858	0.688	0.929	1.47	1.31	2.12	1.70	0.386	1.11
Lu	0.125	0.120	0.113	0.136	0.114	0.080	0.107	0.197	0.162	0.262	0.215	0.045	0.127
Y	10.0	8.93	8.66	9.94	12.3	6.37	12.1	15.7	13.8	21.8	16.2	3.98	15.3
ΣREE	55.2	49.5	47.9	59.0	64.2	46.0	62.7	72.1	60.9	71.4	55.6	30.5	92.3
LREE	48.7	43.7	42.0	52.3	57.0	41.4	55.4	61.2	51.5	57.1	44.7	27.8	82.4
HREE	6.57	5.80	5.86	6.73	7.17	4.56	7.31	10.9	9.42	14.3	10.9	2.70	9.95
LREE/HREE	7.41	7.53	7.17	7.77	7.95	9.09	7.58	5.62	5.46	4.00	4.11	10.32	8.28

Table 3. *Cont.*

La/Yb_N	6.31	6.74	5.84	7.02	6.30	6.56	5.45	3.48	3.09	1.94	1.94	8.10	6.51
La/Sm_N	2.73	2.85	3.11	2.45	1.57	2.40	1.51	1.31	1.31	0.88	1.01	2.32	1.37
Eu/Eu*	0.48	0.62	0.55	0.57	0.51	0.66	0.48	0.41	0.47	0.41	0.41	0.48	0.42
δCe	1.12	1.11	1.12	1.17	1.16	1.22	1.21	1.19	1.26	1.19	1.14	1.15	1.21

Sample	Hydrothermal dolomite	Average	Hydrothermal barite									Average
	BBJ-15-2-9		BBJ-16	DY29-1	BBJ-13	BBJ-14	BBJ-1	BBJ-4	DY26-3	DY2-1	DY13	
La	6.28	6.22	15.4	7.9	8.33	7.47	4.7	4.85	12.1	4.59	5.57	7.88
Ce	25.8	21.8	27.9	23	28.3	16.5	10.9	11.4	33.3	10.7	20.4	20.3
Pr	4.31	3.18	3.25	3.83	4.72	2.08	1.7	1.82	5.69	1.9	4.35	3.26
Nd	19.6	14.01	12.3	18.4	21.3	8.19	8.27	8.92	29.3	10.8	24.3	15.8
Sm	4.7	3.13	2.05	3.74	4.69	1.44	1.89	2.06	6.47	2.74	6.55	3.51
Eu	0.692	0.426	2.22	7	2.63	2.59	6.28	6.78	12.6	8.35	6.65	6.12
Gd	3.65	2.48	2.46	4.7	5.07	2.6	3.53	3.46	7.47	4.06	6.88	4.47
Tb	0.628	0.387	0.169	0.281	0.618	0.153	0.181	0.206	0.451	0.289	0.884	0.359
Dy	4.07	2.55	0.705	1.23	3.76	0.794	1.01	1.2	2.19	1.86	5.96	2.08
Ho	0.82	0.504	0.124	0.215	0.752	0.153	0.197	0.233	0.375	0.362	1.19	0.4
Er	2.29	1.39	0.333	0.546	1.98	0.397	0.53	0.612	0.914	0.836	2.71	0.984
Tm	0.344	0.201	0.042	0.068	0.257	0.048	0.074	0.087	0.121	0.112	0.341	0.128
Yb	2.12	1.3	0.228	0.39	1.49	0.274	0.431	0.503	0.667	0.61	1.87	0.718
Lu	0.253	0.172	0.033	0.058	0.194	0.038	0.064	0.073	0.103	0.084	0.248	0.099
Y	21.4	13.9	5.12	8.94	22	5.86	7.51	8.01	15	13.3	35.1	13.4
ΣREE	75.5	57.7	67.2	71.4	84.1	42.7	39.8	42.2	112	47.3	87.9	66
LREE	61.3	48.7	63.1	63.9	70	38.3	33.7	35.8	99.4	39.1	67.8	56.8
HREE	14.2	8.98	4.09	7.49	14.1	4.46	6.02	6.37	12.3	8.21	20.1	9.24
LREE/HREE	4.33	5.87	15.4	8.53	4.96	8.59	5.61	5.62	8.09	4.76	3.38	7.22
$(La/Yb)_N$	2	3.96	45.5	13.7	3.77	18.38	7.35	6.5	12.2	5.07	2.01	12.7
$(La/Sm)_N$	0.84	1.38	4.73	1.33	1.12	3.26	1.56	1.48	1.17	1.05	0.54	1.8
Eu/Eu*	0.49	0.46	3.02	5.1	1.64	4.04	7.33	7.71	5.53	7.65	3.01	5
δCe	1.16	1.18	0.91	1.01	1.08	1	0.94	0.93	0.97	0.88	0.95	0.96

(Note: data of BBJ13-2-8, -9 and -10 are from [14], Standardized REE data for chondrites are from [46]; $\delta Ce = Ce_N/(La_N \times Pr_N)^{1/2}$, $Eu/Eu^* = Eu_N/(Sm_N \times Gd_N)^{1/2}$).

Hydrothermal calcite and dolomite are also rich in LREE ($\sum$LREE/$\sum$HREE = 4.0–24.6, La/Yb$_N$ = 3.20–24.78 and $\sum$LREE/$\sum$HREE = 2.98–10.3, La/Yb$_N$ = 1.11–10.0, respectively), with changed Eu and Ce anomalies (Eu/Eu* = 0.29–1.13 and Ce/Ce* = 0.99–1.08) (Figure 6b,c) and negative Eu (Eu/Eu* = 0.33–0.66) and positive Ce anomalies (Ce/Ce* = 1.11–1.26), respectively. Hydrothermal barite also has the characteristics of LREE enrichment ($\sum$LREE/$\sum$HREE = 3.38–15.42, La/Yb$_N$ = 2.01–45.5) (Figure 6d) but with significantly positive Eu (Eu/Eu* = 1.64–7.71) and insignificant Ce anomalies (Ce/Ce* = 0.88–1.08) (Table 3).

6. Discussion

6.1. The Nature and Source of Ore-Forming Fluids

Previous studies have suggested that the trace element characteristics of hydrothermal gangue minerals (i.e., calcite, dolomite and barite) can indicate the nature and source of ore-forming fluids [50–52]. The evidence of mineralogy (Figure 5) suggests that these gangue minerals formed in the ore-forming stage, so they are hydrothermal minerals. Their trace elements suggest that Sr, Ba, Zn, Fe, Pb and Ge are enriched in the hydrothermal fluids. This is also confirmed by the form of mineral assemblage, such as Sr-rich barite, Ge-rich sphalerite, pyrite and galena (Figure 5). In addition, barite contains variable contents of other trace elements but generally has a similar variation trend, such as systematic enrichment of large ionic lithophile elements and depletion of high-field-strength elements (Figure 7). This is consistent with the basin brine characteristics of MVT deposits (Tables 2 and 3 [53]). Furthermore, the high-temperature metallogenic elements (such as Co, Ni, Mo and Bi) show obviously depleted characteristics (Figure 7), indicating that these deposits have nothing to do with high-temperature magmatism. Therefore, the nature of ore-forming fluids is low-temperature basin brine.

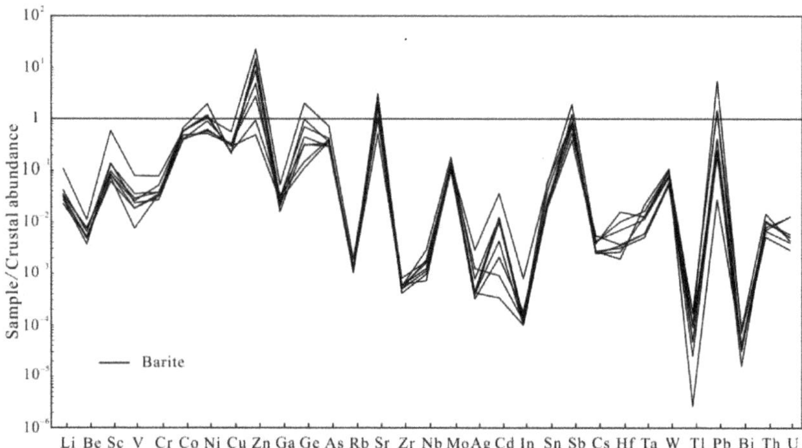

Figure 7. The enrichment degree of trace elements in barite relative to crustal abundances for the Banbianjie deposit (crustal abundances after [1]).

Rare earth elements (REEs) record important information about hydrothermal fluids [54–61]. For example, the Tb/Ca-Tb/La diagram (Figure 8) can effectively determine the genesis of calcite [62]. The calcite from the Banbianjie deposit was plotted into the hydrothermal genesis zone (since the difference in the contents between Ca and REE is more than 5 orders of magnitude, the author directly uses the theoretical values of Ca when calculating the Yb/Ca ratios) in the plot of Tb/Ca-Tb/La (Figure 8), indicating hydrothermal genesis, which is consistent with the actual geology. It is obvious that calcite coexists with sulfide minerals in mineral assemblages (Figure 5). Hence, the Banbianjie deposit may not be genetically related to magmatism or sedimentation.

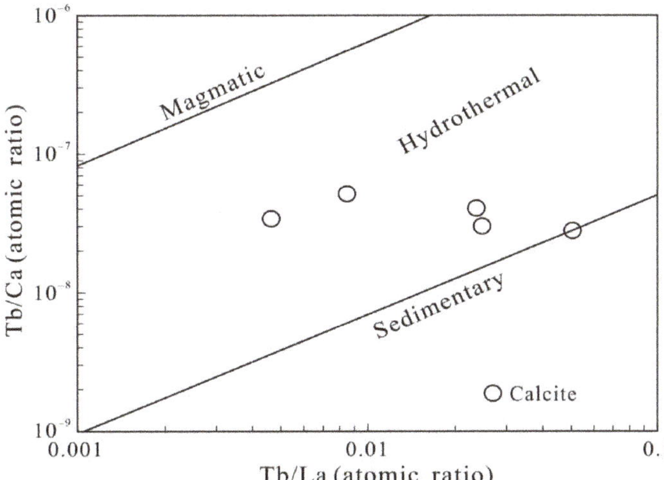

Figure 8. Tb/Ca-Tb/La diagram of calcites from Banbianjie Ge-Zn deposit (after [62]).

The La/Ho-Y/Ho diagram (Figure 9a) can identify the cognate of minerals [63]. The hydrothermal dolomite, calcite and barite from the Banbianjie deposit are distributed horizontally in the diagram of La/Ho-Y/Ho (Figure 9a), suggesting that they are the evolution products of the same fluids. This is also supported by the mineral assemblages, dolomite and calcite, which coexist with barite (Figure 5).

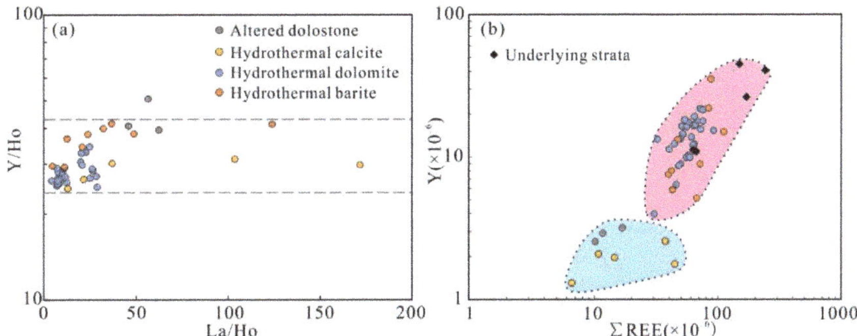

Figure 9. La/Ho vs. Y/Ho (**a**) and Y vs. ΣREE (**b**), the underlying strata data are from [64] diagrams of the Banbianjie Ge-Zn deposit.

On the chondrite-standardized distribution diagram of REE (Figure 6), dolomite, calcite and barite have the characteristics of LREE enrichment, which further suggests that there is an intrinsic genesis relationship between them, although the ore-forming fluids were not directly or wholly contributed by the wall rocks.

In the Banbianjie deposit, the $\sum$REE contents of hydrothermal dolomite and barite are significantly higher than those of altered dolostone and hydrothermal calcite (Table 3). As the $\sum$REE contents of quartz and sulfide minerals are negligible [12,63–65], so the $\sum$REE contents of hydrothermal dolomite and barite can approximately represent the $\sum$REE contents of the fluids ($\sum$REE$_{\text{dolomite+barite}}$ ≈ $\sum$REE$_{\text{fluids}}$). In Figure 6, the $\sum$REE contents of dolomite and barite are significantly higher than those of both altered dolostone and carbonate rocks in Eastern China. Therefore, the REEs in the ore-forming fluids cannot be completely contributed by ore-bearing carbonate rocks [56], but it cannot be ruled out that some REEs are inherited from wall rocks during the water/rock (W/R) interaction between ore-forming fluids and wall rocks.

Y and REEs are close in radius, and both are considered to have similar geochemical natures. Therefore, using the variation trend between Y and REE is one of the effective methods to determine the source of fluids [63–65]. In the diagram of Y-$\sum$REE (Figure 9b), the hydrothermal dolomite, calcite and barite are mainly distributed in two concentrated zones. The hydrothermal dolomite and barite that represent the $\sum$REE contents of the ore-forming fluids are located between altered dolostone and the underlying strata [64]. This indicates that REEs in the ore-forming fluids were likely to be provided by the wall rocks and underlying strata. The results are consistent with the study on the chemical characteristics of carbonate minerals in the adjacent Zhulingou deposit [6]. In summary, the ore-forming fluids of the Banbianjie deposit that are similar to the basin brine may originate from and/or flow through the underlying strata, and parts of ore-forming materials in the fluids are inherited from wall rocks during the water/rock (W/R) interaction.

6.2. Evolution of Ore-Forming Fluids

The trace elements of altered dolostone and hydrothermal dolomite and calcite show that they have different ratios of Fe/Mn, Sr/Ba and Th/U (Table 1). For example, the Fe/Mn ratios (1.56–2.79) of altered dolostone are higher than those of hydrothermal calcite and dolomite (0.01–0.83). This suggests that the initial ore-forming fluids have high Fe/Mn ratios. In contrast, the Sr/Ba ratios of hydrothermal calcite and dolomite (8–634) are higher than those of altered dolostone (6–17). This indicates that the Sr/Ba ratios are increased during the mineralization process. In addition, The Th/U ratios of altered dolostone from the Banbianjie deposit are between 0.29 and 0.88, which are quite different from those of hydrothermal calcite and dolomite (0.26–241) and barite (2–13). This implies that the Th/U ratios of hydrothermal fluids have an evolutionary trend of increasing first and then decreasing.

The REE^{3+} ion has different geochemical characteristics from those of the Eu^{2+} and Ce^{4+} ions [66–69]. Therefore, the Eu^{2+} and Ce^{4+} ions are separated from REE^{3+} in the change process of fluid physical and chemical environments, resulting in positive or negative Eu and Ce anomalies, which is called the oxidation reduction mode of REEs [69,70]. Hence, europium (Eu) and cerium (Ce) anomalies are often used to discuss the change in physical–chemical condition during the evolution of hydrothermal fluids [68–78].

There are obviously negative Eu and Ce anomalies in altered dolostone (Figure 6a) and obviously negative Eu and positive Ce anomalies of hydrothermal calcite and dolomite (Figure 6c), as the ore-forming fluids originated from and/or flowed through the underlying strata and interacted with wall rocks. Hence, the ore-forming fluids were acidic reduced during the early mineralization stage, which is consistent with the fluids' characteristics reflected in the presence of the acicular marcasite (T < 240 °C, pH < 5, Figure 5l, [38]). Moreover, barite has significant positive Eu anomalies (Figure 6d), and the late stage of the Banbianjie deposit is an open environment, suggesting that the late ore-forming fluids were oxidized. Therefore, the evolution of ore-forming fluids experienced a shift from reduction to oxidation during the Ge-Zn mineralization.

6.3. Ore Genesis

There are great controversies on the ore genesis of the Banbianjie Ge-Zn deposit, including: (i) sedimentary exhalative (SEDEX) deposit [9]; (ii) syn-sedimentary modified stratigraphically controlled deposit [38]; and (iii) the Mississippi Valley-type (MVT) deposit [13,14].

According to the geological survey, the Banbianjie Ge-Zn deposit has veined and stockwork ores with metasomatism (Figure 5) and developed the veined ore body that crosscuts the wall rocks (Figure 4b). These structures and textures of ores and hydrothermal minerals suggest the characteristics of epigenetic mineralization. In addition, more than 200 Pb-Zn deposits have been found in the Western Hunan–Eastern Guizhou Pb-Zn metallogenic belt, which are mainly hosted by carbonate rocks of the Cambrian to Devonian, and their ore-forming ages are 349–477 Ma (concentrated in 410–450 Ma) [6,10,31]. Hence, this deposit cannot be a SEDEX deposit. The genesis of hydrothermal calcite also excludes the genetic connection between the deposit and magmatism or sedimentation (Figure 8).

The REE patterns and La/Yb$_N$ ratios of barite can be used to distinguish the ore genesis [48]. The La/Yb$_N$ values of barite from the Banbianjie deposit are similar to those of the Lehong MVT Pb-Zn deposit in Yunnan province, suggesting that their ore genesis is similar, and both belong to MVT deposits [25]. In addition, barite of the Banbianjie deposit has significantly different REE characteristics from those of the Zhenning–Ziyun Devonian large-scale hot-water sedimentary barite deposits (Figure 6d, [48]), the Guangxi hot-water sedimentary barite deposits [79] and magmatic genesis barite deposits [80]. This shows that the Banbianjie deposit has nothing to do with hot-water precipitation and high-temperature magmatism.

The geological characteristics of most Pb-Zn deposits in the Western Hunan–Eastern Guizhou Pb-Zn metallogenic belt are very similar to those of MVT deposits: (i) all deposits occurred in carbonate sequences; (ii) all deposits are controlled by structures; (iii) with obvious epigenetic metallogenic characteristics; (iv) Pb-Zn mineralization has no relationship with magmatism in time and space; (v) the ore bodies occur mainly as strata bound; and (vi) the ore-forming fluids are medium–low temperatures (120–220 °C) and medium–high salinities (8–23 wt.% NaCl equiv.) [4,12–14,18,81,82]. Hence, they can be classified as MVT deposits.

7. Conclusions

1. The trace elements of hydrothermal minerals indicate that the ore-forming fluids are rich in Sr, Ba, Zn, Fe, Pb and Ge, which are similar to those of basin brine. The REE characteristics of hydrothermal minerals suggest that the ore-forming fluids of the Banbianjie deposit were likely derived from and/or flowed through the underlying strata and interacted with wall rocks.
2. The Eu and Ce anomalies of the gangue minerals from the Banbianjie deposit and the mineral assemblages imply that the evolution of ore-forming fluids experienced a shift from reduction to oxidation during Ge-Zn mineralization.
3. The ore deposit geology, mineralogy and trace element geochemistry of the gangue minerals imply that the ore genesis of the Banbianjie deposit is best classified as an MVT deposit.

Author Contributions: Investigation, formal analysis, writing—original draft, Y.-L.A.; methodology, supervision, writing—review and editing, J.-X.Z.; writing—review and editing, Q.-T.M., G.-T.S. and Z.-M.Y. All authors have read and agreed to the published version of the manuscript.

Funding: This research was funded by the National Natural Science Foundation of China [grant numbers U1812402 and 42172082], Yunnan University Scientific Research Start-up Project [grant number YJRC4201804] and the Major collaborative innovation projects for prospecting breakthrough strategic action of Guizhou Province, China ((2022) ZD004).

Data Availability Statement: The data presented in this study are available on reasonable request from the corresponding authors.

Acknowledgments: The field work was supported and assisted by workers of the Nos. 104 and 109 Geological Teams of Guizhou Bureau of Geology and Mineral Exploration & Development. The experimental process was guided and assisted by the staff of the State Key Laboratory of Ore Deposit Geochemistry, Institute of Geochemistry, Chinese Academy of Sciences. We would like to express our heartfelt thanks to them!

Conflicts of Interest: The authors declare no conflict of interest.

References

1. Taylor, S.R.; McLennan, S.M. The geochemical evolution of the continental crust. *Rev. Geophys.* **1995**, *33*, 241–265. [CrossRef]
2. Tu, G.Z.; Gao, Z.M.; Hu, R.Z.; Li, C.Y.; Zhao, Z.H.; Zhang, B.G. *Geochemistry and Metallogenic Mechanism of Disperse Elements*; Geology Press: Beijing, China, 2004. (In Chinese)
3. Wen, H.J.; Zhu, C.W.; Du, S.J.; Fan, Y.; Luo, C.G. Gallium (Ga), germanium (Ge), thallium (Tl) and cadmium (Cd) resources in China. *Sci. Bull.* **2020**, *65*, 3688–3699. (In Chinese with English Abstract) [CrossRef]
4. Ye, L.; Cook, N.J.; Liu, T.G.; Ciobanu, C.L.; Gao, W.; Yang, Y.L. The Niujiaotang Cd-rich zinc deposit, Duyun, Guizhou province, southwest China: Ore genesis and mechanisms of cadmium concentration. *Miner. Deposita* **2012**, *47*, 683–700. [CrossRef]
5. Luo, K.; Cugerone, A.; Zhou, M.F.; Zhou, J.X.; Sun, G.T.; Xu, J.; He, K.J.; Lu, M.D. Germanium enrichment in sphalerite with acicular and euhedral textures: An example from the Zhulingou carbonate-hosted Zn(-Ge) deposit, South China. *Miner. Deposita* **2022**, *57*, 1343–1365. [CrossRef]
6. Yang, Z.M.; Zhou, J.X.; Lou, K.; Yang, D.Z.; Yu, J.; Zhou, F.C. Mineralogy and mineral chemistry of carbonates from the Zhulingou Ge-Zn deposit in Guizhou Province and its geological significance. *Acta Petrol. Sin.* **2021**, *37*, 2743–2760. (In Chinese with English Abstract)
7. Zhou, J.X.; Yang, Z.M.; An, Y.L.; Luo, K.; Liu, C.X.; Ju, Y.W. An evolving MVT hydrothermal system: Insights from the Niujiaotang Cd-Zn ore field, SW China. *J. Asian Earth Sci.* **2022**, *237*, 105357. [CrossRef]
8. Wen, H.J.; Zhou, Z.B.; Zhu, C.W.; Luo, C.G.; Wang, D.Z.; Du, S.J.; Li, X.F.; Chen, M.H.; Li, H.Y. Critical scientific issues of super-enrichment of dispersed metals. *Acta. Petrol. Sin.* **2019**, *35*, 327–3291. (In Chinese with English Abstract)
9. Chen, G.Y.; An, Q.; Wang, M. Geological Characteristics and Genesis of the Banbianjie Type Lead-Zinc Deposits in Southern Guizhou Province. *Acta Geosci. Sin.* **2006**, *27*, 570–576. (In Chinese with English Abstract)
10. Ye, L.; Pang, Z.P.; Li, C.Y.; Liu, T.G.; Xia, B. Isotopic Geochemical Characters in Niujiaotang Cd Rich Zinc Deposit, Duyun, Guizhou. *J. Mineral. Petrol.* **2005**, *25*, 70–74. (In Chinese with English Abstract)
11. Zhou, J.X.; Huang, Z.L.; Zhou, M.F.; Li, X.B.; Jin, Z.G. Constraints of C-O-S-Pb isotope compositions and Rb-Sr isotopic age on the origin of the Tianqiao carbonate-hosted Pb-Zn deposit, SW China. *Ore Geol. Rev.* **2013**, *53*, 77–92. [CrossRef]
12. Li, K.; Tang, C.Y.; Liu, J.S.; Cai, Y.X.; Liu, F. Sources of metallogenic materials of Xiunao Pb-Zn deposit in eastern Guizhou: Constraints from REE and C, O, S, Pb isotope geochemistry. *J. Guilin Univ. Technol.* **2018**, *38*, 365–376. (In Chinese with English Abstract)
13. Meng, Q.T.; Zhou, J.X.; Sun, G.T.; Zhao, Z.; An, Q.; Yang, X.Y.; Lu, M.D.; Xiao, K.; Xu, L. Geochemical characteristics and ore prospecting progress of the Banbianjie Zn deposit in Guiding City, Guizhou Province, China. *Acta Mineral. Sin.* **2022**, *42*, 51–58. (In Chinese with English Abstract)
14. An, Y.L.; Luo, K.; Zhou, J.X.; Nguyen, A.; Lu, M.D.; Meng, Q.T.; An, Q. Origin of the Devonian carbonate-hosted Banbianjie Ge-Zn deposit, Guizhou Province, South China: Geological, mineralogical and geochemical constraints. *Ore Geol. Rev.* **2022**, *142*, 104696. [CrossRef]
15. Zhou, J.X.; Huang, Z.L.; Zhou, G.F.; Jin, Z.G.; Li, X.B.; Ding, W.; Gu, J. Sources of the Ore Metals of the Tianqiao Pb-Zn Deposit in Northwestern Guizhou Province: Constraints form S, Pb Isotope and REE Geochemistry. *Geol. Rev.* **2010**, *56*, 513–524. (In Chinese with English Abstract)
16. Zhou, J.X.; Huang, Z.L.; Zhou, G.F.; Zeng, Q.S. C, O Isotope and REE Geochemistry of the Hydrothermal Calcites from the Tianqiao Pb-Zn Ore Deposit in NW Guizhou Province, China. *Geotecton. Metallog.* **2012**, *36*, 93–101. (In Chinese with English Abstract)
17. Zhou, J.X.; Luo, K.; Li, B.; Huang, Z.L.; Yan, Z.F. Geological and isotopic constraints on the origin of the Anle carbonate-hosted Zn-Pb deposit in northwestern Yunnan Province, SW China. *Ore Geol. Rev.* **2016**, *74*, 88–100. [CrossRef]
18. Zhou, J.X.; Xiang, Z.Z.; Zhou, M.F.; Feng, Y.X.; Luo, K.; Huang, Z.L.; Wu, T. The giant Upper Yangtze Pb-Zn province in SW China: Reviews, new advances and a new genetic model. *J. Asian Earth Sci.* **2018**, *154*, 280–315. [CrossRef]
19. Zhang, C.Q.; Mao, J.W.; Wu, S.P.; Li, H.M.; Liu, F.; Guo, B.J.; Gao, D.R. Distribution, characteristics and genesis of Mississippi Valley-Type Pb-Zn deposits in Sichuan-Yunnan-Guizhou area. *Miner. Deposits* **2005**, *24*, 336–348. (In Chinese with English Abstract).
20. Tang, Y.Y.; Zhang, K.X.; Tian, Y.J.; Zhang, J.W.; Huang, Z.L.; Wu, T. REE compositions of calcites from Pb-Zn deposits in the eastern Guizhou and their metallogenic implications. *Acta Mineral. Sin.* **2020**, *40*, 356–366. (In Chinese with English Abstract)
21. Elderfield, H.; Upstill-Goddard, R.; Sholkovitz, E.R. The rare earth elements in rivers, estuaries, and coastal seas and their significance to the composition of ocean waters. *Geochim. Cosmochim. Acta* **1990**, *54*, 971–991. [CrossRef]
22. Johannesson, K.H.; Stetzenbach, K.J.; Hodge, V.F. Rare earth elements as geochemical tracers of regional groundwater mixing. *Geochim. Cosmochim. Acta* **1997**, *61*, 3605–3618. [CrossRef]
23. Debruyne, D.; Hulsbosch, N.; Muchez, P. Unraveling rare earth element signatures in hydrothermal carbonate minerals using a source–sink system. *Ore Geol. Rev.* **2016**, *72*, 232–252. [CrossRef]
24. Sun, G.T.; Shen, N.P.; Su, W.C.; Feng, Y.X.; Zhao, J.X.; Peng, J.T.; Dong, W.D.; Zhao, H. Characteristics and Implication of Trace Elements and Sr-Nd Isotope Geochemistry of Calcites from the Miaolong Au-Sb Deposit, Guizhou Province, China. *Acta Mineral. Sin.* **2016**, *36*, 404–412. (In Chinese with English Abstract)
25. Zhao, D.; Han, R.S.; Wang, J.S.; Ren, T. REE Geochemical Characteristics in Lehong Large Pb-Zn Deposit, Northeastern Yunnan Province, China. *Acta Mineral. Sin.* **2017**, *37*, 588–595. (In Chinese with English Abstract)

26. Li, X.G.; Yang, K.G.; Hu, X.Y.; Dai, C.G.; Zhang, H. Formation and evolution of the Kaili-Sandu fault in East Guizhou, China. *J. Chengdu Univ. Technol. Sci. Technol. Ed.* **2012**, *39*, 18–26. (In Chinese with English Abstract)

27. Liu, Y.L.; Yang, K.G.; Deng, X. Activities History of Zhenyuan-Guiyang Fault Belts and Constraint on the Evolution of Centre Guizhou Uplift. *Bull. Geol. Sci. Technol.* **2009**, *28*, 41–47. (In Chinese with English Abstract)

28. Deng, X.; Yang, K.G.; Liu, Y.L.; She, Z.B. Characteristics and tectonic evolution of Qianzhong Uplift. *Earth Sci. Front.* **2010**, *17*, 79–89. (In Chinese with English Abstract)

29. Yang, K.G.; Li, X.G.; Dai, C.G.; Zhang, H.; Zhou, Q. Analysis of the origin of trough-like folds in Southeast Guizhou. *Earth Sci. Front.* **2012**, *19*, 53–60. (In Chinese with English Abstract)

30. Dai, C.G.; Wang, M.; Chen, J.S.; Wang, X.H. Tectonic Movement Characteristic and Its Geological Significance of Guizhou. *Guizhou Geol.* **2013**, *30*, 119–124. (In Chinese with English Abstract)

31. Ye, L.; Liu, T.G.; Shao, S.X. Geochemistry of mineralizing fluid of Cd-rich zinc deposit: Taking Niujiaotang Cd-rich zinc deposit, Duyun, Guizhou for example. *Geochimica* **2000**, *29*, 597–603. (In Chinese with English Abstract)

32. Zhao, Z.; Bao, G.P.; Qian, Z.K.; Huang, L.; Lu, M.D.; Xu, L. Geochemical Characteristics of C-O Isotopes and REE in the Hydrothermal Calcite from the Shuanglongquan Pb-Zn Deposit, Southern Guizhou, China. *Acta Mineral. Sin.* **2018**, *38*, 627–636. (In Chinese with English Abstract)

33. Cui, M.; Tang, L.J.; Guo, T.L.; Ning, F.; Tian, H.Q.; Hu, D.F. Structural Style and Thrust Breakthrough Model of Fold in Southeast Guizhou. *Earth Sci. (Wuhan China)* **2009**, *34*, 907–913. (In Chinese with English Abstract)

34. Dai, C.G.; Chen, J.S.; Lu, D.B.; Ma, H.Z.; Wang, X.H. Appearance and geologic significance of Caledonian Movement in southeastern Guizhou, China and its adjacent area. *Geol. Bull. China* **2010**, *29*, 530–534. (In Chinese with English Abstract)

35. Xu, Z.Y.; Yao, G.S.; Guo, Q.X.; Chen, Z.L.; Dong, Y.; Wang, P.W.; Ma, L.Q. Genetic Interpretation about Geotectonics and Structural Transfiguration of the Southern Guizhou Depression. *Geotecton. Metallog.* **2010**, *34*, 20–31. (In Chinese with English Abstract)

36. Luo, K.; Zhou, J.X.; Huang, Z.L.; Caulfield, J.; Zhao, J.X.; Feng, Y.X.; Ouyang, H. New insights into the evolution of Mississippi Valley-Type hydrothermal system: A case study of the Wusihe Pb-Zn deposit, South China, using quartz in-situ trace elements and sulfides in situ S-Pb isotopes. *Am. Mineral.* **2020**, *105*, 35–51. [CrossRef]

37. Zhou, J.X.; Meng, Q.T.; Ren, H.Z.; Sun, G.T.; Zhang, Z.J.; An, Q.; Zhou, C.X. A super-large co-(associative) Ge deposit was discovered in the Huangsi anticline area of Guizhou. *Geotecton. Metallog.* **2020**, *44*, 1025–1026. (In Chinese with English Abstract)

38. Zuo, J.L. Geological characteristics and ore-controlling factors of the Banbianjie Pb-Zn deposit in Guiding County, Guizhou Province. *Land Resour. South. China* **2013**, *5*, 39–41. (In Chinese)

39. Mo, L.L. Geological Characteristics and Origin of Banbianjie Zinc Deposit in Guiding County, Guizhou Province. *West-China Explor. Eng.* **2020**, *32*, 125–127+132. (In Chinese with English Abstract)

40. Qi, L.; Hu, J.; Gregoire, D.C. Determination of trace elements in granites by inductively coupled plasma mass spectrometry. *Talanta* **2000**, *51*, 507–513.

41. Raczek, I.; Stoll, B.; Hofmann, A.W.; Jochum, K.P. High-Precision Trace Element Data for the USGS Reference Materials BCR-1, BCR-2, BHVO-1, BHVO-2, AGV-1, AGV-2, DTS-1, DTS-2, GSP-1 and GSP-2 by ID-TIMS and MIC-SSMS. *Geostand. Newsl.* **2001**, *25*, 77–86. [CrossRef]

42. Cotta, A.J.B.; Enzweiler, J. Classical and new procedures of whole rock dissolution for trace element determination by ICP-MS. *Geostand. Geoanal. Res.* **2012**, *36*, 27–50. [CrossRef]

43. Chen, D.F.; Huang, Y.Y.; Yuan, X.L.; Cathles, L.M. Seep carbonates and preserved methane oxidizing archaea and sulfate reducing bacteria fossils suggest recent gas venting on the seafloor in the Northeastern South China Sea. *Mar. Petrol. Geol.* **2005**, *22*, 613–621. [CrossRef]

44. Chen, L.; Liu, Y.S.; Hu, Z.C.; Gao, S.; Zong, K.Q.; Chen, H.H. Accurate determinations of fifty-four major and trace elements in carbonate by LA-ICP-MS using normalization strategy of bulk components as 100%. *Chem. Geol.* **2011**, *284*, 283–295. [CrossRef]

45. Jochum, K.P.; Weis, U.; Stoll, B.; Kuzmin, D.; Yang, Q.C.; Raczek, I.; Jacob, D.; Stracke, A.; Birbaum, K.; Frick, D.A.; et al. Determination of reference values for NIST SRM 610-617 glasses following ISO guidelines. *Geostand. Geoanal. Res.* **2011**, *35*, 397–429. [CrossRef]

46. Jochum, K.P.; Scholz, D.; Stoll, B.; Weis, U.; Wilson, S.A.; Yang, Q.Y.; Schwalb, A.; Börner, N.; Jacob, D.E.; Andreae, M.O. Accurate trace element analysis of speleothems and biogenic calcium carbonates by LA-ICP-MS. *Chem. Geol.* **2012**, *318*, 31–44. [CrossRef]

47. Boynton, W.V. *Cosmochemistry of the Rare Earth Elements: Meteorite Studies Developments in Geochemistry*; Elsevier: Amsterdam, The Netherlands, 1984; pp. 63–114.

48. Yan, M.C.; Chi, Q.H. *Geochemical Composition of Crust and Rocks in Eastern China*; Science Press: Beijing, China, 1997; pp. 1–292. (In Chinese)

49. Gao, J.B.; Yang, R.D.; Tao, P.; Cheng, W.; Zheng, L.L. Study on the Genesis of Devonian Magnesian Siderite Deposits in the Northwestern Guizhou Province. *Geol. Rev.* **2015**, *61*, 1305–1320. (In Chinese with English Abstract)

50. Shannon, R.D. Revised effective ionic radii and systematic studies of interatomic distances in halides and chalcogenides. *Acta Crystallogr. A* **1976**, *32*, 751–767. [CrossRef]

51. Liu, Y.J.; Cao, L.M.; Li, Z.L. *Elemental Geochemistry*; Science Press: Beijing, China, 1984; pp. 360–420. (In Chinese)

52. Tang, Y.Y.; Bi, X.W.; He, L.P.; Wu, L.Y.; Feng, C.X.; Zou, Z.C.; Tao, Y.; Hu, R.Z. Geochemical characteristics of trace elements, fluid inclusions and carbon-oxygen isotopes of calcites in the Jinding Zn-Pb deposit, Lanping, China. *Acta Petrol. Sin.* **2011**, *27*, 2635–2645. (In Chinese with English Abstract)

53. Leach, D.L.; Sangster, D.F.; Kelley, K.D.; Ross, R.L.; Garven, G.; Allen, C.R.; Gutzmer, J.; Walters, S. Sediment-hosted lead-zinc deposits: A global perspective. *Econ. Geol.* **2005**, *100*, 561–607.

54. Yang, Q.K.; Meng, X.J.; Guo, F.S.; Zhou, W.P.; Sun, Q.Z. Characteristics of Trace Elements in Gangue Minerals of the Xiangshan Uranium Polymetallic Deposit, Jiangxi, and its Geological Significance. *Bull. Mineral. Petrol. Geochem.* **2014**, *33*, 457–465+483. (In Chinese with English Abstract)

55. Michard, A. Rare earth element systematics in hydrothermal fluids. *Geochim. Cosmochim. Acta* **1989**, *53*, 745–750. [CrossRef]

56. Lottermoser, B.G. Rare earth elements and hydrothermal ore formation processes. *Ore Geol. Rev.* **1992**, *7*, 25–41. [CrossRef]

57. Bau, M.; Dulski, P. Comparing yttrium and rare earths in hydrothermal fluids from the Mid-Atlantic Ridge: Implications for Y and REE behavior during near-vent mixing and for the Y/Ho ratio of Proterozoic seawater. *Chem. Geol.* **1999**, *155*, 77–90. [CrossRef]

58. Wang, G.Z.; Hu, R.Z.; Liu, Y.; Sun, G.S.; Sun, W.C.; Liu, H. REE Geochemical Characteristic from Fluorite in Qinglong Antimony Deposit, South-Western Guizhou. *J. Mineral. Petrol.* **2003**, *23*, 62–65. (In Chinese with English Abstract)

59. Zhang, Y.; Xia, Y.; Wang, Z.P.; Yan, B.W.; Fu, Z.K.; Chen, M. REE and stable isotope geochemical characteristics of Bojitian gold deposit, Guizhou Province. *Earth Sci. Front.* **2010**, *17*, 385–395. (In Chinese with English Abstract)

60. Zhou, J.X.; Huang, Z.L.; Zhou, G.F.; Li, X.B.; Ding, W.; Bao, G.P. The trace elements and rare earth elements geochemistry of sulfide minerals of the Tianqiao Pb-Zn ore deposit, Guizhou Province, China. *Acta Geol. Sin.* **2011**, *85*, 189–199.

61. Möller, P.; Parekh, P.P.; Schneider, H.J. The application of Tb/Ca-Tb/La abundance ratios to problems of fluorspar genesis. *Miner. Deposita* **1976**, *11*, 111–116. [CrossRef]

62. Bau, M.; Dulski, P. Comparative study of yttrium and rare-earth element behaviors in fluorine-rich hydrothermal fluids. *Contrib. Mineral. Petr.* **1995**, *119*, 213–223. [CrossRef]

63. Li, H.M.; Shen, Y.C.; Mao, J.W.; Liu, T.B.; Zhu, H.P. REE features of quartz and pyrite and their fluid inclusions: An example of Jiaojia-type gold deposits, northwestern Jiaodong peninsula. *Acta Petrol. Sin.* **2003**, *19*, 267–274. (In Chinese with English Abstract)

64. Wei, H.T.; Shao, Y.J.; Ye, Z.; Xiong, Y.Q.; Zhou, H.D.; Xie, Y.L. REE and Sr isotope geochemistry of gangue calcites from Huayuan Pb-Zn orefield in western Hunan, China. *Chin. J. Nonferrous Met.* **2017**, *27*, 2329–2339. (In Chinese with English Abstract)

65. Liu, S.W.; Shi, S.; Li, R.X.; Gao, Y.B.; Liu, L.F.; Duan, L.Z.; Chen, B.Z.; Zhang, S.N. REE geochemistry of Mayuan Pb-Zn deposit on northern margin of Yangtze Plate. *Miner. Deposits* **2013**, *32*, 979–988. (In Chinese with English Abstract)

66. Chen, Y.J.; Fu, S.G. Variation of REE patterns in early Precambrian sediments: Theoretical study and evidence from the southern margin of the northern China craton. *Chin. Sci. Bull.* **1991**, *36*, 1100–1104.

67. Chen, Y.J.; Zhao, Y.C. Geochemical characteristics and evolution of REE in the Early Precambrian sediments: Evidences from the southern margin of the North China craton. *Episodes J. Int Geosci.* **1997**, *20*, 109–116.

68. Ma, Y.J.; Liu, C.Q. Trace element geochemistry during weathering as exemplified by the weathered crust of granite, Longnan, Jiangxi. *Sci. Bull.* **1999**, *44*, 2260–2263. [CrossRef]

69. Rimstidt, J.D.; Balog, A.; Webb, J. Distribution of trace elements between carbonate minerals and aqueous solutions. *Geochim. Cosmochim. Acta* **1998**, *62*, 1851–1863. [CrossRef]

70. Bau, M.; Möller, P. Rare earth element fractionation in meta-morphogenic hydrothermal calcite, magnesite and siderite. *Miner. Petrol.* **1992**, *45*, 231–246. [CrossRef]

71. He, S.; Xia, Y.; Xiao, J.; Gregory, D.; Xie, Z.J.; Tan, Q.P.; Yang, H.Y.; Guo, H.Y.; Wu, S.W.; Gong, X.X. Geochemistry of REY-Enriched Phosphorites in Zhijin Region, Guizhou Province, SW China: Insight into the Origin of REY. *Minerals* **2022**, *12*, 408. [CrossRef]

72. Cherniak, D.J.; Zhang, X.Y.; Wayne, N.K.; Watson, E.B. Sr, Y and REE diffusion in fluorite. *Chem. Geol.* **2001**, *181*, 99–111. [CrossRef]

73. Bau, M. Rare-earth element mobility during hydrothermal and metamorphic fluid-rock interaction and the significance of the oxidation state of europium. *Chem. Geol.* **1991**, *93*, 219–230. [CrossRef]

74. Cocherie, A.; Calvez, J.Y.; Oudin-Dunlop, E. Hydrothermal activity as recorded by Red Sea sediments: Sr-Nd isotopes and REE signatures. *Mar. Geol.* **1994**, *118*, 291–302. [CrossRef]

75. Mills, R.A.; Elderfield, H. Rare earth element geochemistry of hydrothermal deposits from the active TAG Mound, 26°N Mid-Atlantic Ridge. *Geochim. Cosmochim. Acta* **1995**, *59*, 3511–3524. [CrossRef]

76. Zhong, S.; Mucci, A. Partitioning of rare earth elements (REEs) between calcite and seawater solutions at 25°C and 1 atm, and high dissolved REE concentrations. *Geochim. Cosmochim. Acta* **1995**, *59*, 443–453. [CrossRef]

77. Peng, J.T.; Hu, R.Z.; Qi, L.; Zhao, J.H.; Fu, Y.Z. REE Distribution Pattern for the Hydrothermal Calcites from the Xikuangshan Antimony Deposit and Its Constraining Factors. *Geol. Rev.* **2004**, *50*, 25–32. (In Chinese with English Abstract)

78. Murowchick, J.B.; Barnes, H.L. Marcasite precipitation from hydrothermal solutions. *Geochim. Cosmochim. Acta* **1986**, *50*, 2615–2629. [CrossRef]

79. Wang, M.Y.; Xi, C.Z.; Li, Y. 2008. Geological and geochemical characteristics of the barite rocks in the hydrothermal sedimentary deposit in Guangxi. *Min. Resour. Geol.* **2008**, *22*, 335–341. (In Chinese with English Abstract)

80. Niu, H.C.; Chen, F.R.; Lin, M.Q. REE Geochemistry of Magmatic genetic Barite and Fluorite. *Acta Mineral. Sin.* **1996**, *16*, 328–388. (In Chinese with English Abstract)

81. Yang, D.Z.; Zhou, J.X.; Luo, K.; Yu, J.; Zhou, Z.H. New discovery and research value of Zhulingou zinc deposit in Guiding, Guizhou. *Bull. Mineral. Petrol. Geochem.* **2020**, *39*, 344–345. (In Chinese with English Abstract)
82. Zhou, J.X.; Wang, X.C.; Wilde, S.A.; Luo, K.; Huang, Z.L.; Wu, T.; Jin, Z.G. New insights into the metallogeny of MVT Zn-Pb deposits: A case study from the Nayongzhi in South China, using field data, fluid compositions, and in situ S-Pb isotopes. *Am. Mineral.* **2018**, *103*, 91–108. [CrossRef]

Article

Zircon and Garnet U–Pb Ages of the Longwan Skarn Pb–Zn Deposit in Guangxi Province, China and Their Geological Significance

Xuejiao Zhang [1,2], Wei Ding [1,2,*], Liyan Ma [1,2], Wei Fu [1,2,3], Xijun Liu [1,2] and Saisai Li [1,2]

[1] College of Earth Sciences, Guilin University of Technology, Guilin 541004, China
[2] Guangxi Key Laboratory of Hidden Metallic Ore Deposits Exploration, Guilin University of Technology, Guilin 541004, China
[3] Collaborative Innovation Center for Exploration of Nonferrous Metal Deposits and Efficient Utilization of Resources by the Province and Ministry, Guilin University of Technology, Guilin 541004, China
* Correspondence: dingwei@glut.edu.cn

Abstract: Garnet is the most common alteration mineral in skarn-type deposits, and the geochronological research on it can limit the mineralization age. The Longwan Pb–Zn deposit, situated within the Fozichong Pb–Zn ore field in Guangxi, lacks precise geochronological data, limiting the in-depth comprehension of its genesis and tectonic setting. This study employs LA-ICP-MS U–Pb dating of garnets developed in the skarn orebody and zircons in the associated granitic porphyry to determine the deposit's mineralization age. Backscatter electron images and electron probe microanalysis reveal common zonation characteristics in garnets from the Longwan Pb–Zn deposit, with dominant end-member compositions of Andradite and Grossular. The values of U concentrations range from 1.8 ppm to 3.7 ppm, and a garnet U–Pb age of 102.6 ± 1.9 Ma was obtained, consistent with the zircon U–Pb age of 102.1 ± 1.2 Ma from the granite porphyry within the deposit. The Longwan Pb–Zn deposit formed during the late Early Cretaceous as a skarn deposit resulting from contact metasomatism between the granite porphyry and the host rock. The deposit likely formed in response to the Neo-Tethys plate subducting beneath the South China continent during the Cretaceous, followed by a retreat during the Late Cretaceous. The Cenxi-Bobai Fault experienced reactivation under the extensional tectonic regime induced by the Neo-Tethys Ocean's retreat, leading to a series of magmatic activities along the NE-trending direction within the Fault. The Longwan Pb–Zn deposit formed during the processes of magma emplacement and contact metasomatic reactions with the country rock.

Keywords: garnet; granite porphyry; ore-forming age; tectonic setting

Citation: Zhang, X.; Ding, W.; Ma, L.; Fu, W.; Liu, X.; Li, S. Zircon and Garnet U–Pb Ages of the Longwan Skarn Pb–Zn Deposit in Guangxi Province, China and Their Geological Significance. *Minerals* **2023**, *13*, 644. https://doi.org/10.3390/min13050644

Academic Editor: Maria Boni

Received: 24 March 2023
Revised: 25 April 2023
Accepted: 2 May 2023
Published: 6 May 2023

1. Introduction

Geochronological investigations play a pivotal role in deciphering the genesis and mineralization processes of ore deposits. With the advancements in analytical techniques, various dating methods have been extensively applied in metallic ore deposit studies, such as ^{40}Ar–^{39}Ar, U–Pb, Sm–Nd, Rb–Sr, and Re–Os methods [1]. Garnet, a typical alteration mineral in the skarn deposits, has been widely utilized for tracing the geochemical evolution of hydrothermal fluids [2–4]. Recent research highlights the higher U–Pb closure temperature system (>850 °C) in garnets compared to other minerals [5], and the elevated levels of U and Pb in Andradite, fulfilling the LA-ICP-MS elemental testing requirements [6]. Consequently, garnets have garnered significant attention as subjects for geochronological research in the skarn deposits [1,7–11].

The Longwan lead–zinc mine, a large-scale deposit within the Fozichong lead–zinc ore field in the southeastern Pb–Zn–Ag–Au polymetallic metallogenic belt, has a cumulative proven metal content of 357,000 tons of lead and 546,000 tons of zinc with average grades of 2.39% Pb and 3.66% Zn. Previous investigations on the Fozichong ore field, including the Longwan Pb–Zn deposit, sought to constrain the mineralization age through dating intrusive bodies [12]. However, due to the spatial relationship between the intrusive body and the ore body, especially when multiple magmatic events occur in the mining area, it is difficult to accurately determine the genetic relationship between the intrusive body and the skarn [13]. As a result, a direct ore-forming age for the Longwan Pb–Zn deposit has not been obtained, which hinders the understanding of the deposit's formation mechanism and its related tectonic and geodynamic background. Furthermore, the Longwan lead–zinc deposit lies at the transition zone between the South China-Pacific tectonic domain and the Tethyan tectonic domain. Its mineralization dynamics have been attributed to the westward subduction of the Pacific plate [14]; however other scholars propose that contemporaneous plutons near the Longwan deposit formed due to the northward subduction of the Neo-Tethys [15]. Abundant garnets are present in the skarn orebodies of the Longwan lead–zinc deposit. This study, based on the comprehensive geological features of the deposit, aims to determine the mineralization age by dating garnets in the Longwan skarn deposit and zircons in the associated granitic porphyry, providing robust evidence for deposit petrogenesis research. Integrating the mineralization age with previous tectonic evolution studies, this paper delves deeper into the mineralization dynamics background of the Longwan Pb–Zn deposit.

2. Geological Background

2.1. Regional Geological Features

The Longwan lead–zinc deposit is situated in southeastern Guangxi Province, China, and lies on the Cenxi-Bobai Fault zone in the southwest section of the Qin-Hang suture zone [16]. The deposit is situated at the junction of the South China Pacific tectonic domain and the Tethys tectonic domain. Cretaceous magmatic rocks are extensively distributed throughout the region (Figure 1a). The northeast-trending Cenxi-Bobai Fault zone represents a significant geological structure in southeastern Guangxi (Figure 1b). This deep fault zone is likely to have originated during the Sinian period (a geological time period, that spans from 850 Ma to 542 Ma) and has played a crucial role in controlling the distribution of rocks and ores, displaying multiple stages of activity that have induced various magmatic and tectonic events [14,17]. The late Yanshanian intrusive rocks in the region predominantly occur along the Cenxi-Bobai Fault zone, with granites being the primary rock type. They appear in the form of small plutons, and dykes (Figure 1c). The geological evolution of this area has experienced multiple tectonic events, including the Caledonian orogeny, the Hercynian-Indosinian collision orogeny, and the Yanshanian-Himalayan intraplate extension [18]. These tectonic events, coupled with intense magmatic activity, regional metamorphism, and migmatization, have formed the current complex geological structure [19] and led to multi-stage mineralization events [20,21]. Intermediate-acid magmatic rocks and volcanic rocks of the Yanshan period are primarily distributed along the NE–SW trending Cenxi-Bobai Fault zone, constituting a multiphase magmatic belt.

The geology of this region, with the exception of the Permian and Triassic systems, displays exposures spanning from the Cambrian to Cretaceous periods. Among these strata, the Lower Silurian and the Upper Ordovician are the most significant ore-bearing units in the Fozichong mining district [16]. The Ordovician strata primarily consist of sandstone, limestone, and siltstone, with slate being less common, and are predominantly distributed in the central and eastern parts of the mining area. The Silurian strata are characterized by dark gray slate interbedded with sandstone, siltstone, mudstone with limestone, and fine-grained sandstone, mainly found in the western part of the mining area. The Devonian strata have isolated remnants in the southwestern portion of the study area, featuring the lower section with sandstone interbedded with thin layers of limestone and the upper section composed of marble [19]. The mining area showcases an abundance of folds and fault structures, with the NE–NNE trending thrust-fold belt serving as the primary structural framework [14,22]. Intense magmatic activity within the Fozichong ore field is marked by a diverse range of lithologies and complex lithofacies characteristics. The late Hercynian and Yanshanian (a geological event that occurred during the Mesozoic era, mainly affecting eastern China) intrusive acidic and intermediate-acidic magmatic rocks, such as biotite monzonitic granite and granodiorite, are dominant [14,23–26]. These rocks provide ideal subjects for studying and refining the genetic mechanisms and tectonic background of Late Mesozoic magmatic activity in South China.

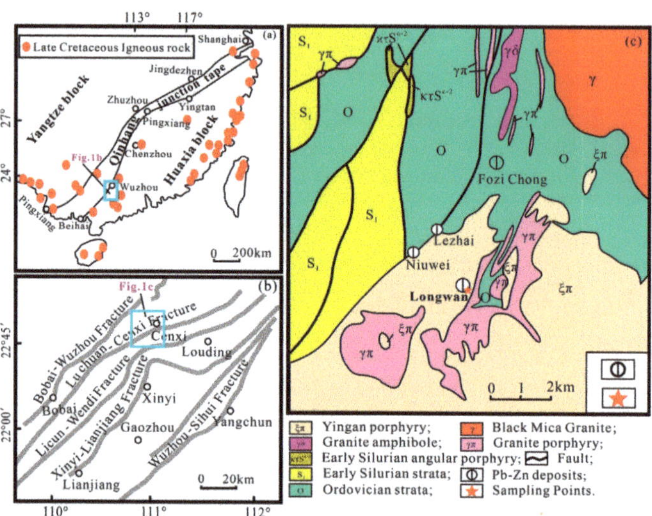

Figure 1. Geotectonic position (**a**) and regional structural domain map (**b,c**) of the Fozichong geological map from field mapping. (**a**) Modified after [27]; (**b**) modified after [27]; and (**c**) modified after [12].

2.2. Geological Characteristics of the Ore Deposit

The Longwan Pb–Zn deposit is situated within the Fozichong ore field, which hosts numerous Pb–Zn deposits, including the Fozichong, Niuwei, Longwan and Lezhai Pb–Zn deposits, and other mining sites.

The deposit is located at the southern margin of the Fozichong mining district and the southeastern block of the Niwei Fault. The Longwan mining area is relatively small, with the surface predominantly covered by extensive, thick volcanic lava layers. The exposed strata mainly consist of Lower Paleozoic Ordovician-Silurian systems. The ore bodies occur within the upper section of the Upper Ordovician Formation (O_3l^2) characterized by muddy sandstone, sandstone, and inter-bedded marble layers. The ore bodies primarily exhibit strata-like and lens-like morphologies [14]. The lithology mainly comprises low-grade metamorphic sandstone, mudstone, siltstone, and limestone [22]. The total exposed strata thickness is 2115 m, with a general strike direction of NNE 20° to 30° and a dip angle of 60° to 70°. The overall inclination direction is NWW. The mining area's structure is dominated by faults, exhibiting multidirectional and composite features [28], primarily controlled by the Xinsheng-Longwan backfold and the Fenghuangchong-Lingjiao (overturned) syncline. NE–NNE trending faults are the main ore-controlling structures, forming a distinct NNE-oriented structural belt [22]. The mining area has experienced frequent and long-lasting magmatic activity, resulting in the widespread distribution and diverse types of magmatic rocks, which mainly occur as stocks and dikelets. The granitic porphyry within the study area primarily intrudes the strata and earlier magmatic rocks in a vein-like manner, displaying grayish-white to grayish-black colors, porphyritic textures, and blocky structures (Figure 2). Phenocrysts mainly consist of quartz, feldspar, and hornblende, with grain sizes ranging from 2 to 8 mm and accounting for 30% to 45% of the total rock volume. Quartz and feldspar phenocrysts exhibit high euhedrality, with some feldspar phenocrysts partially altered by later hydrothermal events. The matrix predominantly comprises fine-grained textures, and accessory minerals include apatite and zircon.

Figure 2. Photographs and photomicrographs of Granitic Porphyry. (**a,b**) Longwan granite porphyry hand specimen; (**c,d**) microscopic photograph of granite porphyry. Qzt—Quartz, Amp—Amphibole, Pl—Plagioclase, Py—Pyrite.

The ore body's occurrence is generally consistent with the surrounding rock's occurrence and exhibits a strong association with intrusive bodies and veins. The ore body is characterized by significant thickness, high grades, and frequent limestone inclusions [28]. The mineral assemblage within the ore is relatively simple, with the primary metallic minerals being sphalerite, galena, pyrite, pyrrhotite, chalcopyrite, and magnetite. Non-metallic minerals include quartz, calcite, garnet, chlorite, and glauconite. The ore's structural types are relatively uncomplicated, encompassing euhedral to subhedral granular structures, xenomorphic structures, brecciated structures, replacement structures, and dissolution-residual structures. The ore textures mainly consist of vein-like, disseminated, and blocky structures (Figure 3). The alteration types of the surrounding rocks in the mining area primarily involve skarnization, silicification, and marblization.

Based on the analysis of metal and gangue mineral assemblages, structural features, and interrelationships, the ore deposit can be broadly divided into two mineralization stages and four mineralization phases (Figure 4). (1) Sedimentary-diagenetic stage (initial ore-bearing formation period): During the Early Paleozoic, southeastern Guangxi experienced a tensional fault tectonic environment, which led to the formation of a pyrite-bearing calcareous mudstone. Pyrite commonly exhibits euhedral crystals, predominantly with cubic faces, medium-to-coarse-grained structure, euhedral granular texture, and sparse disseminated structure, displaying sedimentary ore characteristics [29]. (2) Hydrothermal mineralization stage, further subdivided into four mineralization phases: ① Skarn stage: Characterized by weak mineralization, marly limestone underwent hydrothermal alteration, forming skarn-type minerals, such as diopside, tremolite, and garnet, accompanied by a small amount of quartz. ② Early sulfide stage: The primary ore body consists of metallic and gangue minerals. ③ Late sulfide stage: The metallic minerals are primarily pyrite, with minor amounts of galena and sphalerite, and the grains are coarser than those in the earlier stage. ④ Calcite stage: Thin carbonate veins are observed crosscutting the blocky ore, composed of calcite, and minor fine-grained pyrite.

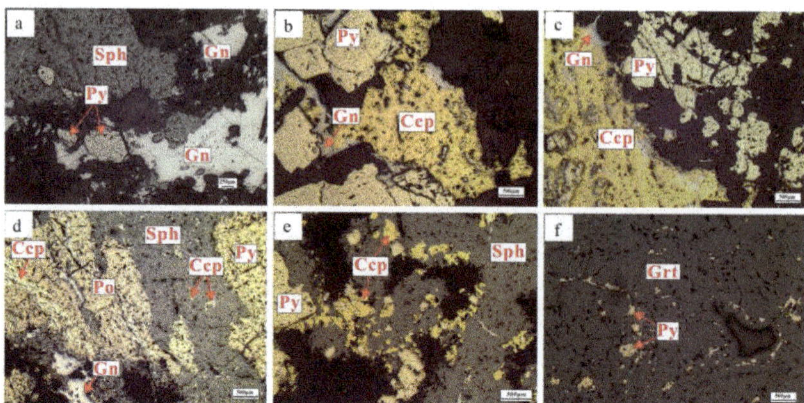

Figure 3. Reflected light photomicrographs of Longwan skarn deposit. (**a**) Galena replaces early pyrite and is later replaced by galena itself; (**b**) pyrite appears as euhedral to subhedral granular, while chalcopyrite is anhedral; (**c**) chalcopyrite and galena coexist, filling the spaces around euhedral pyrite and gangue minerals; (**d**) sphalerite and galena filled and replaced the early chalcopyrite, pyrite, and Pyrrhotite; (**e**) chalcopyrite and sphalerite coexist, filling the spaces around pyrite; (**f**) pyrite and chalcopyrite fill the gaps within garnet crystals. Sph—Sphalerite, Py—Pyrite, Gn—Galena, Ccp—Chalcopyrite, Grt—Garnet, Po—Pyrrhotite.

Mineralization period	Deposition period	Hydrothermal mineralization period			
Ore formation stage		Skarn stage	Early sulfide stage	Late Sulfide Stage	Calcite stage
Pyrite					
Galena					
Sphalerite					
Chalcopyrit					
Magnetic					
Translucent					
Tremolite					
Quartz					
Calcite					
Epidote					
Chlorite					

Figure 4. Main mineral formation sequence chart of the Longwan Pb–Zn deposit.

3. Sample Collection and Analytical Methods

3.1. Sample Collection and Description

The granitic porphyry sample (LW-16) utilized for zircon U–Pb dating was obtained from the middle section of 280 (at an elevation of 280 m), along the 60th exploration line. The porphyritic samples are located near the skarn and exhibit a distinct porphyritic texture, making them good representatives of granitic porphyry. This grayish-white rock primarily comprises quartz, plagioclase, and hornblende, with minor chlorite and actinolite as alteration minerals. Under cathodoluminescence, zircon grains within the granitic porphyry display well-defined crystal morphologies, mainly occurring in prismatic and short prismatic shapes. These grains are colorless and transparent and reveal prominent zonal structures in cathodoluminescence images.

Samples (LW-08 and LW-12) used for garnet major element analysis and U–Pb dating were also collected from the Longwan deposit along the 60th exploration line. Both samples represent coarse-grained magnetite–sphalerite garnet skarn (Figure 5) and are products of the alteration stage. The garnets are well-preserved and have not undergone significant post-depositional alteration. The mineral assemblage predominantly consists of garnet, sphalerite, galena, and magnetite. Garnet is yellowish-brown in color, and under microscopic examination, both LW-08 and LW-12 display inconspicuous zoning, lacking evident growth zones in their core regions. They exhibit notable alteration and fracturing, with minor oscillatory zoning along the edges, presenting euhedral to subhedral structures and grain sizes ranging from 1 to 2 cm. Metallic sulfides can be observed within garnet fractures and interstitial spaces between minerals, coexisting with magnetite. Additionally, garnet is altered by later-stage minerals such as diopside and actinolite (Figure 5).

Figure 5. Garnet stone hand specimen and reflected light photomicrographs of Longwan lead–zinc deposit. (**a**,**b**) Garnet specimens; (**c**) sphalerite replaced garnet; (**d**) sphalerite-replaced garnet; (**e**) galena occurs as veinlets filling the fractures within garnets; (**f**) pyrrhotite replacing sphalerite, with chalcopyrite scattered throughout the sphalerite. Sph–Sphalerite, Gn–Galena, Grt–Garnet, Ccp–Chalcopyrite, Po–Pyrrhotite.

3.2. Analytical Methods

Garnet samples for this study were prepared by directly cutting hand specimens into polished thin sections. Following identification under a microscope, representative, unaltered garnet samples devoid of inclusions were chosen for electron probe microanalysis (EPMA) and LA-ICP-MS U–Pb dating. EPMA of garnets' chemical composition was performed at the Guangxi Key Laboratory of Hidden Metallic Ore Deposits Exploration of the Guilin University of Technology using a JEOL JXA-8230 M Electron Microprobe Analyzer (Japan Electron Optics Laboratory, Tokyo, Japan). We utilized its WDS (wavelength dispersive X-ray spectrometer) function. The instrument settings included an accelerating voltage of 15 kV, a beam current of 20 nA, and a beam diameter of 5 μm. Characteristic peak measurement time for elements was set at 10 s with a background collection time of 5 s. Major elements analyzed comprised SiO_2, Al_2O_3, FeO, MgO, MnO, TiO_2, CaO, Na_2O, K_2O, and P_2O_5, with a detection limit of 0.01% for oxides [30]. Calibrations were conducted using silicate mineral and oxide standards from the American SPI company.

The garnet LA-ICP-MS U–Pb dating test analysis in this study was conducted on an Analytik-Jena PlasmaQuant MS quadrupole ICP-MS at Beijing Yandu Zhongzhi Test Technology Co., Ltd., using a 193 nm NWR193 Ar-F excimer laser. The MALI grandite (U–Pb TIMS = 202.0 $\pm$ 1.2 Ma) [31] was employed as the primary standard to calibrate the U–Pb geochronology of garnet. Instrument drift, mass bias, and fractionation of the U–Pb ratio were corrected using a standard-sample bracketing method. Trace element concentrations of garnet were quantified using SRM610 as an external standard and ^{29}Si as the internal standard element, assuming a stoichiometric garnet composition. Each analysis on the garnet began with a 15 s blank gas measurement, followed by an additional 40 s of analysis time when the laser was activated. The laser was operated at 10 Hz, with the beam size set at 55 µm and a density of approximately 3 J/cm^2. A flow of He carrier gas at a rate of 0.6 L/min carried particles ablated by the laser out of the chamber to be mixed with Ar gas and conveyed to the plasma torch. The isotopes measured included ^{27}Al, ^{29}Si, ^{44}Ca, ^{206}Pb, ^{207}Pb, ^{208}Pb, ^{232}Th, ^{235}U, and ^{238}U, with longer counting times for Pb isotopes compared to other elements. Raw data were corrected offline using ICP-MS-DataCal software (Version 10.9) [32,33] and ZSkits software [34]. Common Pb was corrected using the ^{207}Pb-based correction method outlined in reference [35]. A total of 29 sample points were analyzed for LW-08, with each analysis taking 3 min. The detailed analytical method adhered to the procedures outlined in [36]. Isoplot 4.15 was utilized to calculate U–Pb ages and obtain the lower intercept ages in the Tera-Wasserburg diagram, which can be used as the formation time of common-lead-bearing minerals [33,36].

Zircon selection and cathodoluminescence (CL) imaging were completed at Nanjing Hongchuang Geological Exploration Technology Service Co., Ltd. (Najing, China) and LA-ICP-MS U–Pb dating analysis was conducted at the Guangxi Key Laboratory of Hidden Metallic Ore Deposits Exploration of the Guilin University of Technology. The LA-ICP-MS system consisted of a laser and an Agilent 7900 ICP-MS instrument (Agilent, USA). The experimental parameters included a laser beam diameter of 32 µm, a laser frequency of 6 Hz, and an ablation time of 40 s. The standard zircon TEMORA (^{206}Pb/^{238}U age of 416.75 $\pm$ 0.24 Ma) [37] was utilized as an external standard for zircon U–Pb age analysis. The standard was measured every 5–6 samples to calibrate isotopic fractionation effects in zircon. The GJ-1 zircon standard was interspersed among test samples to verify the accuracy of the analytical method. During testing, the ^{206}Pb/^{238}U weighted average age for the GJ-1 zircon standard was 603.8 $\pm$ 7.3 Ma (2σ, MSWD = 0.51, n = 5) (MSWD-mean square of weighted deviates), consistent with the recommended value of 605.4 $\pm$ 3.0 Ma (2σ) [38] within the error range. Along with obtaining the U and Pb isotopic ratios, trace element data for zircon were also acquired simultaneously. Zircon trace element concentrations and U–Pb age corrections were performed using the ICPMSDataCal (Version 12.2) program developed by Liu Yongsheng at the China University of Geosciences [39]. Age calculations employed standard zircon TEMORA as an external standard for isotopic ratio fractionation correction, while ^{29}Si was utilized as an internal standard and NIST610 as an external standard for correcting zircon trace element concentrations. Due to the influence of radiogenic Pb isotopes, ^{206}Pb/^{238}U was used for ages < 1000 Ma. Most of the measured analysis points had ^{206}Pb/^{204}Pb >1000, and no common Pb correction was applied. ^{204}Pb was detected by an ion counter, and analysis points with abnormally high ^{204}Pb content, potentially influenced by inclusions or other common Pb sources, were excluded from calculations. Sample age data U–Pb concordia diagrams and age distribution diagrams were generated using Ludwig's Isoplot program.

4. Results

4.1. Major Element Compositions of Garnets

Major element analyses of garnet samples LW-08 and LW-12 from the Longwan Pb–Zn deposit are presented in Table 1. Garnets are primarily composed of andradite and grossular solid solution series (Figure 6). Both samples exhibit similar major element compositions. In sample LW-08, SiO$_2$ content ranges from 32.95% to 36.09%, CaO from

34.99% to 36.25%, Al_2O_3 from 0 to 0.92%, FeO from 26.75% to 30.01%, MnO from 1.06% to 1.39%, and MgO from 0% to 0.04%. The calculated end-member compositions for this garnet are predominantly andradite (96.13% to 100.11%) and grossular (6.51% to 16.09%), with minor contributions from spessartine (0.00% to 0.15%) and almandine (2.57% to 3.36%). In sample LW-12, SiO_2 content ranges from 33.53% to 35.60%, CaO from 35.26% to 36.47%, Al_2O_3 from 0.09% to 1.97%, FeO from 26.07% to 28.82%, MnO from 0.89% to 1.26%, and MgO from 0.01% to 0.16%. The calculated end-member compositions for this garnet are predominantly andradite (91.97% to 99.81%) and grossular (8.28% to 19.11%), with minor contributions from spessartine (0.07% to 0.68%) and almandine (2.25% to 3.10%). SiO_2 (32.95% to 36.09%) and CaO (34.99% to 36.47%) content ranges in all samples exhibit limited variation, while FeO (26.07% to 30.01%) and Al_2O_3 (0 to 1.97%) content show slight variation and display a negative correlation (Figure 7).

Table 1. Electron probe major elements data of representative garnet in the Longwan deposit.

Component	LW-08			LW-12		
	Max	Min	Average (n = 21)	Max	Min	Average (n = 10)
			wt.%			
SiO_2	36.09	32.95	34.81	35.60	33.53	34.70
Al_2O_3	0.92	0.00	0.19	1.97	0.09	0.58
FeO	30.01	26.75	28.34	28.82	26.07	27.83
MnO	1.39	1.06	1.19	1.26	0.89	1.10
MgO	0.04	0.00	0.02	0.16	0.01	0.09
CaO	36.25	34.99	35.51	36.47	35.26	35.85
Total	104.69	95.75	100.06	104.27	95.85	100.15
			Calculated from 12 oxygen atoms			
Si	3.18	3.07	3.15	3.16	3.07	3.12
Al	0.10	0.00	0.02	0.21	0.01	0.07
Ti	0.01	0.00	0.00	0.01	0.00	0.00
Fe^{3+}	2.50	2.22	2.29	2.41	2.13	2.28
Fe^{2+}	2.22	2.06	2.14	2.17	1.95	2.09
Mn	0.11	0.08	0.09	0.10	0.07	0.08
Mg	0.00	0.00	0.00	0.02	0.00	0.01
Ca	3.56	3.37	3.46	3.46	3.45	3.46
Total	11.68	10.08	11.15	11.53	10.68	11.12
Andradite	100.11	96.13	99.15	99.81	91.97	97.61
Grossular	16.09	6.51	10.17	19.11	8.28	12.92
Pyrope	0.15	0.00	0.08	0.68	0.07	0.42
Spessartine	3.36	2.57	2.90	3.10	2.25	2.64

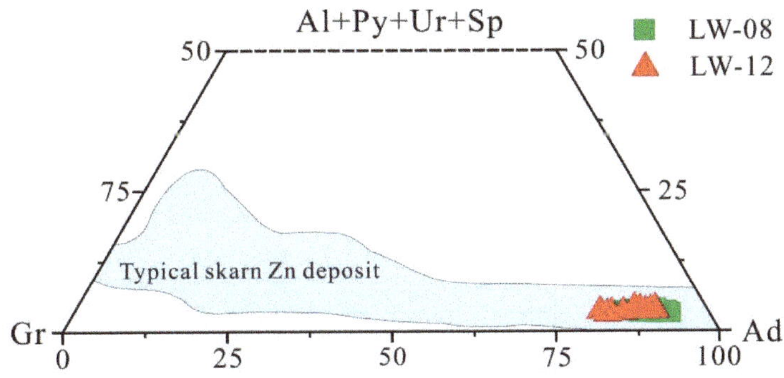

Figure 6. Triangular distribution of garnet samples in the Longwan Pb–Zn deposit (Modified after [40]).

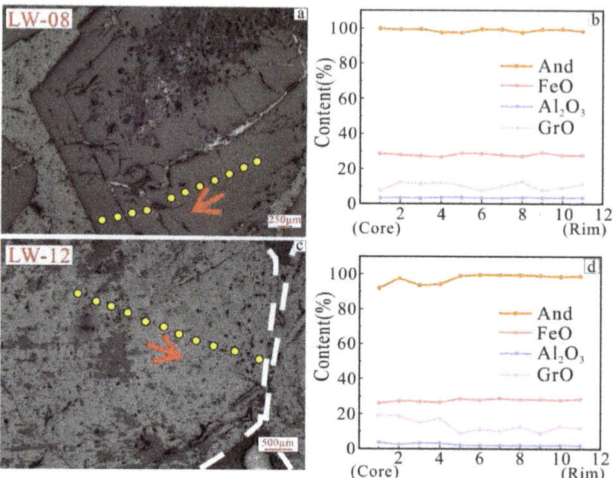

Figure 7. Relative proportions of Adr–Gro and Al_2O_3–TFeO profiles of the garnet particle composition in the ring zone of the Longwan Pb–Zn deposit. (**a,b**) the variation of the ring-band composition of garnet particles in LW-08; (**c,d**) the variation of the ring-band composition of garnet particles in LW-12. Additionally, Gro are abbreviations for Andalusite and Grossular.

4.2. LA-ICP-MS U–Pb Isotope Analysis Results of Garnet

Garnets from the Longwan Pb–Zn deposit, as revealed by backscattered images and electron probe analyses, are brown minerals with fracture structures in samples LW-08 and LW-12. Metallic sulfides are observed within garnet fractures and interstitial spaces between minerals. The garnets selected for LA-ICP-MS U–Pb dating mainly come from the LW-08 skarn sample, which features distinct zoning at the edges. The experimental data show that U and Th concentrations in garnets vary considerably, with U concentrations ranging from 0.49 ppm to 3.34 ppm and Th concentrations from 0 to 0.037 ppm (Table 2). The average U concentration in garnets is 1.56×10^{-6}, with andradite as the dominant component followed by grossular. The $^{207}Pb/^{206}Pb$ and $^{238}U/^{206}Pb$ isotope ratios of the samples are plotted on a Concordia diagram, with 29 data points distributed evenly on or near the Concordia curve (Figure 8). The lower intercept age is determined to be 102.4 ± 2.0 Ma (MSWD = 0.74, $n = 29$).

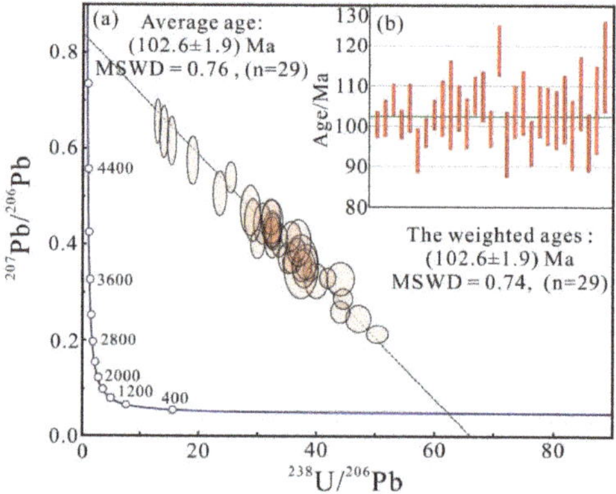

Figure 8. U–Pb age of garnet in skarn of the Longwan Pb–Zn deposit. (**a**) Tera-Wasserburg diagram and (**b**) weighted average diagram of ^{207}Pb-corrected $^{206}Pb/^{238}U$ age for Longwan garnet.

Table 2. Garnet LA-ICP-MS U–Pb isotope analysis results of the Longwan Pb–Zn Deposit.

Spot No.	ppm		Isotope Ratio						Age/Ma					
	Th	U	$^{207}Pb/^{206}Pb$	1σ	$^{207}Pb/^{235}U$	1σ	$^{206}Pb/^{238}U$	1σ	$^{207}Pb/^{206}Pb$	1σ	$^{207}Pb/^{235}U$	1σ	$^{206}Pb/^{238}U$	1σ
LM-08-01	0.002	1.41	0.325	0.02	1.12	0.060	0.03	0.0008	3591.4	113	764.50	29	159.80	5
LM-08-02	0.002	0.88	0.471	0.03	2.27	0.114	0.03	0.0014	4154.2	110	1204.30-	36	221.70	9
LM-08-05	0.001	1.33	0.373	0.02	1.46	0.080	0.03	0.0009	3801.6	86	914.70	33	180.90	6
LM-08-06	0.002	1.38	0.426	0.02	1.82	0.068	0.03	0.0009	4004.3	77	1054.40	24	197.10	5
LM-08-07	0.002	1.40	0.428	0.03	1.82	0.075	0.03	0.0008	4010.7	90	1052.30	27	195.70	5
LM-08-08	0.002	1.40	0.401	0.02	1.85	0.078	0.03	0.0009	3912.2	73	1064.50	28	212.60	6
LM-08-09	0.001	1.45	0.433	0.03	1.85	0.079	0.03	0.0010	4029	86	1063.10	28	196.50	6
LM-08-10	0.001	1.41	0.418	0.02	1.62	0.073	0.03	0.0009	3976.2	73	979.10	28	178.90	5
LM-08-11	0.002	1.38	0.329	0.02	1.03	0.055	0.02	0.0008	3612.6	106	720.40	27	145.10	5
LM-08-13	0.001	1.35	0.410	0.02	1.69	0.074	0.03	0.0008	3944.7	85	1003.20	28	189.60	5
LM-08-14	0.001	1.36	0.364	0.02	1.32	0.069	0.03	0.0009	3768.1	102	853.70	30	167.00	5
LM-08-16	0.002	0.96	0.354	0.04	1.31	0.124	0.03	0.0014	3726.2	188	851.10	54	170.80	9
LM-08-17	0.000	0.99	0.404	0.03	1.52	0.092	0.03	0.0010	3925.5	121	936.70	37	172.90	7
LM-08-18	0.006	1.33	0.452	0.03	2.15	0.105	0.03	0.0011	4091.1	97	1165.40	34	218.80	7
LM-08-19	0.001	1.35	0.351	0.03	1.28	0.069	0.03	0.0009	3712.4	117	837.90	31	168.50	6
LM-08-21	0.002	1.08	0.416	0.03	1.77	0.089	0.03	0.0009	3967.7	102	1035.90	33	196.40	6
LM-08-23	0.003	1.14	0.449	0.03	1.92	0.089	0.03	0.0012	4080.8	101	1086.90	31	196.70	7
LM-08-28	0.028	0.78	0.444	0.03	2.00	0.105	0.03	0.0011	4066.5	95	1115.30	36	207.00	7
LM-08-39	0.013	0.77	0.508	0.03	3.00	0.136	0.04	0.0014	4263.2	87	1406.50	35	270.20	9
LM-08-40	0.008	3.34	0.364	0.01	1.40	0.060	0.03	0.0008	3765.2	62	890.40	25	178.00	5
LM-08-41	0.021	2.92	0.246	0.02	0.72	0.054	0.02	0.0006	3159.4	120	551.60	32	135.70	4
LM-08-42	0.014	3.02	0.288	0.01	0.89	0.035	0.02	0.0005	3404.8	75	648.20	19	143.60	3
LM-08-43	0.015	1.87	0.458	0.02	1.97	0.107	0.03	0.0010	4111.6	79	1103.80	37	197.60	7
LM-08-44	0.037	2.39	0.260	0.01	0.82	0.035	0.02	0.0006	3248	90	606.40	20	145.10	4
LM-08-45	0.002	3.26	0.345	0.01	1.22	0.047	0.03	0.0005	3685.8	60	811.60	22	163.70	3
LM-08-46	0.005	1.66	0.356	0.02	1.32	0.067	0.03	0.0008	3734.4	91	852.50	29	170.30	5
LM-08-47	0.002	2.16	0.389	0.02	1.46	0.057	0.03	0.0007	3867.6	62	915.80	23	173.60	4
LM 08 49	0.008	3.05	0.331	0.01	1.09	0.041	0.02	0.0005	3621.1	65	749.30	20	152.40	3
LM-08-50	0.016	2.94	0.214	0.01	0.59	0.031	0.02	0.0005	2936.6	92	469.00	20	127.00	3

4.3. LA-ICP-MS U–Pb Dating Results of Zircons

Zircon crystals in the granitic porphyry of the Longwan Pb–Zn deposit are well-formed, mainly exhibiting elongated and short prismatic shapes, and are colorless and transparent. Cathodoluminescence images show distinct zoning in the zircons. The LA-ICP-MS U–Pb dating results for zircons (Table 3) reveal U concentrations ranging from 950 ppm to 1760 ppm and Th concentrations from 179.1 ppm to 949.6 ppm. The Th/U ratios of zircons mainly vary between 0.17 and 0.61 ppm (average 0.40). Excluding a few samples, the Th/U ratios are all greater than 0.3, indicating a magmatic origin for the zircons. In this study, zircons from the granitic porphyry in close contact with the skarn were obtained through 20 data points. These points are all located on or near the Concordia curve on the Concordia diagram, yielding a weighted average age of 102.1 ± 1.2 Ma (MSWD = 0.26, n = 20) (Figure 9).

Table 3. LA-ICP-MS U–Pb zircon analysis data of granite-porphyry in the Longwan Pb–Zn Deposit.

Spot No.	ppm		Th/U	Isotope Ratio						Age/Ma					
	Th	U		$^{207}Pb/^{206}Pb$	1σ	$^{207}Pb/^{235}U$	1σ	$^{206}Pb/^{238}U$	1σ	$^{207}Pb/^{206}Pb$	1σ	$^{207}Pb/^{235}U$	1σ	$^{206}Pb/^{238}U$	1σ
LW-16-01	794.54	1645.89	0.48	0.0521	0.0027	0.1148	0.0062	0.0159	0.0004	300	149	110	6	102	2
LW-16-03	466.87	1273.65	0.37	0.0470	0.0023	0.1010	0.0048	0.0158	0.0003	56	111	98	4	101	2
LW-16-04	613.44	1760.37	0.35	0.0492	0.0021	0.1102	0.0054	0.0162	0.0004	167	98	106	5	104	3
LW-16-05	599.73	1556.85	0.39	0.0492	0.0021	0.1075	0.0048	0.0161	0.0004	167	102	104	4	103	3
LW-16-07	323.25	1055.83	0.31	0.0473	0.0029	0.1030	0.0070	0.0156	0.0004	65	146	100	6	100	3
LW-16-09	484.09	1356.38	0.36	0.0481	0.0023	0.1088	0.0063	0.0162	0.0006	106	107	105	6	104	4
LW-16-11	838.38	1608.91	0.52	0.0472	0.0021	0.1021	0.0054	0.0160	0.0006	58	104	99	5	102	4
LW-16-12	386.89	1274.58	0.30	0.0463	0.0020	0.1014	0.0050	0.0163	0.0006	9	113	98	5	104	4
LW-16-14	688.09	1760.77	0.39	0.0491	0.0021	0.1075	0.0050	0.0157	0.0003	154	103	104	5	100	2
LW-16-16	561.01	1685.58	0.33	0.0479	0.0023	0.1068	0.0053	0.0161	0.0004	98	107	103	5	103	2
LW-16-18	605.52	1100.01	0.55	0.0477	0.0026	0.1015	0.0054	0.0160	0.0005	83	126	98	5	102	3
LW-16-19	652.70	1654.61	0.39	0.0472	0.0022	0.1003	0.0047	0.0156	0.0004	58	107	97	4	100	2
LW-16-20	179.10	1074.91	0.17	0.0439	0.0021	0.0966	0.0054	0.0160	0.0005	-	-	94	5	102	3
LW-16-21	581.77	950.49	0.61	0.0489	0.0031	0.1064	0.0069	0.0162	0.0006	146	141	103	6	104	4
LW-16-22	568.28	1246.50	0.46	0.0470	0.0025	0.1033	0.0063	0.0162	0.0005	56	113	100	6	103	3
LW-16-23	236.28	995.77	0.24	0.0478	0.0027	0.1050	0.0062	0.0161	0.0006	100	120	101	6	103	4
LW-16-24	641.59	1702.33	0.38	0.0505	0.0023	0.1093	0.0055	0.0160	0.0006	220	107	105	5	102	4
LW-16-25	394.91	841.16	0.47	0.0545	0.0033	0.1173	0.0068	0.0163	0.0006	391	135	113	6	104	4
LW-16-27	399.93	1259.69	0.32	0.0464	0.0023	0.0979	0.0048	0.0159	0.0004	17	115	95	4	102	3
LW-16-31	949.58	1544.68	0.61	0.0510	0.0028	0.1105	0.0058	0.0161	0.0003	243	126	106	5	103	2

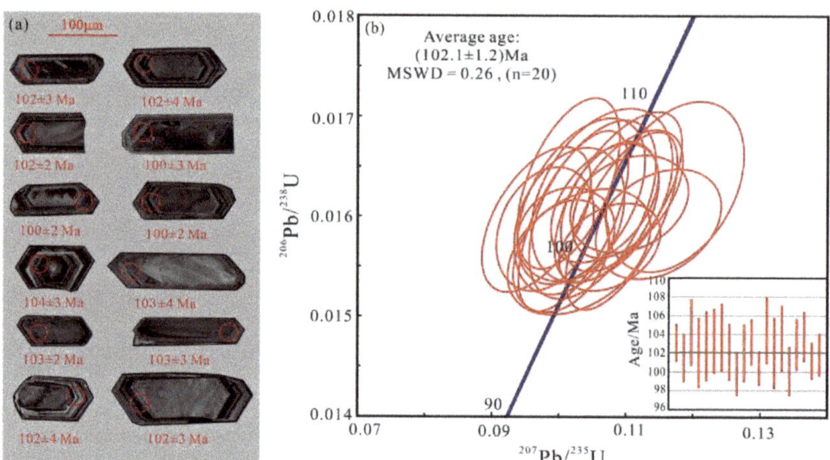

Figure 9. Zircon U–Pb ages of granitic porphyry in the Longwan Pb–Zn deposit. (**a**) Representative cathodoluminescent images of analyzed zircon spots and (**b**) Zircon U–Pb Concordia diagram for granitic porphyry, also shown is the weighted mean ages.

5. Discussion

5.1. Formation Age of the Longwan Pb–Zn Deposit

The Longwan Pb–Zn deposit is an important deposit within the Fozichong mining district, with great exploration potential. Previous studies have carried out K-Ar and Rb-Sr dating on Longwan granites (Table 4); however, these methods can be easily influenced by subsequent magmatic and tectonic activities, possibly affecting dating results [39]. Zircon U–Pb dating from previous research suggests an age of 104.2 Ma for the Longwan monzogranite porphyry. However, in the skarn deposits, intrusive bodies often do not spatially contact the skarns, rendering the age of the intrusion insufficient for directly indicating the mineralization age. Garnet is a common alteration mineral in skarn deposits, characterized by high closure temperature (>850 °C), stability, and limited influence by subsequent magmatic-hydrothermal and tectonic activities [5]. Garnet can better constrain the timing of metamorphic events and subsequently the formation age of skarn deposits, making it an ideal mineral for dating skarn deposits.

Table 4. Age table of the Longwan Pb–Zn deposit.

Minerals	Methods	Age	Source
Potassium feldspar in the Longwan rock mass	K-Ar	75.5 Ma	[23]
Longwan Erchang granite porphyry whole rock	Rb-Sr isochrones	127.6 Ma	[29]
Longwan Erchang granite porphyry	LA-ICP-MS U–Pb	104.2 ± 1.5 Ma	[13]
Guyi Erchang granite porphyry	LA-ICP-MS U–Pb	105.1 ± 1.7 Ma	[13]
Hesan Second-length granite porphyry	LA-ICP-MS U–Pb	105.2 ± 0.5 Ma	[13]
Longwan granite porphyry	LA-ICP-MS U–Pb	102.1 ± 1.2 Ma	This study
Longwan Garnet	LA-ICP-MS U–Pb	102.6 ± 1.9 Ma	This study

Garnet can be classified into grossular ($Ca_3Al_2Si_3O_{12}$), andradite ($Ca_3Fe_2Si_3O_{12}$), almandine ($Fe_3Al_2Si_3O_{12}$), pyrope ($Mg_3Al_2Si_3O_{12}$), and spessartine ($Mn_3Al_2Si_3O_{12}$), based on differences in primary chemical compositions. The most commonly occurring garnet in skarn deposits is the solid solution of grossular and andradite. In the Longwan Pb–Zn deposit, the main garnet series is andradite-grossular (Figure 6). Ore minerals such as galena, sphalerite, and chalcopyrite in the Longwan Pb–Zn deposit fill the interiors or interstices between garnet mineral grains (Figure 3f), indicating a close genetic relationship and formed during the same fluid evolution process. Experimental studies have shown that only grossular garnets possess a highly homogeneous U distribution, and they contain higher U concentrations than other garnet types [41]. U–Pb isotope analysis results show that the LW-08 garnet samples possess high U concentrations (0.49 ppm to 3.34 ppm), meeting the requirements for garnet U–Pb isotope dating methods. Consequently, the U–Pb isotope dating results from sample LW-08, collected from the skarn within the Longwan Pb–Zn deposit, can represent the mineralization age of the deposit.

The obtained garnet U–Pb age of 102.6 ± 1.9 Ma in this study is consistent with the zircon U–Pb age of 102.1 ± 1.2 Ma from the granitic porphyry within the Longwan deposit and the previously reported zircon U–Pb age of 104.2 ± 1.5 Ma for the Longwan monzogranite porphyry [12]. This consistency suggests that the garnet U–Pb age obtained in this study is reliable and represents the mineralization age of the Longwan Pb–Zn deposit. In recent years, the method of using garnet for geochronological studies of skarn deposits has gained increasing recognition [42,43] and has been more widely applied [9,11], particularly in areas with multiple episodes of skarn activity and a lack of exposed intrusive bodies, where it has demonstrated good application results [10]. During the Late Early Cretaceous, the Longwan Pb–Zn deposit formed through contact metasomatic reactions that occurred during the intrusion of granitic porphyry into the Silurian limestone. The ore-forming process was a continuous event in the temporal scale, closely associated with magmatic evolution. Therefore, this study provides important geochronological evidence for determining the mineralization age and genesis of the Longwan Pb–Zn deposit.

5.2. Tectonic Setting of the Longwan Pb–Zn Deposit

The extensive distribution and diverse types of Cretaceous magmatic rocks in southeastern Guangxi establish this region as an ideal location for investigating the genesis and tectonic setting of Late Mesozoic magmatic activities in South China. The area lies within the transitional zone between the South China Pacific tectonic domain and the Tethys tectonic domain, situated at the southwestern end of the significant multi-metallic mineralization belt of South China, the Qin-Hang suture belt. Late Mesozoic volcanic rocks, predominantly from the Yanshanian period, are exposed in this region. These magmatic rocks are primarily acidic to intermediate in composition, encompassing biotite granites, granitic porphyries, and granite-porphyries, among others. The zircon and garnet U–Pb ages obtained in this study for the Longwan granitic porphyry and skarn are 102.1 ± 1.2 Ma and 102.6 ± 1.9 Ma, respectively, representing the formation age of the Longwan Pb–Zn deposit. These ages are consistent with the formation times of other intrusions in the southeastern Guangxi region, such as the Xinzhoutang intrusion (98.5~100.5 Ma), Daye volcanic rocks (99.2 Ma) [44], Liuwang quartz porphyry (98.72 ± 0.64 Ma) [15], Luchuan monzogranite (107.6 ± 1.2 Ma) [45], and Guantian granite (98 Ma) [46]. All these rock bodies formed during the Late Early Cretaceous to Early Late Cretaceous period (<110 Ma).

Previous research indicates that during the Late Cretaceous, the Pacific Plate underwent northwestward subduction. Some scholars argue that, between the Early and Late Cretaceous transition (107–86 Ma), the South China Block experienced an extensional phase due to the southeastward rollback of the Pacific Plate [47–49]. The magmatic rocks' ages reveal a progressively younger trend towards the southeast coast, where the magmatic belt exhibits a NE orientation [24]. Consequently, the genesis of the NE-trending magmatic rocks along the southeast coast is thought to be associated with the southeastward rollback of the Pacific Plate [50,51]. The Longwan Pb–Zn deposit is situated relatively far from the NE-trending magmatic belt along the southeast coast, suggesting it may not be influenced by the Pacific tectonic domain. In southern South China, an EW-trending magmatic belt spans Guangdong and Guangxi provinces, extending westward to Yunnan. The ages of these rocks predominantly range from 110 to 80 Ma, with no apparent trend of decreasing age towards the southeast coast [50–52]. Numerous magmatic rocks, including the Longwan granitic porphyry with ages younger than 110 Ma, are positioned within the EW-trending magmatic belt in southern South China. Observations of the spatial distribution of

Cretaceous magmatic rocks in the South China Block reveal that the EW-trending magmatic belt from the Middle-Late Cretaceous (younger than 110 Ma) intersects at a large angle with the contemporaneous NE-trending magmatic belt along the southeast coast [51]. The South China Block's tectonic environment during the Cretaceous underwent multiple episodes of extension and compression [53–59]. Based on the reconstruction of the Pacific Plate's drift history, earlier researchers have proposed that the genesis of the EW-trending granites in South China during the Late Cretaceous is related to the northward subduction of the Neo-Tethys Ocean [51]. They have proposed a model involving northward subduction of the Neo-Tethys plate, ridge subduction, and slab rollback [51,60]. Other scholars have also suggested that the formation of granites and andesites in southeastern Guangxi around 100 Ma is connected to the northward subduction of the Neo-Tethys plate [45,59,61]. The Shilu intrusion (103.9 Ma) within the Yangchun Basin displays adakitic features and may have been formed through partial melting of the slab during the northward subduction of the Neo-Tethys Ocean [51]. Sr-Nd isotopic analysis of the monzogranitic porphyries from the Longwan Pb–Zn deposit has demonstrated that their source region is crustal [14], differing from the Shilu intrusion's origin and not being a product of oceanic crust partial melting.

At approximately 100 Ma, magmatic activity was present from the northeast to the southwest along the southern segment of the Qin-Hang metallogenic belt. In the Huaiji Basin, located about 100 km northeast of Longwan, 102 Ma volcanic rocks have been identified [62]. The Fozichong granitic porphyry near Longwan has a formation age of 100 Ma, the Jinzhu dacite is dated at 98.1 Ma, the Liuwang quartz porphyry 98 Ma is situated 50 km southwest of Longwan [16], and the Xinzhoutang andesite, 150 km southwest of Longwan, has an age range of 98.5–100.5 Ma [47]. These magmatic rocks are predominantly distributed along a NE–SW axis along the Cenxi-Bobai Fault, suggesting extensive magmatic activity around 100 Ma during the mid-Cretaceous along this fault. Evidence such as the deceleration of the Indian Plate's drift rate, the emergence of adakitic rocks, and plate reconstructions point to the Neo-Tethys ridge subducting beneath southern South China around 100 Ma [51,60]. With the increase in subduction angle, slab rollback transpired, establishing an NS-trending extensional tectonic regime in southern South China. This incited the activation of the southern section of the Qin-Hang metallogenic belt, giving rise to a series of magmatic events. Concurrently, as the slab retreated southward, a pattern of progressively younger magmatic evolution from north to south materialized, aligning with the age variation trend of volcanic rocks from the Huaiji Basin (102 Ma) to the Daban Basin (94 Ma) and the Shuiwen Basin (82 Ma) [62]. The Longwan Pb–Zn deposit is situated within the Cenxi-Bobai Fault zone, in the southwestern segment of the Qin-Hang metallogenic belt [16,63]. The Late Cretaceous granitic porphyry associated with the Longwan Pb–Zn deposit may have formed during the reactivation of the NE-trending Cenxi-Bobai Fault under the context of the Neo-Tethys plate rollback. This process facilitated the generation of a series of NE-trending Pb–Zn deposits in the southwestern section of the Qin-Hang metallogenic belt, including the Longwan, Jilongding, Longjing, and Dongtao Pb–Zn deposits (Figure 10).

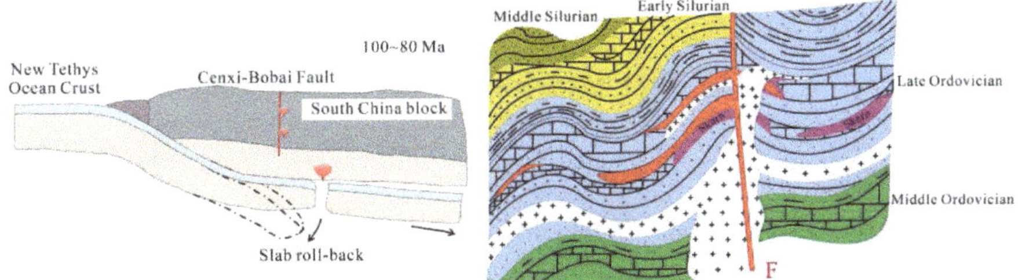

Figure 10. Conceptual diagram illustrating the tectonic and ore-forming evolution of the Longwan lead–zinc deposit during the early Late Cretaceous (~100–80 Ma).

6. Conclusions

By conducting U–Pb dating on garnet from skarn within the Longwan Pb–Zn deposit and zircon from granitic porphyry, we obtained a garnet U–Pb age of 102.6 ± 1.9 Ma, representing the mineralization epoch of the Longwan Pb–Zn deposit. This age signifies that the formation of the Longwan Pb–Zn deposit involved a continuous process of magmatic evolution and mineralization. The consistency between U–Pb ages of garnet in the skarn and zircon in the granitic porphyry validates the accuracy of garnet dating in constraining the mineralization age of skarn-type deposits. Considering previous research, this study suggests that the geodynamic setting of the Longwan Pb–Zn deposit is associated with the northward subduction of the Neo-Tethys Ocean. The extensive magmatic activity along the southern segment of the Qin-Hang metallogenic belt around 100 Ma during the mid-Cretaceous correlates with the geological process of the Neo-Tethys ridge subducting beneath southern South China at approximately 100 Ma, followed by the subsequent slab rollback, which led to an extensional tectonic background.

Author Contributions: Investigation, formal analysis, writing—original draft, X.Z.; methodology, supervision, writing—review and editing, W.D.; writing—review and editing, L.M., W.F., X.L. and S.L. All authors have read and agreed to the published version of the manuscript.

Funding: This research was funded by the National Natural Science Foundation of China (92162218, 42262026), the Guangxi Natural Science Foundation (2022GXNSFBA035588), and the Guangxi Science Innovation Base Construction Foundation (GuikeZY21195031).

Data Availability Statement: The data presented in this study are available on reasonable request from the corresponding authors.

Acknowledgments: We express our sincere gratitude to the geologists of the Longwan Pb–Zn Deposit for their valuable support and assistance during the fieldwork. Special thanks go to Qijun Yang and Jinbao Yang from Guilin University of Technology for their insightful comments and suggestions on the initial manuscript. We are also grateful to Hongxia Yu and Zhenglin Li from the Guangxi Key Laboratory of Hidden Metallic Ore Deposits Exploration at Guilin University of Technology for their expert guidance and help during the experimental process. Finally, we are thankful for the constructive feedback provided by two anonymous reviewers, which has greatly improved the quality of our manuscript. This is a contribution to Guangxi Key Mineral Resources Deep Exploration Talent Highland.

Conflicts of Interest: The authors declare no conflict of interest.

References

1. Zhang, Y.; Shao, Y.J.; Zhang, R.Q.; Li, D.F.; Liu, Z.F.; Chen, H.Y. Dating ore deposit using garnet U–Pb geochronology: Example from the Xinqiao Cu-S-Fe-Au deposit, Eastern China. *Minerals* **2018**, *8*, 31. [CrossRef]
2. Jamtveit, B.; Ragnarsdottir, K.V.; Wood, B.J. On the origin of zoned grossular-andradite garnets in hydrothermal systems. *Eur. J.* **1995**, *7*, 1399–1410. [CrossRef]
3. Gaspar, M.; Knaack, C.; Meinert, L.D.; Moretti, R. REE in skarn systems: A LA-ICP-MS study of garnets from the Crown Jewel gold deposit. *Geochim. Cosmochim. Acta* **2008**, *72*, 185–205. [CrossRef]

4. Baghban, S.; Hosseinzadeh, M.R.; Moayyed, M.; Mokhtari, M.A.A.; Gregory, D.D.; Mahmou-di, N.H. Chemical composition andevolution of the garnets in the Astamal Fe-LREE distal skarn deposit, Qara-Dagh-Sabalan metallogenic belt, Lesser Caucasus, NW Iran. *Ore Geol. Rev.* **2016**, *78*, 166–175. [CrossRef]

5. Mezger, K.; Hanson, G.N.; Bohlen, S.R. U–Pb systematics of garnet: Dating the growth of garnet in the late Archean Pikwitonei granulite domain at Cauchon and Natawahunan Lakes, Manitoba, Canada. *Contrib. Miner. Petrol.* **1989**, *101*, 136–148. [CrossRef]

6. Gevedon, M.; Semen, S.; Barnes, J.D.; Lackey, J.S.; Stockli, D.F. Unraveling histories of hydrothermal systems via U–Pb laser ablation dating of skarn garnet. Earth Planet. Sci. Lett. **2018**, *498*, 237–246.

7. Li, D.F.; Tan, C.Y.; Miao, F.Y.; Liu, Q.F.; Zhang, Y.; Sun, X.M. Initiation of Zn-Pb mineralization in the Pingbao Pb–Zn skarn district, South China: Constraints from U–Pb dating of grossular-rich garnet. *Ore Geol. Rev.* **2019**, *107*, 587–599. [CrossRef]

8. Tang, Y.W.; Gao, J.F.; Lan, T.G.; Cui, K.; Han, J.J.; Zhang, X.; Chen, Y.W.; Chen, Y.H. In situ low-U garnet U–Pb dating by LA-SF-ICP-MS and its application in constraining the origin of Anji skarn system combined with Ar-Ar dating and Pb isotopes. *Ore Geol. Rev.* **2021**, *130*, 103970. [CrossRef]

9. Deng, X.D.; Li, J.W.; Luo, T.; Wang, H.Q. Dating magmatic and hydrothermal processes using andradite-rich garnet U–Pb geochronometry. *Contrib. Miner. Petrol.* **2017**, *172*, 71–82. [CrossRef]

10. Reinhardt, N.; Gerdes, A.; Beranoaguirre, A.; Frenzel, M.; Meinert, L.D.; Gutzmer, J.; Burisch, M. Timing of magmatic-hydrothermal activity in the Variscan Orogenic Belt: LA-ICP-MS U–Pb geochronology of skarn-related garnet from the Schwarzenberg District, Erzgebirge. *Miner. Depos.* **2022**, *57*, 1071–1087. [CrossRef]

11. Zang, Z.J.; Dong, L.L.; Liu, W.; Zhao, H.; Wang, X.S.; Cai, K.D.; Wan, B. Garnet U–Pb and O isotopic deter-minations reveal a shear-zone induced hydrothermal system. *Sci. Rep.* **2019**, *9*, 10382. [CrossRef] [PubMed]

12. Yang, Q.J.; Qin, Y.; Wang, T.S.; Zhang, Q.W. Geochronology and Geochemistry Characteristics of Monzo-granitic porphyry from Fozichong ore field Guangxi Province and their geological implication. *J. Jilin Univ. (Earth Sci. Ed.)* **2017**, *47*, 760–774, (In Chinese with English Abstract).

13. Maleki, S.; Alirezaei, S.; Corfuc, F. Dating of Oligocene granitoids in the KhakSorkh area, Central Urumieh-Dokhtar arc, Iran, and a genetic linkage with the associated skarn iron deposit. *J. Asian Earth Sci.* **2019**, *182*, 103930. [CrossRef]

14. Fu, W.; Chai, M.C.; Yang, Q.J.; Wei, L.M.; Huang, X.R.; Feng, J.P. Genesis of the Fozichong Pb–Zn polyme-tallic deposit: Constraints from fluid inclusions and H-O-S-Pb isotopic evidences. *Acta Petrol. Sin.* **2013**, *29*, 4136–4150, (In Chinese with English Abstract).

15. Liu, X.Y. Petrogenesis of Late Mesozoic Granites in Southeastern Guangxi and Their Geodynamic Implications. Master's Thesis, Guiling University of Technology, Guilin, China, 2022.

16. Yu, P.P.; Zheng, Y.; Huang, X.; Wang, C.M. Stratabound skarn Pb–Zn mineralization in the Yunkai Domain (South China): The Fozichong case. *Ore Geol. Rev.* **2020**, *125*, 103673. [CrossRef]

17. Yu, P.P.; Zheng, Y.; Zhou, Y.Z.; Chen, B.H.; Niu, J.; Yang, W. Zircon U–Pb geochronology and geochemistry of the metabasite and gabbro: Implications for the Neoproterozoic and Paleozoic tectonic settings of the Qinzhou Bay-Hangzhou Bay suture zone, South China. *Geol. J.* **2018**, *53*, 2219–2239. [CrossRef]

18. Wang, Y.J.; Fan, W.M.; Zhang, G.W.; Zhang, Y.H. Phanerozoic tectonics of the South ChinaBlock: Key observations and controversies. *Gondwana Res.* **2013**, *23*, 1273–1305. [CrossRef]

19. Cheng, S.B.; Fu, J.M.; Ma, L.Y.; Chen, X.Q.; Lu, Y.Y. Zircon SHRIMP U–Pb Dating of Dachong Granodioritic Stock and Its geological Significancal in Fozichong Deposit, Guangxi Province. Geol. Miner. *Resour. South China* **2012**, *28*, 315–320, (In Chinese with English Abstract).

20. Hu, R.Z.; Zhou, M.F. Multiple Mesozoic mineralization events in South China: An introduction to the thematic issue. *Miner. Depos.* **2012**, *47*, 579–588. [CrossRef]

21. Mao, J.W.; Cheng, Y.B.; Chen, M.H.; Franco, P. Major types and time-space distribution of Mesozoic ore deposits in South China and their geodynamic settings. *Miner. Depos.* **2013**, *48*, 267–294.

22. Lei, L.Q. Age and Geochemical character of magmatites in Fozichong Pb–Zn (Ag) ore field, Guangxi, China. *Acta Petrol. Sin.* **1995**, *11*, 77–82, (In Chinese with English Abstract).

23. Wang, C.M.; Deng, J.; Carraza, E.J.M.; Lai, X.R. Nature, diversity and temporalspatial distributions of sediment-hosted Pb–Zn deposits in China. *Ore Geol. Rev.* **2014**, *56*, 327–351. [CrossRef]

24. Li, X.H.; Zhou, H.W.; Liu, Y.; Lee, C.Y.; Sun, M.; Chen, C.H. Shoshonitic intrusive suite in SE Guangxi: Petrology and geochemistry. *Chin. Sci. Bull.* **2000**, *45*, 653–659. [CrossRef]

25. Zheng, W.; Mao, J.W.; Pirajno, F.; Zhao, H.J.; Zhao, C.S.; Mao, Z.H.; Wang, Y.J. Geochronology and geochemistry of the Shilu Cu-Mo deposit in the Yunkai area, Guangdong province, South China and its implication. *Ore Geol. Rev.* **2015**, *67*, 382–398. [CrossRef]

26. Wang, Y.J.; Fan, W.M.; Cawood, P.A.; Ji, S.C.; Peng, T.P.; Chen, X.Y. Indosinian high-strain deformation for the Yunkaidashan tectonic belt, South China: Kinematics and $^{40}Ar/^{39}Ar$ geochronological constraints. *Tectonics* **2007**, *26*, 1–21. [CrossRef]

27. Liu, X.Y. The Subduction of the Tethyan Ocean Beneath the South China Continent in Late Cretaceous: Evidences from the Liuwang Quartz Porphyry in the Southeastern Guangxi. *Bull. Mineral. Petro. Geochem.* **2022**, *41*, 374–387, (In Chinese with English Abstract).

28. Zhai, L.N.; Wang, J.H.; Wei, C.S.; Jiang, H.; Cai, J.H. Researching on the Age of rock-forming and metallogenic epoch of Fozichong ore field in Guangxi. *Geol. Miner. Resour. South. China* **2008**, *95*, 46–49, (In Chinese with English Abstract).

29. He, X.R.; Chen, A.B.; Xie, X.M. Ore Minerals Fabric and Mineralization Stage Division of Pb–Zn Depositin Longwan Guangxi. *Henan Sci.* **2014**, *32*, 1571–1575, (In Chinese with English Abstract).
30. Fan, X.; Wang, X.; Lü, X.; Wei, W.; Chen, W.; Yang, Q. Garnet composition as an indicator of skarn formation: LA-ICP-MS and EPMA studies on oscillatory zoned garnets from the Haobugao skarn deposit, Inner Mongolia. *China. Geol. J.* **2018**, *54*, 1976–1992. [CrossRef]
31. Seman, S.; Stockli, D.F.; McLean, N.M. U–Pb geochronology of grossular-andradite garnet. *Chem. Geol.* **2017**, *460*, 106–116. [CrossRef]
32. Liu, Y.S.; Hu, Z.C.; Gao, S.; Gunther, D.; Xu, J.; Gao, C.G.; Chen, H.H. In situ analysis of major and trace elements of anhydrous minerals by LA-ICP-MS without applying an internal standard. *Chem. Geol.* **2008**, *257*, 34–43. [CrossRef]
33. Liu, Y.S.; Gao, S.; Hu, Z.C.; Gao, C.G.; Zong, K.Q.; Wang, D. Continental and Oceanic Crust Recycling-induced Melt-Peridotite Interactions in the Trans-North China Orogen: U–Pb Dating, Hf Isotopes and Trace Elements in Zircons from Mantle Xenoliths. *J. Petro* **2010**, *51*, 537–571. [CrossRef]
34. Cai, P.R.; Wang, T.; Wang, Z.Q.; Li, L.M.; Jia, J.L.; Wang, M.Q. Geochronology and geochemistry of late Paleozoic volcanic rocks from eastern Inner Mongolia, NE China: Implications for igneous petrogenesis, tectonic setting, and geodynamic evolution of the south-eastern Central Asian Orogenic Belt. *Lithos* **2020**, *362–363*, 105480. [CrossRef]
35. Chew, D.M.; Petrus, J.A.; Kamber, B.S. U–Pb LA-ICP-MS dating using accessory mineral standards with variable common Pb. *Chem. Geol.* **2014**, *363*, 185–199. [CrossRef]
36. Chew, D.M.; Sylvester, P.J.; Tubrett, M.N. U–Pb and Th-Pb dating of apatite by LA-ICP-MS. *Chem. Geol.* **2011**, *280*, 200–216. [CrossRef]
37. Black, L.P.; Kamo, S.L.; Allen, C.M.; Aleinikoff, J.N.; Davis, D.W.; Korsch, R.J.; Foudoulis, C. TEMORA 1 a new zircon standard for Phanerozoic U–Pb geochronology. *Chem. Geol.* **2003**, *200*, 155–170. [CrossRef]
38. Luan, Y.; He, K.; Tan, X.J. In situ U–Pb dating and trace element determination of standard zircons by LA-ICP-MS. *Geol. Bull. China* **2019**, *38*, 1206–1218, (In Chinese with English Abstract).
39. Chiaradia, M.; Schaltegger, U.; Spikings, R.; Wotzlaw, J.F.; Ovtcharova, M. How accurately can we date the duration of magmatic hydrothermal events in porphyry systems?—An invited paper. *Econ. Geol.* **2013**, *108*, 565–584. [CrossRef]
40. Meinert, L.D.; Dipple, G.M.; Nicolescu, S. World skarn deposits. In *Economic Geology 100th Anniversary Volume*; Society of Economic Geologists, Inc.: Littleton, CO, USA, 2005; pp. 299–336.
41. Lima, S.M.; Corfu, F.; Neiva, A.M.R.; Ramos, J.M.F. U–Pb ID-TIMS dating applied to U-rich inclusions in garnet. *Amer. Miner.* **2012**, *97*, 800–806. [CrossRef]
42. Wafforn, S.; Seman, S.; Kyle, J.R.; Stockli, D.; Leys, C.; Sonbait, D.; Cloos, M. Andradite garnet U–Pb geochronology of the big Gossan skarn, Ertsberg-Grasberg mining district, Indonesia. *Econ. Geol.* **2018**, *113*, 769–778. [CrossRef]
43. Alexander, E.M.; Alexei, V.I.; Vadim, S.K.; Adam, A.; Tamara, Y.Y.; Timur, V.D. Contact metamorphic and metasomatic processes at the Kharaelakh Intrusion, Oktyabrsk Deposit, Norilsk-Talnakh ore district: Application of LA-ICP-MS dating of perovskite, apatite, garnet, and titanite. *Econ. Geol.* **2020**, *115*, 1213–1226.
44. Li, Y.X.; Kang, Z.Q.; Xu, J.F.; Yang, F.; Liu, D.M.; Shan, C.X. Chronological, Geochemical Characteristics and Geological Significance of Volcanic Rocks in the Late Early Cretaceous in Southeast Guangxi. *Earth Sci.* **2021**, 1–28, (In Chinese with English Abstract). [CrossRef]
45. Liu, Y.; Fang, N.; Qiang, M.; Jia, L.; Song, C. The Cretaceous igneous rocks in southeastern Guangxi and their implication for tectonic environment in southwestern South China Block. *Open. Geosci.* **2020**, *12*, 518–531. [CrossRef]
46. Wang, X.Y. Late Yanshanian Magmatism and W-Mineralization in Yunkai Region, Guangxi Province. Ph.D. Thesis, China University of Geosciences, Wuhan, China, 2017.
47. Li, J.; Zhang, Y.; Dong, S.; Johnston, S.T. Cretaceous tectonic evolution of South China: A preliminary synthesis. *Earth Sci. Rev.* **2014**, *134*, 98–136. [CrossRef]
48. Yang, J.B.; Zhao, Z.D.; Hou, Q.Y.; Niu, Y.L.; Mo, X.X.; Sheng, D.; Wang, L.L. Petrogenesis of Cretaceous (133-84 Ma) intermediate dykes and host granites in southeastern China: Implications for lithospheric extension, continental crustal growth, and geodynamics of Palaeo-Pacific subduction. *Lithos* **2018**, *296–299*, 195–211. [CrossRef]
49. Liu, L.; Xu, X.S.; Xia, Y. Cretaceous Pacific plate movement beneath SE China: Evidence from episodic volcanism and related intrusions. *Tectonophysics* **2014**, *614*, 170–184. [CrossRef]
50. Liu, L.; Xu, X.S.; Xia, Y. Asynchronizing paleo-Pacific slab rollback beneath SE China: Insights from the episodic Late Mesozoic volcanism. *Gondwana Res.* **2016**, *37*, 397–407. [CrossRef]
51. Zhang, L.P.; Hu, Y.B.; Liang, J.L.; Ireland, T.; Chen, Y.L.; Zhang, R.Q.; Sun, S.J.; Sun, W.D. Adakitic rocks associated with the shilu copper-molybdenum deposit in the Yangchun Basin, South China, and their tectonic implications. *Acta Geochim.* **2017**, *36*, 132–150. [CrossRef]
52. Liu, H.; Liao, R.; Zhang, L.; Li, C.; Sun, W. Plate subduction, oxygen fugacity, and mineralization. *J. Oceanol. Limnol.* **2020**, *38*, 64–74. [CrossRef]
53. Li, J.H.; Cawood, P.A.; Ratschbacher, L.; Zhang, Y.Q.; Dong, S.W.; Xin, Y.J.; Yang, H.; Zhang, P.X. Building Southeast China in the late Mesozoic: Insights from alternating episodes of shortening and extension along the Lianhuashan fault zone. *Earth Sci. Rev.* **2020**, *201*, 103056. [CrossRef]

54. Suo, Y.H.; Li, S.Z.; Jin, C.; Zhang, Y.; Zhou, J.; Li, X.Y.; Wang, P.C.; Liu, Z.; Somerville, L. Eastward tectonic migration and transition of the Jurassic-Cretaceous Andeantype continental margin along Southeast China. *Earth Sci. Rev.* **2019**, *196*, 102884. [CrossRef]
55. Li, S.Z.; Suo, Y.H.; Li, X.Y.; Zhou, J.; Santosh, M.; Wang, P.C.; Wang, G.Z.; Guo, L.L.; Yu, S.Y.; Lan, H.Y.; et al. Mesozoic tectono-magmatic response in the East Asian ocean-continent connection zone to subduction of the Paleo-Pacific Plate. *Earth Sci. Rev.* **2019**, *192*, 91–137. [CrossRef]
56. Li, J.H.; Ma, Z.L.; Zhang, Y.Q.; Dong, S.W.; Li, Y.; Lu, M.A.; Tan, J.Q. Tectonic evolution of Cretaceous extensional basins in Zhejiang Province, eastern South China: Structural and geochronological constraints. *Int. Geol. Rev.* **2014**, *56*, 1602–1629. [CrossRef]
57. Zhou, X.M.; Sun, T.; Shen, W.Z.; Shu, L.S.; Niu, Y.L. Petrogenesis of Mesozoic granitoids and volcanic rocks in South China: A response to tectonic evolution. *Episodes* **2006**, *9*, 26–33. [CrossRef]
58. Chu, Y.; Lin, W.; Faure, M.; Xue, Z.H.; Ji, W.B.; Feng, Z.T. Cretaceous episodic extension in the South China Block, East Asia: Evidence from the Yuechengling Massif of central South China. *Tectonics* **2019**, *38*, 3675–3702. [CrossRef]
59. Shu, L.S.; Zhou, X.M.; Deng, P.; Wang, B.; Jiang, S.Y.; Yu, J.H.; Zhao, X.X. Mesozoic tectonic evolution of the Southeast China Block: New insights from basin analysis. *J. Asian Earth Sci.* **2009**, *34*, 376–391. [CrossRef]
60. Sun, W. Initiation and evolution of the South China Sea: An overview. *Acta Geochim.* **2016**, *35*, 215–225. [CrossRef]
61. Zhou, J.X.; Li, W.C. Evolution and metallogeny of the Sanjiang arc-back arc basin system in the Eastern Tethys: An introduction. *J. Asian Earth Sci.* **2021**, *222*, 104961. [CrossRef]
62. Yang, Q.J.; Li, C.C.; Yang, J.B. The geological and geochemical characteristics of volcanic rocks in Shuiwen Basin of Guangxi: Magmatic activity response to Mesozoic strike-slip process in Cenxi-Bobai Fault zone. *Miner. Resour. Geol.* **2022**, *36*, 781–795, (In Chinese with English Abstract).
63. Li, C. The Evolution and Geological Significance of Complex Rock Mass in Xintang City, Cenxi City, Guangxi. Master's Thesis, Guilin University of Technology, Guilin, China, 2018. (In Chinese with English Abstract).

Article

LA-ICP-MS Trace Element Geochemistry of Sphalerite and Metallogenic Constraints: A Case Study from Nanmushu Zn–Pb Deposit in the Mayuan District, Shaanxi Province, China

Junjie Wu [1,2], Huixin Dai [1,*], Yong Cheng [3,4,*], Saihua Xu [1,*], Qi Nie [3], Yiming Wen [3] and Ping Lu [3]

1   Faculty of Land and Resource Engineering, Kunming University of Science and Technology, Kunming 650093, China; starwjj555@126.com
2   Shaanxi Institute of Geology and Mineral Resources Experiment Co., Ltd., Xi'an 710054, China
3   Faculty of Metallurgy and Mining Engineering, Kunming Metallurgy College, Kunming 650033, China; kiki158961326@163.com (Q.N.); kmcymw@163.com (Y.W.); lupingin@163.com (P.L.)
4   School of Earth Sciences, Yunnan University, Kunming 650500, China
*   Correspondence: dhxk_must@163.com (H.D.); cheng_yong1988@163.com (Y.C.); xusaihua18@126.com (S.X.)

Abstract: The Nanmushu Zn–Pb deposit is a large-scale and representative deposit in the Mayuan ore field on the northern margin of the Yangtze Block. This study investigates the trace element geochemistry of sphalerite from this deposit using laser ablation inductively coupled plasma mass spectrometry (LA-ICP-MS). The results show that the main trace elements in sphalerite include various trace elements, such as Mn, Fe, Cu, Ga, Ge, Ag, Cd, Pb, Co, Hg, Tl, In, Sn, and Sb. Among them, Ag, Ge, Cd, and Cu are valuable components that may be recovered during mineral processing or smelting techniques. The histograms, LA-ICP-MS time-resolved depth profiles, and linear scan profiles indicated that most trace elements occur in sphalerite as isomorphs, while partial Pb, Fe, and Ag occur as tiny mineral inclusions. The correlation diagrams of trace elements revealed that Fe^{2+}, Mn^{2+}, Pb^{2+}, and Tl^{3+} can substitute Zn^{2+} in sphalerite through isomorphism. In sphalerite, Cd^{2+} and Hg^{2+} together or Mn^{2+}, Pb^{2+}, and Tl^{3+} together can replace Zn^{2+}, i.e., $((3Mn, 3Pb, 2Tl)^{6+}, 3(Cd, Hg)^{2+}) \leftrightarrow 3Zn^{2+}$. Moreover, there is a mechanism of Ge^{4+} with Cu^+ or Ga^{3+} with Cu_+ replacing Zn^{2+} in the Nanmushu deposit, i.e., $Ge^{4+} + 2Cu^+ \leftrightarrow 3Zn^{2+}$ or $2Ga^{3+} + 2Cu^+ \leftrightarrow 4Zn^{2+}$. Furthermore, the trace element compositions indicate that the Nanmushu Zn mineralization occurred under low-temperature conditions (<200 °C) and should be classified as a Mississippi Valley-type (MVT) deposit. This study provides new insights into the occurrence and substitution mechanisms of trace elements in sphalerite and the metallogenic constraints of the Nanmushu deposit.

Keywords: in-site trace elements; sphalerite; substitution mechanisms; Nanmushu Zn–Pb deposit; Mayuan district; Mississippi Valley-type (MVT) deposit

Citation: Wu, J.; Dai, H.; Cheng, Y.; Xu, S.; Nie, Q.; Wen, Y.; Lu, P. LA-ICP-MS Trace Element Geochemistry of Sphalerite and Metallogenic Constraints: A Case Study from Nanmushu Zn–Pb Deposit in the Mayuan District, Shaanxi Province, China. *Minerals* **2023**, *13*, 793. https://doi.org/10.3390/min13060793

Academic Editor: Maria Boni

Received: 9 May 2023
Revised: 1 June 2023
Accepted: 6 June 2023
Published: 10 June 2023

1. Introduction

Sphalerite is the predominant ore mineral of lead–zinc (Pb–Zn) deposits and hosts various trace elements, such as Cd, Ga, Ge, In, and other dispersed elements [1–6]. These elements have diverse applications in many high-tech fields [7]. For instance, Cd alloys serve as bearing materials for aircraft engines and control rods (neutron absorption) for nuclear reactors, and Cd–Ni batteries are essential for aviation and railways [4]. Ga plays a role In new generation information technology, biotechnology, high-end equipment manufacturing, new energy, and new materials [8]. Ge is employed in information communication, modern aviation, modern military, and new energy sectors [9]. In is used in the electronics industry, aerospace, alloy manufacturing, solar cell new materials, and other fields [10,11]. As strategic resources, Ga, Ge, and In are regarded by many countries as "critical metals" for economic development and the national defence industry [12]. Furthermore, trace elements in sphalerite carry valuable metallogenic information. Some elements constrain the physical and chemical conditions of ore-forming fluids and can provide reliable ore-forming temperature data as well as indicate the origin of ore deposits [1,13,14].

In recent years, significant progress has been made in the exploration of lead and zinc in the northern margin of the Yangtze block, especially the huge resource prospect of the Mayuan zinc and lead (Zn–Pb) ore field, which has attracted wide attention from geologists. The Zn–Pb mineralization belt has a width of 10–200 m at the Mayuan district and extends over 60 km, which can be divided into three Zn–Pb ore zones, including the south, middle, and north ore zones (Figure 1). The Mayuan Zn–Pb ore field contains a metal resource of over 10 Mt Zn (1.05–10.82 wt%) and Pb (0.55–7.54 wt%) which are associated with some beneficial elements, such as Ag (2–35 ppm), Ge (0.002–0.05 wt%), Cd (0.002–0.10 wt%), and Cu (0.03–0.35 wt%) [15,16]. Many studies have been conducted on the Mayuan Zn–Pb ore field, including the geological features of the ore deposit [15,17], the source of ore-forming materials [16,18–20], the ore-forming age [21–23], the relationship between hydrocarbon organic matters and mineralization [21,24–26], etc. Regarding the genesis of the deposit, the initial understanding was that it was an epigenetic medium-low temperature hydrothermal deposit [27] or a syn-sedimentary hydrothermal deposit [28], but later, most scholars agreed that it was a Mississippi Valley-type (MVT) deposit [16,18–23]. However, studies on the associated beneficial elements are still scarce.

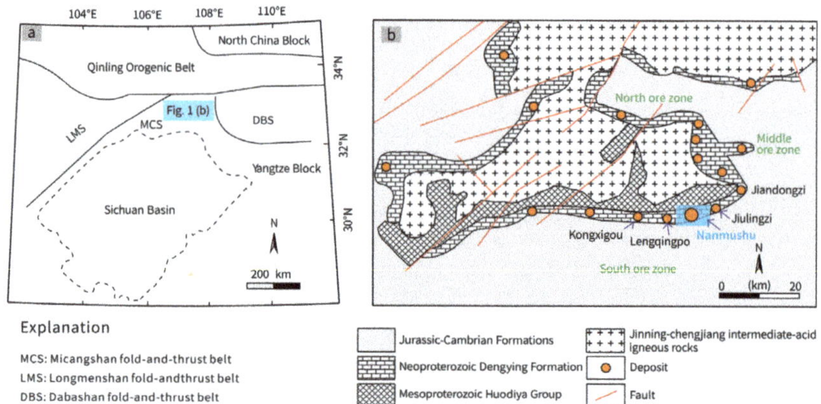

Figure 1. (**a**) Tectonic location of the Mayuan district. (**b**) The distribution of the south, middle, and north ore zones, and major Zn–Pb deposits in the Mayuan district. The figure was modified from reference [20].

The Nanmushu Zn–Pb deposit belongs to the southern ore zone and is the largest deposit in the Mayuan ore field. Previous studies have reported that the associated dispersed elements of the Nanmushu deposit are mainly concentrated in sphalerite [15]. The dissolution method [29] and the in-situ LA-ICP-MS method [19,30,31] show that sphalerite is enriched in the dispersed elements of Cd, Ge, and Ga, while it is relatively depleted in In, Tl, and Se. The occurrence mode and substitution mechanism of major trace elements in sphalerite have not been thoroughly investigated. In this study, the in-situ LA-ICP-MS method was applied to measure the trace elements of sphalerite from the Nanmushu deposit, aiming to discuss the occurrence and substitution mechanism of dispersed elements in sphalerite, estimate and decipher the ore-forming temperature, and provide geochemical constraints on the ore genetic type.

2. Geological Background

2.1. Regional Geology

The study area is located in the south of the Beibadome structure on the northern margin of the Yangtze Block (Figure 1) [20]. The Mayuan district comprises crystalline basement and overlying sedimentary cover. The basement consists of the Huodiya Group of Mesoproterozoic to Neoproterozoic rocks, and its lithology is mainly meso-deep metamorphic volcaniclastic rocks and magmatic rocks from the Chengjiang to Jinning period, which exhibit angular unconformity with the overlying sedimentary cover. The cover is composed of Sinian–Cambrian shallow metamorphic carbonate rocks and clastic rocks [32].

The regional structure is a large EW-striking Beiba dome structure with pre-Sinian basement in the core and Sinian–Cambrian in the wing [19]. The fault structures in the area are mainly developed along the boundary of basement and caprocks, which are mostly located in caprocks and strike in the same direction as caprocks. In addition, there are also faults formed in a later period obliquely crossing and crosscutting the strata [33].

The Beiba dome structure controls the distribution of ore-bearing strata, and the Pb–Zn ore bodies occur in the brecciform dolostone of the Dengying Formation in a zonal pattern around the dome core [34]. There are three Pb–Zn mineralization zones in the north, east, and south of the study area, and nearly 50 Pb–Zn ore bodies have been discovered, mainly in the south and east ore zones [35]. The most important south ore zone extends in the NEE direction, with a length of more than 30 km and a width of 20–200 m. The average grade of Zn and Pb in the ores is 4.02% and 4.16%, respectively [15]. Kongxigou, Lengqingpo, Nanmushu, Jiulingzi, Jiandongzigou, and other deposits are distributed in this ore zone.

2.2. Ore Deposit Geology

The Nanmushu deposit is one of the most typical deposits in the Mayuan lead–zinc ore field (Figure 2). The exposed strata in the deposit area from old to new are the Mesoproterozoic Huodiya Group, the Upper Sinian Dengying Formation, and the Lower Cambrian Guojiaba Formation [23] (Figure 2). Dengying Formation can be divided into upper and lower lithologic members. The lithology of the lower member is sandstone, including pebbly sandstone with thin layer dolostone. The upper member is dolostone, which can be subdivided into four lithologic beds from bottom to top according to the thickness of the sedimentary layer and rock structure: medium to thick layered dolostone, striated dolostone, brecciform dolostone, and laminar dolostone [20].

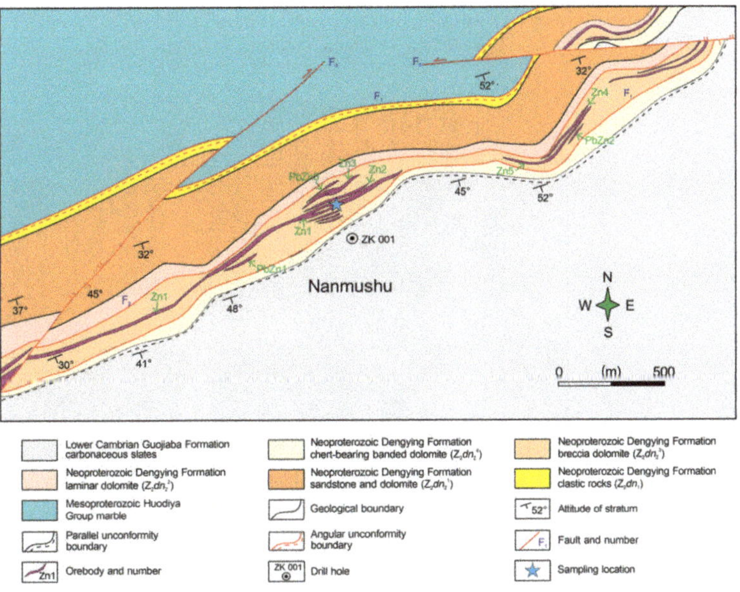

Figure 2. Geological map of the Nanmushu Zn–Pb deposit showing the orebodies are mainly hosted in the third section of Neoproterozoic Dengying Formation breccia dolomite ($Z_2dn_2{}^3$) (modified from reference [34]).

The ore bodies are stratiform or stratiform-like along the brecciform dolostone of the third lithologic layer of the upper member of the Dengying Formation (Figure 3). Five zinc bodies and three lead–zinc bodies have been discovered in the Nanmu deposit with a length of 100–2560 m and a thickness of 0.80–13.14 m [35]. The deposit contains 2.1 Mt Zn with a grade of 1.05–13.09 wt% and 0.1 Mt Pb with a grade of 0.60–4.12 wt% [21,23].

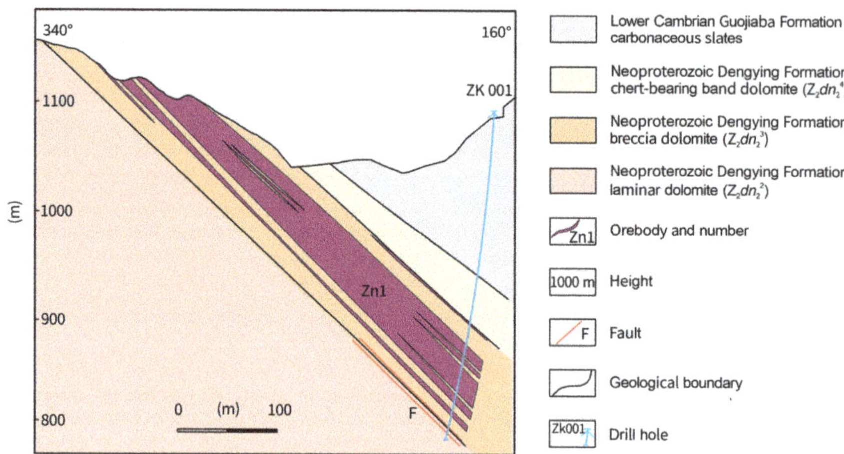

Figure 3. Cross-section of the Nanmushu Zn–Pb deposit (modified from reference [15]).

The ore is dominated by a brecciform structure as well as some stringer disseminated, sparse disseminated, and geode structures. The mineral composition of the ore is relatively simple, and the metal minerals are mainly sphalerite, galena, pyrite, with minor supergene minerals, such as limonite, anglesite, and cerussite. Gangue minerals are mainly dolomite, quartz, bitumen, and minor barite and calcite (Figure 4a–c).

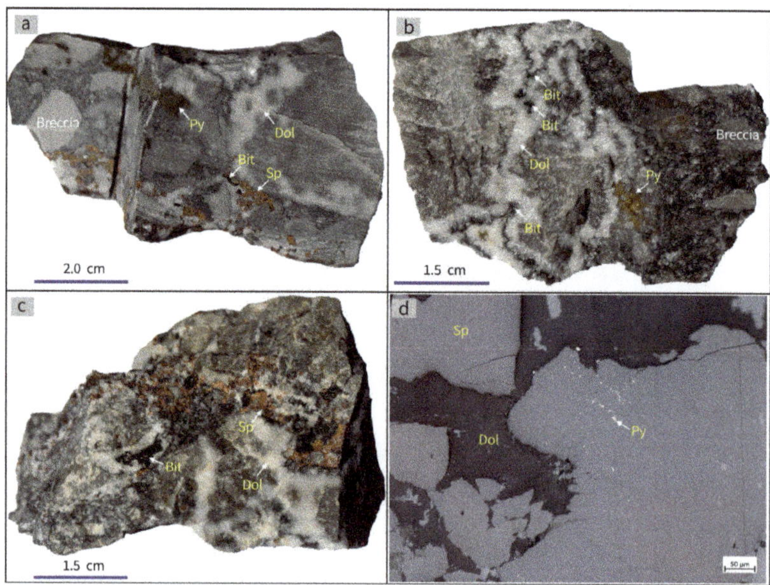

Figure 4. *Cont.*

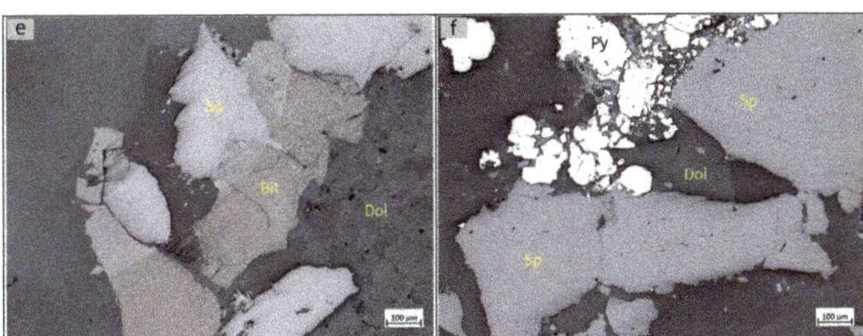

Figure 4. Photographs of the mineralization characteristics of the Nanmushu Zn–Pb deposit. (**a–c**) Breccia-type ore. (**a**) Dolostone breccia cemented by later hydrothermal mineral. (**b**) Bitumen is distributed in the dolomite as a thin film or grain, and pyrite is in clumps. (**c**) Sphalerite is associated with bitumen and dolomite. (**d**) Many pyrite inclusions can be seen in the sphalerite particles. (**e**) There are many solid bitumens distributed in the edge of the sphalerite particles. (**f**) Sphalerite is associated with pyrite. Abbreviations: Py = pyrite; Sp = sphalerite; Bit = bitumen; Dol = dolomite.

As the main ore mineral in the ore, sphalerite is mainly distributed in the cement between dolomite breccia, which is tawny or brown, showing glass luster. Microscopically, they are mostly irregular xenomorphic grains with particle sizes ranging from 0.1 mm to 3.0 mm. Sphalerite often has a paragenetic relationship with pyrite and bitumen and is mostly replaced by smithsonite, dolomite, willemite, and hydrozincite (Figure 4e,f). The wall rock alteration of the deposit is weak, mainly involving silicification, carbonatization, pyritization, bituminization, baritization, etc. [26]. According to the geological features and mineral paragenesis relationships of the deposit, it is inferred that the Nanmushu deposit had experienced pre-ore sedimentary and the diagenetic stage, main-ore hydrothermal ore-forming stage, and post-ore supergene stage [22]. In the sedimentary and diagenetic stage, the fine-grained (10 to 50 μm) euhedral pyrite is disseminated within dolomite. A large amount of sphalerite is formed in the hydrothermal stage, which is the economic mineralization stage. Smithsonite, cerussite, and limonite are the main minerals in the post-ore supergene stage [16,20].

3. Samples and Analytical Methods

In this study, samples were collected from the main ore body of the Nanmushu deposit in the Mayuan lead–zinc ore field. Four representative samples were selected to make laser thin sections for testing. Trace elements of sphalerite were determined using a NWR 193 nm ArF Excimer laser ablation system coupled to an iCAP RQ (ICPMS) at Guangzhou Tuoyan Analytical Technology Co., Ltd., Guangzhou, China.

The LA-ICP-MS was tuned using NIST 610 standard glass to yield low oxide production rates. A carrier gas of 0.7 L/min He was fed into the cup, and the aerosol was subsequently mixed with 0.89 L/min Ar make-up gas. The laser fluence was 3.5 J/cm^2, with a repetition rate of 6 Hz, a 30 μm spot size, and an analysis time of 40 s, followed by a 40 s background measurement. The raw isotope data were reduced using the "trace elements" data reduction scheme (DRS). The DRS runs within the freeware IOLITE package of Paton et al. (2011) [36]. In IOLITE, user-defined time intervals are established for the baseline correction procedure to calculate session-wide baseline-corrected values for each isotope. Blocks of two standards (one NIST 610 and one GSE-2G) and one MASS-1 sulfide standard analyses were followed by five to eight unknown samples. For sphalerite, the following 18 isotopes were measured (with their respective dwell times in milliseconds listed in parentheses): ^{34}S, ^{55}Mn, ^{57}Fe, ^{59}Co, ^{60}Ni, ^{65}Cu, ^{66}Zn, ^{71}Ga, ^{74}Ge, ^{75}As, ^{77}Se, ^{107}Ag, ^{111}Cd, ^{118}Sn, ^{121}Sb, ^{126}Hg, ^{205}Tl, and ^{208}Pb, corresponding to a total dwell time of 180 ms. The MASS-1 sulfide standard [37] was used as the primary standard for calibrating ^{34}S, ^{55}Mn, ^{57}Fe, ^{59}Co, ^{60}Ni, ^{65}Cu, ^{66}Zn, ^{71}Ga, ^{74}Ge, ^{75}As, ^{77}Se, ^{107}Ag, ^{111}Cd, ^{118}Sn, ^{121}Sb,

^{126}Hg, ^{205}Tl, and ^{208}Pb. All the isotopes were corrected with an internal standard of ^{66}Zn (Zn = 65).

In addition, three sphalerite particles were selected for line scanning of trace elements. The laser spot size was 10 μm with a scanning speed of 5 μm/S. The other test parameters were the same as above.

4. Results

In this study, a total of 42 LA-ICP-MS spots analyses were completed on 4 sphalerite samples, and the results are listed in Table 1. It can be seen that the trace elements of sphalerite formed in the main mineralization stage of the Nanmushu deposit are mainly Mn, Fe, Cu, Ga, Ge, Ag, Cd, Pb, Co, Hg, Tl, In, Sn, and Sb, while the enrichment degree of other elements is not significant (Figure 5). Furthermore, LA-ICP-MS linear scanning was performed for three sphalerite particles. The complete linear scanning results are shown in Figure 6.

Table 1. LA-ICP-MS in-situ analysis of trace elements of sphalerite in the Nanmushu deposit (ppm).

Sample ID	Spot No.	Mn	Fe	Cu	Ga	Ge	Ag	Cd	Pb	Co	Hg	Tl	In	Sn	Sb
NMS-2 *n* = 10	NMS-2-1	BLD	1531.93	5191.82	85.95	2024.89	41.82	1008.99	23.59	2.84	51.37	0.03	2.85	4.72	1.96
	NMS-2-2	3.96	2321.67	1366.21	12.61	626.32	58.90	3058.57	243.91	2.23	132.66	0.12	BLD	BLD	16.07
	NMS-2-3	1.39	1786.34	2465.35	9.73	1061.00	47.07	1712.13	116.02	2.62	80.30	0.03	BLD	BLD	5.92
	NMS-2-4	5.11	2194.86	1510.91	12.80	753.26	65.98	2763.97	509.68	2.19	138.82	0.51	BLD	0.80	14.26
	NMS-2-5	69.77	11,629.06	1001.99	0.27	539.84	63.40	1166.09	483.56	1.57	107.47	7.56	0.09	1.57	BLD
	NMS-2-6	36.47	2351.52	1297.71	22.40	698.85	64.30	3385.32	272.11	1.92	133.64	1.40	0.45	2.81	16.37
	NMS-2-7	44.03	10,205.90	108.66	BLD	704.46	11.52	794.14	698.57	1.96	42.20	9.38	BLD	BLD	BLD
	NMS-2-8	78.66	15,171.51	600.38	BLD	713.76	39.94	333.94	767.74	1.97	17.86	11.75	BLD	BLD	BLD
	NMS-2-9	3.86	1525.91	2270.94	62.65	1035.60	36.13	1408.38	22.75	2.96	83.45	0.03	5.94	42.19	2.51
	NMS-2-10	49.94	1982.75	1093.54	40.61	563.08	39.31	3035.20	197.26	2.00	164.52	0.51	0.54	3.91	17.92
	Min.	1.39	1525.91	108.66	0.27	539.84	11.52	333.94	22.75	1.57	17.86	0.03	0.09	0.80	1.96
	Max.	78.66	15,171.51	5191.82	85.95	2024.89	65.98	3385.32	767.74	2.96	164.52	11.75	5.94	42.19	17.92
	Median	36.47	2258.26	1331.96	17.60	709.11	44.45	1560.25	258.01	2.10	95.46	0.51	0.54	3.36	14.26
	Average	32.58	5070.14	1690.75	30.88	872.11	46.84	1866.67	333.52	2.23	95.23	3.13	1.98	9.33	10.71
	S.D.	28.53	4899.40	1341.99	27.96	418.92	16.07	1041.99	254.10	0.42	45.62	4.33	2.21	14.75	6.46
NMS-5 *n* = 9	NMS-5-1	8.29	3980.66	2.78	0.45	BLD	1.19	4626.48	9.28	2.44	212.42	0.01	0.25	0.90	BLD
	NMS-5-2	34.27	1943.87	884.46	20.08	488.06	33.36	2554.53	193.72	2.26	142.49	1.19	0.07	0.85	10.78
	NMS-5-3	2.82	3322.38	4.32	0.31	2.40	0.78	5365.12	7.10	2.41	278.48	BLD	BLD	BLD	0.34
	NMS-5-4	44.14	5574.82	292.06	BLD	296.82	26.10	1321.07	542.55	1.72	42.49	2.15	BLD	BLD	BLD
	NMS-5-5	51.55	5677.35	383.21	BLD	312.78	29.80	1083.50	493.23	1.93	48.98	3.10	BLD	BLD	BLD
	NMS-5-6	2.85	1679.23	974.75	12.63	482.70	35.65	2106.33	224.25	2.48	121.12	0.04	0.02	BLD	13.60
	NMS-5-7	53.07	8194.90	47.47	BLD	27.47	2.34	3089.33	22.33	2.33	71.84	BLD	BLD	BLD	BLD
	NMS-5-8	5.71	1724.06	2669.10	49.38	1153.91	54.11	995.82	148.22	3.00	55.08	0.09	2.61	26.49	0.73
	NMS-5-9	9.42	1996.63	530.07	4.63	270.75	20.37	3029.74	140.93	2.54	175.38	0.17	0.08	1.09	11.64
	Min.	2.82	1679.23	2.78	0.31	2.40	0.78	995.82	7.10	1.72	42.49	0.01	0.02	0.85	0.34
	Max.	53.07	8194.90	2669.10	49.38	1153.91	54.11	5365.12	542.55	3.00	278.48	3.10	2.61	26.49	13.60
	Median	9.42	3322.38	383.21	8.63	304.80	26.10	2554.53	148.22	2.41	121.12	0.17	0.08	0.99	10.78
	Average	23.57	3788.21	643.14	14.58	379.36	22.63	2685.77	197.96	2.34	127.59	0.96	0.60	7.33	7.42
	S.D.	20.56	2153.90	792.28	17.06	337.53	17.30	1447.80	187.15	0.35	77.67	1.15	1.01	11.06	5.70
NMS-7 *n* = 8	NMS-7-1	3.89	2567.59	489.89	15.67	240.82	20.64	3601.98	158.27	2.65	189.52	0.02	BLD	BLD	13.52
	NMS-7-2	60.57	11,731.80	304.14	1.01	401.96	17.93	436.34	499.88	2.12	23.26	7.34	0.03	1.37	BLD
	NMS-7-3	144.33	15,271.94	940.19	0.52	540.53	63.74	424.42	682.66	1.93	87.63	15.05	0.06	2.67	0.34
	NMS-7-4	177.73	14,415.03	1018.49	2.45	567.10	69.51	445.94	656.87	1.94	126.48	21.78	0.47	6.28	0.22
	NMS-7-5	85.35	15,385.92	551.72	BLD	685.86	44.87	276.80	983.69	1.88	14.80	18.52	BLD	BLD	BLD
	NMS-7-6	2.90	3020.49	1322.35	50.91	555.34	16.23	3562.22	200.60	2.34	196.73	0.02	4.02	33.30	14.97
	NMS-7-7	7.75	3257.11	1062.29	2.65	576.17	52.68	3168.06	484.91	2.01	122.25	0.41	0.12	3.17	2.76
	NMS-7-8	10.80	3192.97	978.40	11.08	527.36	45.42	3509.59	424.68	1.87	147.07	0.52	0.41	7.89	4.35
	Min.	2.90	2567.59	304.14	0.52	240.82	16.23	276.80	158.27	1.87	14.80	0.02	0.03	1.37	0.22
	Max.	177.73	15,385.92	1322.35	50.91	685.86	69.51	3601.98	983.69	2.65	196.73	21.78	4.02	33.30	14.97
	Median	35.68	7494.46	959.29	2.65	547.93	45.14	1807.00	492.39	1.97	124.36	3.93	0.26	4.72	3.56
	Average	61.66	8605.36	833.43	12.04	511.89	41.38	1928.17	511.44	2.09	113.47	7.96	0.85	9.11	6.03
	S.D.	62.05	5739.59	304.98	15.74	130.38	19.93	1457.77	257.48	0.60	61.26	8.39	1.32	10.25	5.55

Table 1. *Cont.*

Sample ID	Spot No.	Mn	Fe	Cu	Ga	Ge	Ag	Cd	Pb	Co	Hg	Tl	In	Sn	Sb
NMS-10 *n* = 7	NMS-10-1	3.39	2803.48	1306.69	2.61	688.60	56.52	2932.11	371.69	1.98	145.62	1.70	BLD	BLD	13.84
	NMS-10-2	2.96	2821.35	1656.85	3.97	828.48	55.45	2942.67	411.12	2.09	146.42	0.08	BLD	BLD	14.26
	NMS-10-3	4.38	4790.85	7.90	3.18	1.19	1.29	6286.87	2.85	2.31	321.92	0.02	0.19	6.16	BLD
	NMS-10-4	3.13	2493.68	1232.51	19.92	631.79	44.39	2880.02	316.88	2.05	169.27	0.14	BLD	0.29	13.29
	NMS-10-5	1.36	2322.53	1563.03	21.13	746.85	21.77	2521.65	301.71	2.29	131.45	0.04	BLD	BLD	11.74
	NMS-10-6	1.29	2351.22	1136.79	6.98	584.54	35.70	2632.58	326.53	2.07	125.92	0.02	BLD	BLD	9.29
	NMS-10-7	70.77	2871.43	1514.86	1.93	772.40	63.10	3276.86	317.01	1.91	141.28	2.47	BLD	1.54	7.53
	Min.	1.29	2322.53	7.90	1.93	1.19	1.29	2521.65	2.85	1.91	125.92	0.02	0.19	0.29	7.53
	Max.	70.77	4790.85	1656.85	21.13	828.48	63.10	6286.87	411.12	2.31	321.92	2.47	0.19	6.16	14.26
	Median	3.13	2803.48	1306.69	3.97	688.60	44.39	2932.11	317.01	2.07	145.62	0.08	0.19	1.54	12.51
	Average	12.47	2922.08	1202.66	8.53	607.69	39.75	3353.25	292.54	2.10	168.84	0.64	0.19	2.66	11.66
	S.D.	23.82	791.74	517.84	7.74	259.33	20.40	1218.48	123.45	0.14	63.78	0.94	0.00	2.53	2.48
NMS-14 *n* = 8	NMS-14-1	65.56	14,588.81	179.84	BLD	631.20	15.99	531.72	1235.21	1.90	46.64	9.14	BLD	BLD	BLD
	NMS-14-2	2.71	2828.70	1016.89	22.30	466.81	37.14	3759.95	259.70	2.11	187.21	0.05	0.36	7.69	19.63
	NMS-14-3	15.97	2246.54	1885.02	27.66	802.82	47.10	2584.62	208.92	2.24	155.08	0.37	0.09	0.20	12.55
	NMS-14-4	21.83	4628.39	1071.64	BLD	642.09	238.37	1992.45	873.42	1.41	72.42	0.69	BLD	BLD	BLD
	NMS-14-5	41.35	9110.29	530.27	18.22	447.72	49.44	1096.14	591.05	1.66	96.59	6.01	0.25	0.81	BLD
	NMS-14-6	43.24	10,450.27	89.05	BLD	524.40	10.28	947.50	677.07	1.66	67.70	7.55	BLD	BLD	BLD
	NMS-14-7	49.08	3327.04	681.20	6.47	408.88	53.33	3812.57	237.45	1.63	180.04	8.80	0.33	2.57	19.70
	NMS-14-8	85.86	14,429.99	1021.45	9.05	783.46	123.61	347.06	990.70	1.67	55.62	24.73	0.16	1.84	BLD
	Min.	2.71	2246.54	89.05	6.47	408.88	10.28	347.06	208.92	1.41	46.64	0.05	0.09	0.20	12.55
	Max.	85.86	14,588.81	1885.02	27.66	802.82	238.37	3812.57	1235.21	2.24	187.21	24.73	0.36	7.69	19.70
	Median	42.30	6869.34	849.04	18.22	577.80	48.27	1544.30	634.06	1.67	84.51	6.78	0.25	1.84	19.63
	Average	40.70	7701.25	809.42	16.74	588.42	71.91	1884.00	634.19	1.78	107.66	7.17	0.24	2.62	17.29
	S.D.	25.36	4800.11	538.91	7.96	141.26	70.69	1295.65	358.43	0.26	53.86	7.54	0.10	2.66	3.35

BLD means concentration below the detection limit.

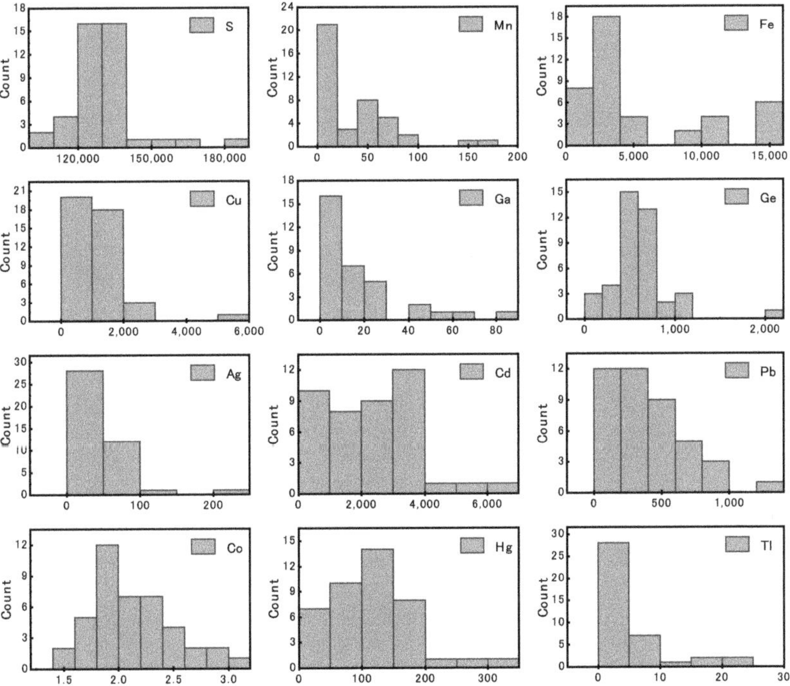

Figure 5. *Cont.*

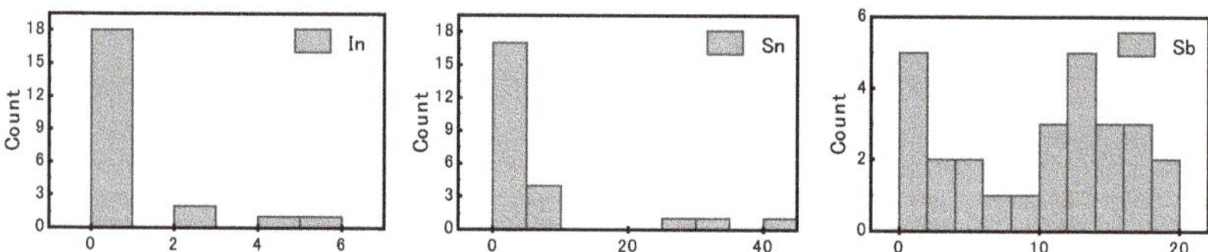

Figure 5. Histogram of trace element composition of sphalerite in Nanmushu deposit.

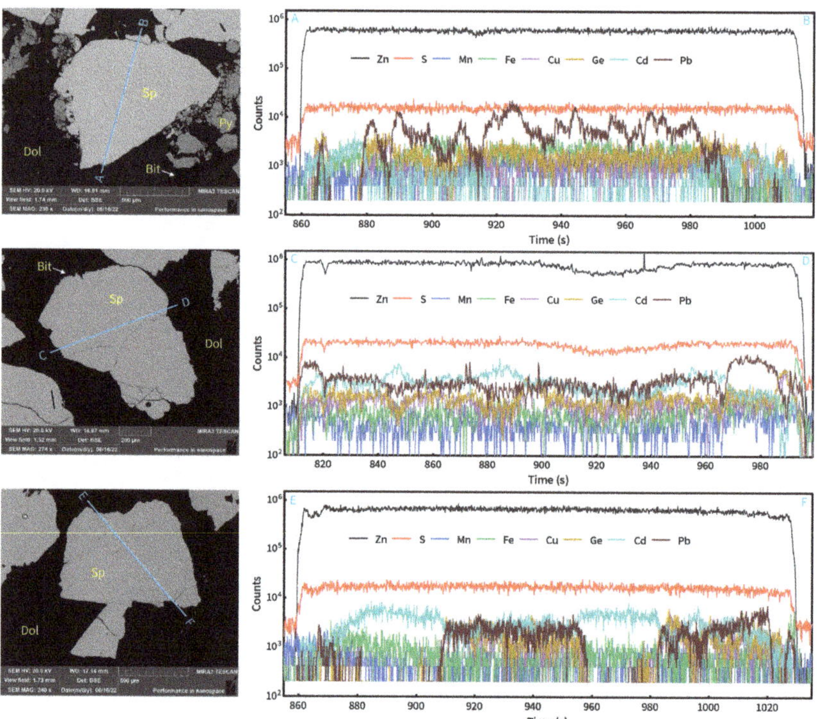

Figure 6. A backscattered electron image of sphalerite and its LA-ICP-MS linear scan profiles. Abbreviations are the same as in Figure 4.

Among these trace elements, Fe concentration was the highest, ranging from 1525.91 ppm to 15,385.92 ppm, with an average of 5741.24 ppm. The following trace elements were Cd and Cu, whose contents varied from 276.80 ppm to 6286.87 ppm (with an average of 2349.37 ppm) and from 2.78 ppm to 5191.82 ppm (with an average of 1123.90 ppm), respectively. Sphalerite also contains abundant trace elements Ge, Pb, and Hg. Their concentrations varied from 1.19 ppm to 2024.89 ppm (with an average of 624.15 ppm), 2.85 ppm to 1235.21 ppm (with an average of 399.27 ppm), and 14.80 ppm to 321.92 ppm (with an average of 122.46 ppm), respectively. The average content of Ag, Mn, Ga, and Sb is less than 50 ppm, whose average contents are 44.20 ppm, 34.43 ppm, 17.04 ppm, and 10.00 ppm, respectively. Besides, sphalerite contains minor trace elements Sn, Tl, Co, and In, the average content of which is less than 10 ppm. Sphalerite from the Nanmushu deposit contains high concentrations of Ag, Ge, Cd, and Cu elements, which exceed the minimum grades required for their extraction and utilization in China. These grades are 2 ppm for Ag, 10 ppm for Ge, 100 ppm for Cd, and 600 ppm for Cu. Therefore, these elements may be recovered through mineral processing or smelting techniques.

5. Discussion

5.1. Occurrence of Trace Elements in Sphalerite

This paper examines the occurrence states of trace elements in sphalerite by using three methods: contents histogram, LA-ICP-MS time-resolved depth profiles, and linear scan profiles [38,39]. The contents histogram shows that Cu, Ge and Pb have a wide range of concentrations in sphalerite, from 2.78 ppm to 5191.82 ppm, from 1.19 ppm to 2024.89 ppm, and from 2.85 ppm to 1235.21 ppm, respectively (Table 1). The LA-ICP-MS linear scan reveals that Pb has a large variation in the three curves (Figure 6). The time-resolved depth profiles indicate that most test spots have flat curves with small fluctuations in trace elements (Figure 7a), while some test spots have large fluctuations in Pb (Figure 7b,e) and Ag (Figure 7e), moderate fluctuations in Fe (Figure 7d), and smooth fluctuations in Ge, Cu, and Mn (Figure 7c,f).

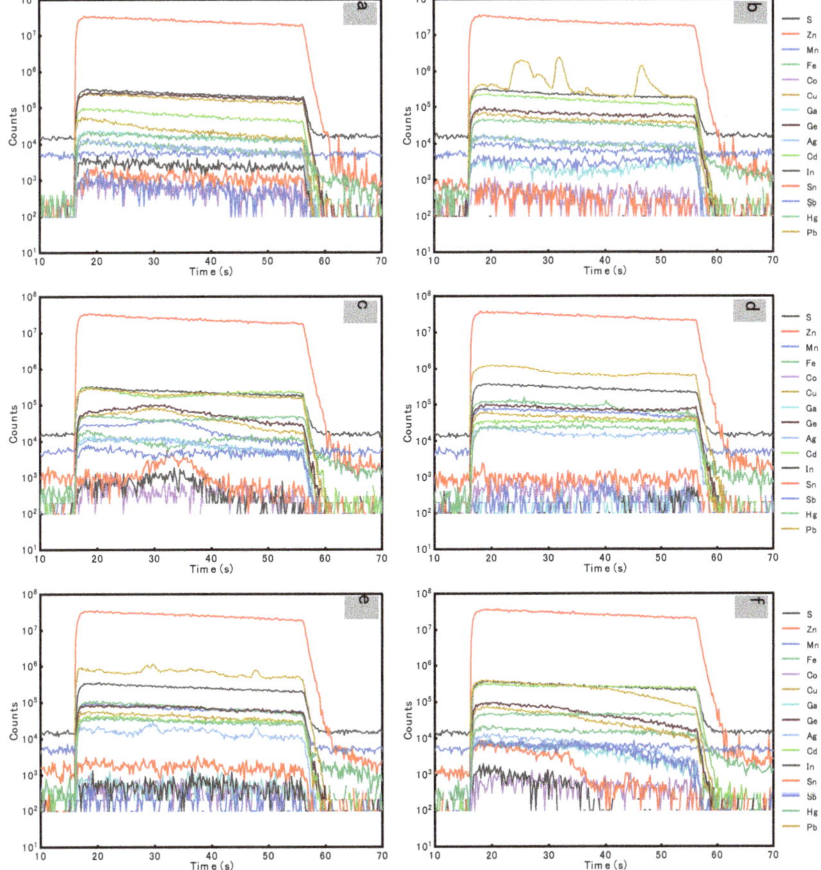

Figure 7. Representative time-resolved depth profiles for selected elements in sphalerite analyzed in this study. (**a**) The stable test signals indicate that trace elements occur as isomorphism in sphalerite; (**b**) The signal curves of Pb show large fluctuation; (**c**) The signal curves of Ge, Cu, and Mn show smooth fluctuation; (**d**) The signal curve of Fe shows moderate fluctuation; (**e**) The signal curves of Pb and Ag show large fluctuation; (**f**) The signal curves of Ge, Cu, and Pb show smooth fluctuation.

The results suggest that Pb, Fe, and Ag have different occurrence states in sphalerite besides isomorphism. The micrograph and BSE electron photograph show that some minerals such as pyrite are included in sphalerite (Figures 4d and 6). The time-resolved depth profiles and linear scan profiles show that Pb, Fe, and Ag have large fluctuations in some test spots (Figure 7b,e), indicating the presence of inclusions. The other trace elements have small variations in their concentration and smooth curves, indicating that they mainly occur as isomorphism in sphalerite particles.

5.2. Substitution Mechanisms of Zn by Trace Elements

5.2.1. Iron Element

Fe is the most enriched trace element in sphalerite from the Nanmushu deposit. The correlation between Fe and other trace elements can reveal whether Fe facilitates the substitution of Zn by other trace elements in sphalerite. The scatter plot shows that Fe has a positive correlation with Mn, Pb, and Tl, with correlation coefficients (R^2) of 0.66, 0.58, and 0.79, respectively (Figure 8a–c). Previous studies have suggested that divalent cations, such as Fe^{2+}, Mn^{2+}, and Cd^{2+}, can directly replace Zn^{2+} in sphalerite because they have similar ion radii and oxidation states in tetrahedral coordination [3,5,40,41]. Pb^{2+} can also enter the sphalerite structure by simple substitution for Zn^{2+} [38,42], while monovalent (Ag^+ and Cu^+) or trivalent (In^{3+}, Ga^{3+}, Fe^{3+}, and Tl^{3+}) cations can enter the sphalerite structure by coupled substitution for Zn^{2+} [43]. Therefore, Fe^{2+} in sphalerite can easily form isomorphism with Mn^{2+}, Pb^{2+}, and Tl^{3+} by substituting Zn^{2+}.

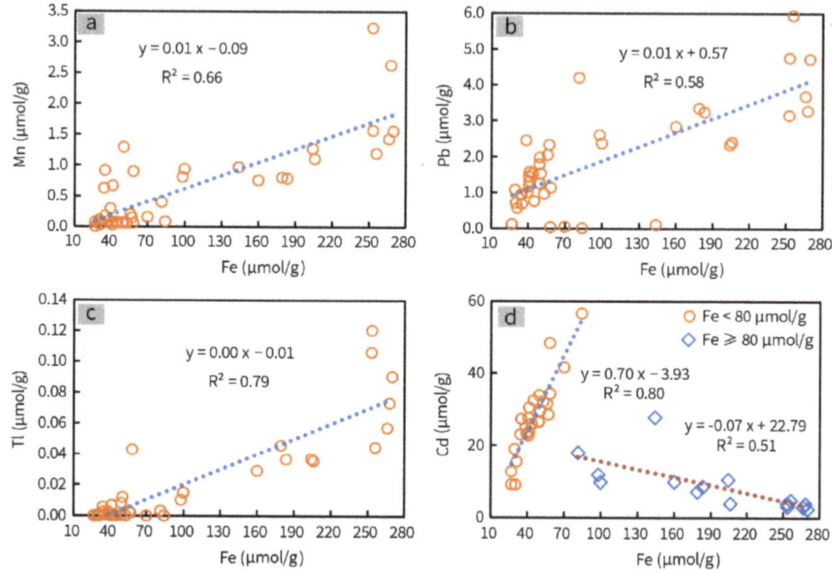

Figure 8. Plots of (**a**) Fe vs. Mn; (**b**) Fe vs. Pb; (**c**) Fe vs. Tl; (**d**) Fe vs. Cd.

The scatter plot also shows that Cd has a different relationship with Fe than other trace elements. Cd has a positive correlation with Fe at lower concentrations ($R^2 = 0.82$), but it has a negative correlation with Fe at higher concentrations ($R^2 = 0.51$) (Figure 8d). This suggests that low concentration Fe occurs as an isomorph in sphalerite and co-substitutes Zn^{2+} with Cd^{2+}, while high concentration Fe may be mainly due to Fe mineral inclusions (Figure 4d), which have low Cd content but high Mn, Pb, and Tl content (Figure 8a–c). Therefore, the occurrence state of Fe in sphalerite affects the substitution of Cd for Zn.

5.2.2. Cadmium Element

Sphalerite is the main carrier of the dispersed element Cd [4,44]. The scatter plot shows that Cd has a good positive correlation with Hg ($R^2 = 0.83$) (Figure 9a), but it has a negative correlation with Pb, Mn, and Tl ($R^2 = 0.44$, 0.35, and 0.39, respectively) (Figure 9b–d). In contrast, Pb and Mn, Pb and Tl, and Mn and Tl have positive correlations with each other ($R^2 = 0.32$, 0.51, and 0.74, respectively) (Figure 9e–g). Overall, (Cd + Hg) and (Mn + Pb + Tl) have a good negative correlation ($R^2 = 0.49$) (Figure 9h). This implies that the trace elements Cd and Hg coexist in sphalerite, or Mn, Pb, and Tl co-substitute Zn in sphalerite; that is, $(2Mn, 2Pb, 3Tl)^{6+}, 3(Cd, Hg)^{2+}) \leftrightarrow 3Zn^{2+}$.

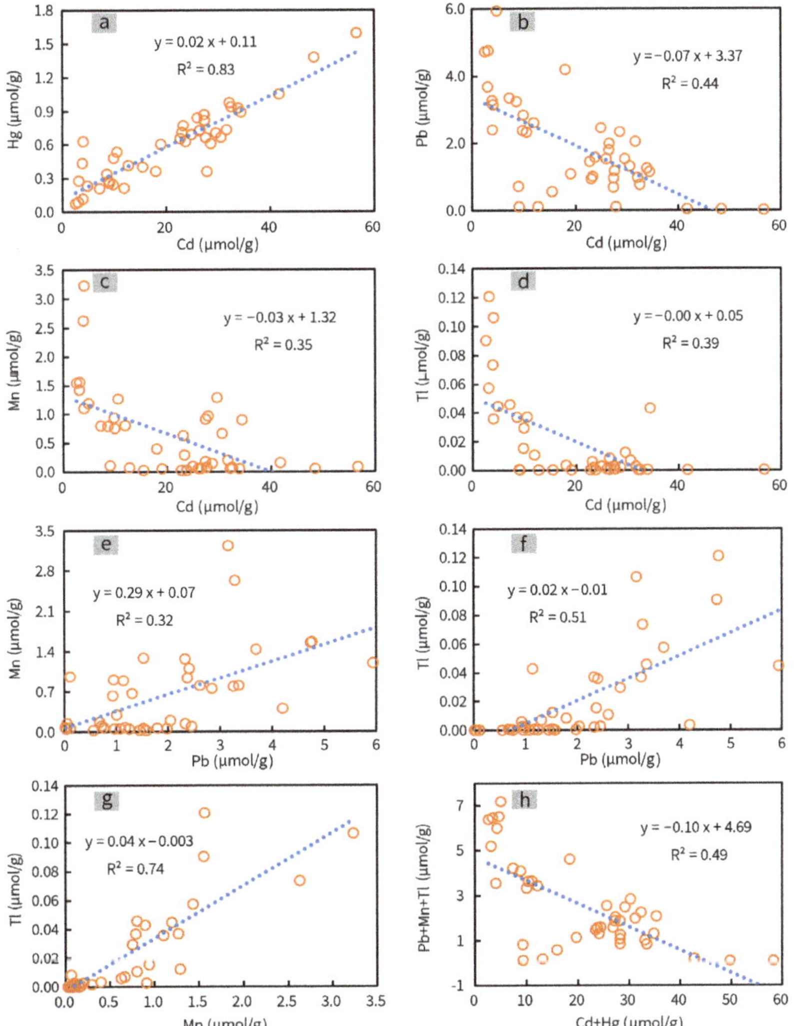

Figure 9. Plots of (**a**) Cd vs. Hg; (**b**) Cd vs. Pb; (**c**) Cd vs. Mn; (**d**) Cd vs. Tl; (**e**) Pb vs. Mn; (**f**) Pb vs. Tl; (**g**) Mn vs. Tl; (**h**) (Cd + Hg) vs. (Pb + Mn + Tl).

5.2.3. Germanium Element

The present study shows that germanium occurs in sphalerite mainly in two forms: independent minerals (mineral inclusions) and isomorphism. Ge minerals are rarely found in sphalerite, and most Ge occurs as isomorphs. However, the substitution mechanism of Ge varies depending on the deposits. Even within the same deposit, Ge may have multiple ways of replacing Zn. For example, Ge can directly replace Zn in sphalerite:

$Ge^{4+} \leftrightarrow 2Zn^{2+}$ or $Ge^{2+} \leftrightarrow Zn^{2+}$ [5,45,46]. In the Saint-Salvy deposit in France, Ge has two other substitution methods: $2Cu^+ + Cu^{2+} + Ge^{4+} \leftrightarrow 4Zn^{2+}$ [47] and $2Ag^+/Cu^+ + Ge^{4+} \leftrightarrow 3Zn^{2+}$ [40]. In the lead–zinc deposits around the Yangtze Block in China, Ge has more diverse substitution methods, such as $2Cu^+ + Ge^{4+} \leftrightarrow 3Zn^{2+}$ [31,48,49], $2Fe^{2+} + Ge^{4+} + \square \leftrightarrow 4Zn^{2+}$ ($\square$ represents vacancy) [50], $nCu^{2+} + Ge^{2+} \leftrightarrow (n + 1) Zn^{2+}$ [51], $Fe + Ge \leftrightarrow 2Zn$ [52], and $Mn^{2+} + Ge^{2+} \leftrightarrow 2(Zn, Cd)^{2+}$ [53] etc.

The scatter plot shows that Ge has a good correlation with Cu in the sphalerite of the Nanmushu deposit ($R^2 = 0.82$) (Figure 10a). Ge and Ga and Ga and Cu also have positive correlations with each other, with R^2 values of 0.47 and 0.56, respectively (Figure 10b,c). Overall, (Ge + Ga) and Cu have a better positive correlation, with R^2 reaching 0.84 (Figure 10d). In minerals, Ge usually has 2 oxidation states of +4 and +2, and Cu usually has 2 oxidation states of +2 and +1. μ-XANES studies indicate that Ge and Cu mainly exist as Ge^{4+} and Cu^+ in sphalerite, rather than +2 valence [54–56]. It has been shown that Ga often replaces Zn^{2+} with +3 valence in sphalerite [5,41,57,58]. Therefore, a possible substitution mechanism of Ge and Ga in the Nanmu deposit is $Ge^{4+} + 2Cu^+ \leftrightarrow 3Zn^{2+}$ or $2Ga^{3+} + 2Cu^+ \leftrightarrow 4Zn^{2+}$.

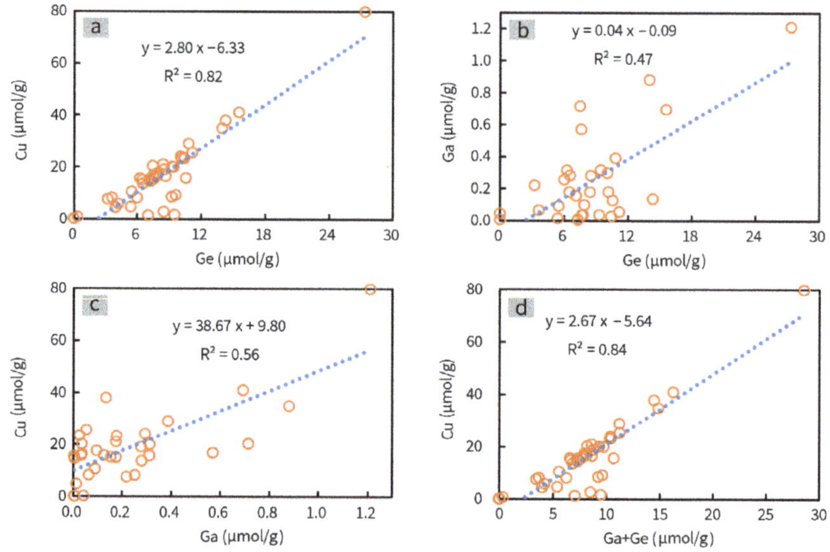

Figure 10. Plots of (**a**) Ge vs. Cu; (**b**) Ge vs. Ga; (**c**) Ga vs. Cu; (**d**) (Ga + Ge) vs. Cu.

5.3. Ore-Forming Temperature

Previous studies have suggested that the trace elements in sphalerite are related to the ore-forming temperature. Since Fe^{2+}, Mn^{2+}, In^{3+}, and Zn^{2+} have very similar ionic radii, and Se, Te, and S have similar geochemical characteristics, they can easily substitute each other in sphalerite under relatively high temperature conditions [3,59]. Dark sphalerite with high concentrations of Fe, Mn, In, Se, and Te is usually formed at higher temperatures with higher In/Ga ratios, while light sphalerite with high concentrations of Cd, Ga, and Ge is often formed at lower temperatures with lower In/Ge ratios [50,60]. The sphalerite in the Nanmushu deposit is dominated by tawny or brown and is characterized by Cd and Ge enrichment with low In/Ge ratios (<0.16), suggesting a lower formation temperature.

Frenzel et al. (2016) established a relationship between the trace element compositions of sphalerite and homogenization temperatures and concluded that Mn, Fe, and In increase while Ga and Ge decrease in sphalerite with increasing temperatures of deposition [1]. A strong correlation between PC 1* and the homogenization temperature, as suggested by Frenzel et al. (2016), shows that this expression can be used as a geothermometer (GGIMFis) and is widely applied in the study of ore-forming temperatures [61–63]. The empirical relationship between PC 1* and the homogenization temperature is as follows:

$$T\ (^{\circ}C) = -(54.4 \pm 7.3) \times PC\ 1^* + (208 \pm 10)$$

The expression of PC 1* is:

$$PC\ 1^* = \ln\left[\left(C_{Ga}^{0.22} \times C_{Ge}^{0.22}\right) / \left(C_{Fe}^{0.37} \times C_{Mn}^{0.20} \times C_{In}^{0.11}\right)\right]$$

with Ga, Ge, In, and Mn concentrations in ppm, while Fe concentration is in wt%.

From this study, the ore-forming temperature of the Nanmushu deposit calculated by the GGIMFis geothermometer ranges from 24 to 199 °C, with an average of 116 °C. As the samples in this study are not extensive, the calculated temperature range is lower than that of previous homogenization temperatures of fluid inclusions in sphalerite, quartz, dolomite, barite, and calcite from the Nanmushu deposit (107–340 °C) [19]. Based on the two geothermometer criteria discussed above, it can be inferred that the Nanmushu deposit originated from a medium-low temperature environment.

5.4. Genesis of Ore Deposit

The trace elements in sphalerite can reflect the genetic type of ore deposits because they are influenced by the physical and chemical conditions of mineralization, the sources of ore-forming materials, the fluid migration and precipitation mechanism, and the types of minerals [3,5,64]. Generally, epithermal deposits have high concentrations of Fe, Cu, Ga, In, Mn, and Sn and low concentrations of Ge and Pb; MVT Pb–Zn deposits have high concentrations of Ag, Ge, and Sb and low concentrations of Co, Cu, Fe, In, and Mn; SEDEX Pb–Zn deposits have high concentrations of Ag, Fe, Pb, and Sb and low concentrations of Cd, Co, Ga, Ge, and Mn; Skarn Pb–Zn deposits have high concentrations of Fe, Mn, Co, and In and low concentrations of Ga, Sb, and Ag; and VMS Pb–Zn deposits have high concentrations of Cd, Fe, Ga, and Mn and low concentrations of Cu, Ge, Pb, Ag, and Sn (Figure 11). The Nanmushu deposit has a trace element composition similar to that of MVT Pb–Zn deposits.

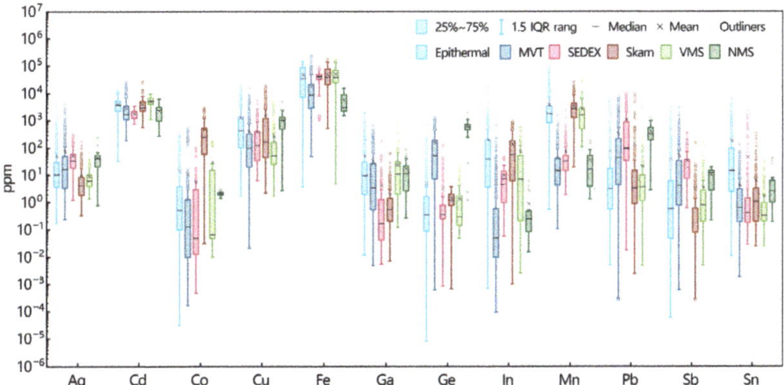

Figure 11. Box plots of trace elements in sphalerite from the Nanmushu deposit and five Pb–Zn deposit types. Epithermal data are collected from references [5,13,14,65–71]; MVT data are collected from references [3,5,18,30,41–43,50,66,72–74]; SEDEX data are collected from references [75–79]; Skarn data are collected from references [3,5,13,14,67,80–82]; and VMS data are collected from references [3,5,66,83,84].

In the lnGa–lnIn diagram of sphalerite trace elements [85,86], the samples from the Nanmushu deposit mostly fall into the field of sedimentary-reworked deposit (Figure 12).

Moreover, the ternary plots of Ag–(Ga + Ge)–(In + Se + Te) (Figure 13a) and Cd–Mn–1000 Ge (Figure 13b) of sphalerite are often used to determine the genesis of lead–zinc deposits [87–89]. In these ternary plots, the Nanmushu sphalerite also falls into the MVT field, which is clearly different from the SEDEX, VMS, and Skarn deposits.

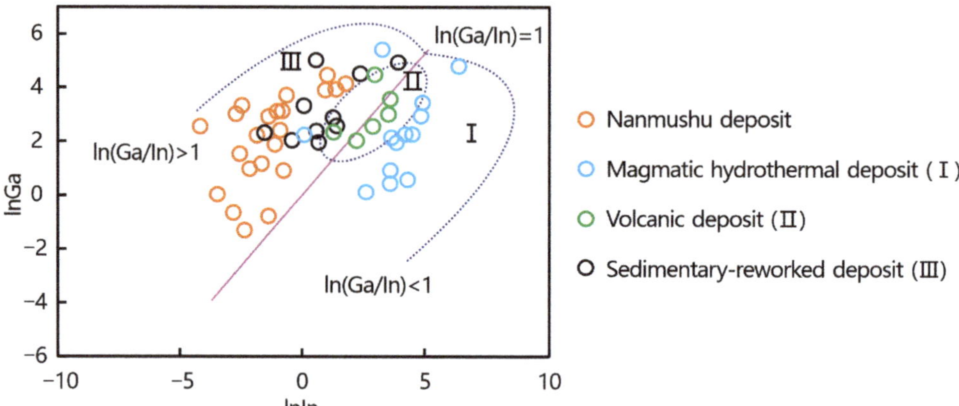

Figure 12. Plot of ln(Ga) vs. ln(In) for sphalerite from different Pb–Zn deposit types (modified after reference [85,86]).

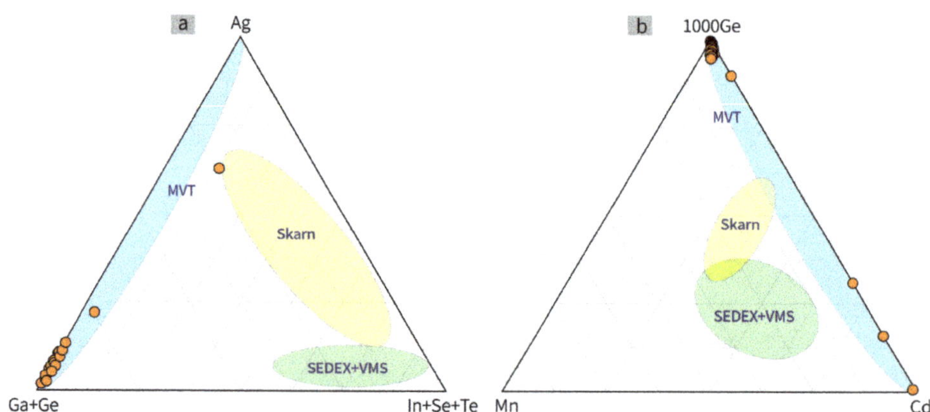

Figure 13. Ternary Ag–(Ga + Ge)–(In + Se + Te) plot (**a**) and Cd–Mn–1000Ge plot (**b**) of sphalerite from the Daliang deposit (modified after reference [87–89]).

In the Ge–In and Ge–Mn relation diagrams of sphalerite trace elements, most samples from the Nanmushu deposit plot into the range of the MVT deposit. However some test points deviate from the MVT field (Figure 14a,b) which is related to the Ge enrichment of sphalerite in the Nanmushu deposit. In the Cd/Fe–Mn and Mn–Fe diagrams, the test points from the Nanmushu deposit almost all plot into the range of the MVT deposit (Figure 14c,d).

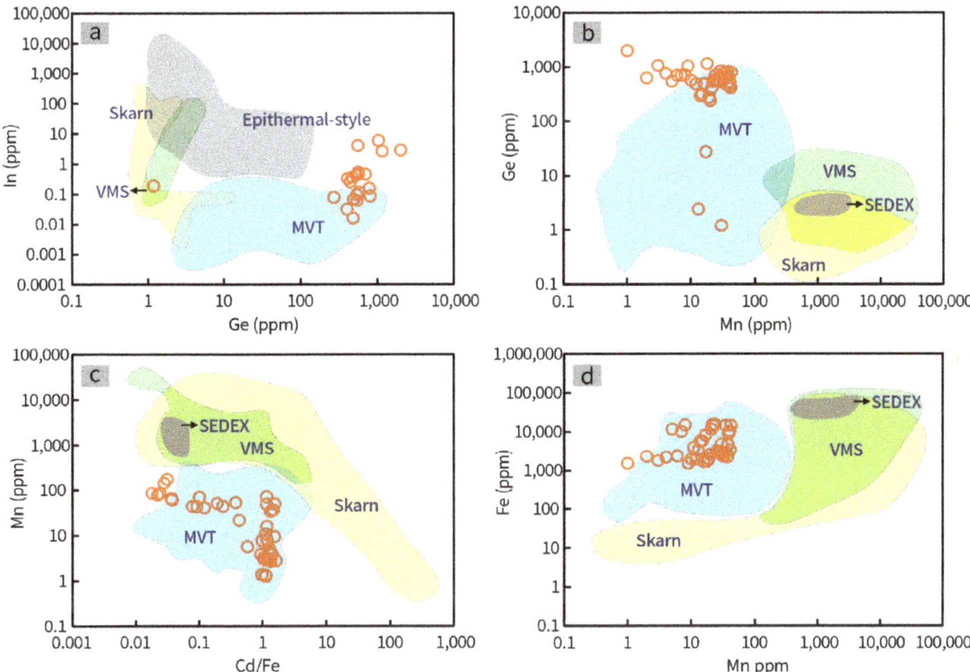

Figure 14. Plots of Ge vs. In. (**a**): Mn vs. Ge; (**b**) Cd/Fe vs. Mn; (**c**) Mn vs. Fe (**d**); (modified after reference [41]).

The ore bodies from the Nanmushu deposit are stratiform or stratiform-like along the brecciform dolostone of the Sinian Dengying Formation. They are related to the paleo oil reservoir but not associated with igneous activity [21,25,26]. Previous studies show that the ore-forming ages are mainly 468–488 Ma [21–23], suggesting that Nanmu belongs to an epigenetic hydrothermal deposit. Furthermore, the ore-forming fluids are basinal brines trapped in a heterogenous fluid system with medium-low temperature and medium-high salinity [19]. Sulfur and lead isotopes indicate that ore-forming sulfur originated from the thermochemical sulfate reduction (TSR) of seawater or evaporitic sourced sulfates and a mixing of lead from the basement and the host rocks [16]. The Cambrian Guojiaba Formation may have provided not only the oil for the paleo oil reservoir but also the ore metals for the Pb–Zn mineralization [25]. As reducing agents of TSR, reaction hydrocarbon organic matter plays an important role in mineralization [21,26]. This study and previous data revealed that the features of Nanmushu are consistent with those of MVT Pb–Zn deposit [90,91]. Therefore, we concluded that Nanmushu belongs to an MVT deposit.

6. Conclusions

(1) Sphalerite from the Nanmushu deposit contains various trace elements, such as Mn, Fe, Cu, Ga, Ge, Ag, Cd, Pb, Co, Hg, Tl, In, Sn, and Sb. Among them, Ag, Ge, Cd, and Cu are valuable components that may be recovered during mineral processing or smelting techniques.

(2) Most of the trace elements in sphalerite occur as isomorphs, while a small amount of Pb, Fe, and Ag occur as tiny mineral inclusions. Zn^{2+} can be easily substituted by Fe^{2+}, Mn^{2+}, Pb^{2+}, and Tl^{3+} in the form of isomorphism, Cd^{2+} and Hg^{2+} together, or Mn^{2+}, Pb^{2+}, and Tl^{3+} together to replace Zn^{2+}; that is, $(3Mn, 3Pb, 2Tl)^{6+}$, $3(Cd, Hg)^{2+}$) $\leftrightarrow 3Zn^{2+}$. Cu^+ exhibits a tendency to combine with Ge^{4+} or Ga^{3+} to substitute for Zn^{2+}; that is, $Ge^{4+} + 2Cu^+ \leftrightarrow 3Zn^{2+}$ or $2Ga^{3+} + 2Cu^+ \leftrightarrow 4Zn^{2+}$.

(3) Trace element compositions of the Nanmushu sphalerite indicate that Zn mineralization occurred under low-temperature conditions (<200 °C). Based on previous studies, we comprehensively believe that Nanmushu is an MVT deposit.

Author Contributions: Conceptualization, H.D., Y.C. and S.X.; investigation, J.W., Y.C., S.X. and Q.N.; data curation, J.W.; writing—original draft preparation, J.W.; writing—review and editing, H.D., Y.C., S.X., Y.W. and P.L.; supervision, H.D. and Y.C.; funding acquisition, H.D. and Y.C. All authors have read and agreed to the published version of the manuscript.

Funding: This research was funded by the National Natural Science Foundation of China [grant number 51764023], the Research Fund of the Yunnan Provincial Department of Education [grant number 2022J1304], and the Kunming University of Science and Technology 2020 Top Innovative Talents Program for postgraduates- [grant number CA22369M111A].

Data Availability Statement: The data presented in this study are available on reasonable request from the corresponding authors.

Acknowledgments: We would like to express special thanks to Jiaxi Zhou for the invitation. The authors thank the three anonymous reviewers for their valuable comments and suggestions for improving the manuscript.

Conflicts of Interest: The authors declare no conflict of interest.

References

1. Frenzel, M.; Hirsch, T.; Gutzmer, J. Gallium, germanium, indium, and other trace and minor elements in sphalerite as a function of deposit type—A meta-analysis. *Ore Geol. Rev.* **2016**, *76*, 52–78. [CrossRef]
2. Luo, K.; Cugerone, A.; Zhou, M.; Zhou, J.; Sun, G.; Xu, J.; He, K.; Lu, M. Germanium enrichment in sphalerite with acicular and euhedral textures: An example from the Zhulingou carbonate-hosted Zn(-Ge) deposit, South China. *Miner. Deposita* **2022**, *57*, 1343–1365. [CrossRef]
3. Ye, L.; Cook, N.J.; Ciobanu, C.L.; Liu, Y.; Zhang, Q.; Liu, T.; Gao, W.; Yang, Y.; Danyushevskiy, L. Trace and minor elements in sphalerite from base metal deposits in South China: A LA-ICPMS study. *Ore Geol. Rev.* **2011**, *39*, 188–217. [CrossRef]
4. Liu, Y.C.; Hou, Z.Q.; Yue, L.L.; Ma, W.; Tang, B.L. Critical metals in sediment-hosted Pb-Zn deposits in China. *Chin. Sci. Bull.* **2022**, *67*, 406–424. [CrossRef]
5. Cook, N.J.; Ciobanu, C.L.; Pring, A.; Skinner, W.; Shimizu, M.; Danyushevsky, L.; Saini-eidukat, B.; Melcher, F. Trace and minor elements in sphalerite: A LA-ICPMS study. *Geochim. Cosmochim. Acta* **2009**, *73*, 4761–4791. [CrossRef]
6. Sun, G.; Zhou, J. Application of Machine Learning Algorithms to Classification of Pb–Zn Deposit Types Using LA–ICP–MS Data of Sphalerite. *Minerals* **2022**, *12*, 1293. [CrossRef]
7. Sahlström, F.; Arribas, A.; Dirks, P.; Corral, I.; Chang, Z. Mineralogical Distribution of Germanium, Gallium and Indium at the Mt Carlton High-Sulfidation Epithermal Deposit, NE Australia, and Comparison with Similar Deposits Worldwide. *Minerals* **2017**, *7*, 213. [CrossRef]
8. Jaskula, B. Gallium. In *US Geological Survey Mineral Commodity Summaries*; US Geological Survey: Reston, VA, USA, 2017; pp. 64–65. Available online: https://minerals.usgs.gov/minerals/pubs/commodity/gallium/mcs-2017-galli.pdf (accessed on 15 September 2017).
9. Guberman, D. Germanium. In *US Geological Survey Mineral Commodity Summaries*; US Geological Survey: Reston, VA, USA, 2017; pp. 70–71. Available online: https://minerals.usgs.gov/minerals/pubs/commodity/germanium/mcs-2017-germa.pdf (accessed on 15 September 2017).
10. Li, X.F.; Zhu, Y.T.; Xu, J. Indium as a critical mineral: A research progress report. *Chin. Sci. Bull.* **2020**, *65*, 3678–3687. [CrossRef]
11. Werner, T.; Mudd, G.M.; Jowitt, S.M. The world's by-product and critical metal resources part III: A global assessment of indium. *Ore Geol. Rev.* **2017**, *86*, 939–956. [CrossRef]
12. Hayes, S.M.; Mccullough, E.A. Critical minerals: A review of elemental trends in comprehensive criticality studies. *Resour. Policy* **2018**, *59*, 192–199. [CrossRef]
13. Zhang, J.; Shao, Y.; Liu, Z.; Chen, K. Sphalerite as a record of metallogenic information using multivariate statistical analysis: Constraints from trace element geochemistry. *J. Geochem. Explor.* **2022**, *232*, 106883. [CrossRef]
14. Xing, B.; Mao, J.; Xiao, X.; Liu, H.; Zhang, C.; Guo, S.; Li, H.; Huang, W.; Lai, C. Metallogenic discrimination by sphalerite trace element geochemistry: An example from the Fengyan Zn-Pb deposit in central Fujian, SE China. *Ore Geol. Rev.* **2022**, *141*, 104651. [CrossRef]
15. Hou, M.T.; Wang, D.G.; Deng, S.B.; Yang, Z.R. Geology and genesis of the Mayuan lead-zinc mineralization belt in the Shaanxi Province. *Northwest. Geol* **2007**, *40*, 42–60. (In Chinese with English Abstract)

16. Xiong, S.; Jiang, S.; Ma, Y.; Liu, T.; Zhao, K.; Jiang, M.; Zhao, H. Ore genesis of Kongxigou and Nanmushu Zn-Pb deposits hosted in Neoproterozoic carbonates, Yangtze Block, SW China: Constraints from sulfide chemistry, fluid inclusions, and in situ S-Pb isotope analyses. *Precambrian Res.* **2019**, *333*, 105405. [CrossRef]
17. Wang, X.H.; Xue, C.J.; Li, Z.M.; Li, Q.; Yang, R.J. Geological and geochemical characteristics of Mayuan Pb-Zn ore deposit on northern margin of Yangtze landmass. *Miner. Depos.* **2008**, *27*, 37–48. (In Chinese with English Abstract)
18. Gao, Y.B.; Li, K.; Qian, B.; Li, W.Y.; Zheng, M.C.; Zhang, C.G. Trace elements, S, Pb, He, Ar and C isotopes of sphalerite in the Mayuan Pb-Zn deposit, at the northern margin of the Yangtze Plate, China. *Acta Petrol. Sin.* **2016**, *32*, 251–263. (In Chinese with English Abstract)
19. Xiong, S.-F.; Gong, Y.-J.; Yao, S.-Z.; Shen, C.-B.; Ge, X.; Jiang, S.-Y. Nature and Evolution of the Ore-Forming Fluids from Nanmushu Carbonate-Hosted Zn-Pb Deposit in the Mayuan District, Shaanxi Province, Southwest China. *Geofluids* **2017**, *2017*, 2410504. [CrossRef]
20. Wang, H.; Wang, J.B.; Zhu, X.Y.; Cheng, X.Y.; Jiang, B.B. Discussion of ore-forming material source and sulfur and lead isotopic geochemistry of Mayuan Pb-Zn deposit in Shaanxi. *Miner. Resour. Geol.* **2017**, *31*, 241–247. (In Chinese with English Abstract)
21. Wang, G.; Huang, Z.; Zhao, F.; Li, N.; Fu, Y. The relationship between hydrocarbon accumulation and Mississippi Valley-type Pb-Zn mineralization of the Mayuan metallogenic belt, the northern Yangtze block, SW China: Evidence from ore geology and Rb-Sr isotopic dating. *Resour. Geol.* **2020**, *70*, 188–203. [CrossRef]
22. Song, Z.; Chen, C.; Yang, Y.; Zhang, Y.; Yin, L.; Li, H. Mineralization age and sources of ore-forming material of the Nanmushu Zn-Pb deposit in the Micangshan Tectonic Belt at the northern margin of the Yangtze Craton, China: Constraints from Rb-Sr dating and Sr-Pb isotopes. *Resour. Geol.* **2020**, *70*, 273–295. [CrossRef]
23. Li, H.M.; Chen, Y.C.; Wang, D.H.; Li, H.Q. Geochemistry and mineralization age of the Mayuan zinc deposit, Nanzheng, southern Shaanxi, China. *Geol. Bull. China* **2007**, *26*, 546–552. (In Chinese with English Abstract)
24. Hou, M.T. The relationship between organic matter and lead-zinc mineralization in the Mayuan lead-zinc deposit, Shaanxi Province. *Geol. China* **2009**, *36*, 861–870. (In Chinese with English Abstract)
25. Huang, Z.; Wang, G.; Li, N.; Fu, Y.; Lei, Q.; Mao, X. Genetic link between Mississippi Valley-Type (MVT) Zn-Pb mineralization and hydrocarbon accumulation in the Nanmushu, northern margin of Sichuan Basin, SW China. *Geochemistry* **2021**, *81*, 125805. [CrossRef]
26. Gao, Y.B.; Li, K.; Zhang, J.W.; Guo, W.; Han, Y.X.; Liu, T.H.; Chao, Y.Y. Geochemical characteristics of organic matters and metallogenesis of Mayuan Pb-Zn deposit in the northern margin of Yangtze Plate, China. *J. Earch Sci. Environ.* **2021**, *43*, 276–290. (In Chinese with English Abstract)
27. Chen, B.Y. Study on Geological Characteristics of Pb-Zn Deposits in Mayuan, Nanzheng County. Master's Thesis, Chang'an University, Xi'an, China, 2013. (In Chinese with English Abstract)
28. Qi, W.; Hou, M.T.; Wang, K.M.; Wang, D.G. A large-scale stratabound lead-zinc metallogenic belt discovered in the Mayuan area. *Geol. Bull. China* **2004**, *23*, 1139–1142. (In Chinese with English Abstract)
29. Li, H.M.; Wang, D.H.; Zhang, C.Q.; Chen, Y.C.; Li, L.X. Characteristics of trace and rare earth elements in minerals from some typical lead-zinc deposits of Shaanxi Province. *Miner. Depos.* **2009**, *28*, 434–448. (In Chinese with English Abstract)
30. Hu, P.; Wu, Y.; Zhang, C.Q.; Hu, M.Y. Trace and minor elements in sphalerite from the Mayuan lead-zinc deposit, northern margin of the Yangtze Plate: Implications from LA-ICP-MS analysis. *Acta Miner. Sin.* **2014**, *34*, 461–468. (In Chinese with English Abstract)
31. Wu, Y.; Kong, Z.G.; Chen, M.H.; Zhang, C.Q.; Cao, L.; Tang, Y.J.; Yuan, X.; Zhang, P. Trace elements in sphalerites from the Mississippi Valley-type lead-zinc deposits around the margins of Yangtze Block and its geological implications: A LA-ICPMS study. *Acta Petrol. Sin.* **2019**, *35*, 3443–3460. (In Chinese with English Abstract)
32. Liu, S.W.; Li, R.X.; Liu, Y.H.; Zeng, R. Geochemical characteristics and metallogenic mechanism of the Mayuan Pb-Zn deposit on the northern margin of Yangtze Plate. *Acta Geosci. Sin.* **2016**, *37*, 101–110. (In Chinese with English Abstract)
33. Song, Z.J.; Chen, C.H.; Zhang, Y.; Yin, L.; Li, H.Z.; Huang, X.D. Metallogenic sources of Nanmushu Pb-Zn deposit in Mayuan area, Shaanxi Province: Constraint from hydrogen and oxygen isotopes and trace elements. *Miner. Depos.* **2018**, *37*, 985–1000. (In Chinese with English Abstract)
34. Han, Y.X.; Liu, Y.H.; Liu, S.W.; Lei, W.S.; Li, Z.; Li, Y.T.; Li, X. Origin of the breccia and metallogenic geological background of Mayuan Pb-Zn deposit. *Earth Sci. Front.* **2016**, *23*, 94–101. (In Chinese with English Abstract)
35. Hou, M.T.; Wang, D.G.; Yang, Z.R.; Gao, J. Geological characteristics of lead-zinc mineralized zones in the Mayuan area, Shaanxi, and their ore prospects. *Geol. China* **2007**, *34*, 101–109. (In Chinese with English Abstract)
36. Paton, C.; Hellstrom, J.; Paul, B.; Woodhead, J.; Hergt, J. Iolite: Freeware for the visualisation and processing of mass spectrometric data. *J. Anal. Atom. Spectrom.* **2011**, *26*, 2508. [CrossRef]
37. Wilson, S.A.; Ridley, I.; Koenig, A.E. Development of sulfide calibration standards for the laser ablation inductively-coupled plasma mass spectrometry technique. *J. Anal. Atom. Spectrom.* **2002**, *17*, 406–409. [CrossRef]
38. Han, R.; Qin, K.; Groves, D.I.; Hui, K.; Li, Z.; Zou, X.; Li, G.; Su, S. Ore-formation at the Halasheng Ag-Pb-Zn deposit, northeast Inner Mongolia as revealed by trace-element and sulfur isotope compositions of ore-related sulfides. *Ore Geol. Rev.* **2022**, *144*, 104853. [CrossRef]

39. Erlandsson, V.B.; Wallner, D.; Ellmies, R.; Raith, J.G.; Melcher, F. Trace element composition of base metal sulfides from the sediment-hosted Dolostone Ore Formation (DOF) CuCo deposit in northwestern Namibia: Implications for ore genesis. *J. Geochem. Explor.* **2022**, *243*, 107105. [CrossRef]

40. Belissont, R.; Boiron, M.; Luais, B.; Cathelineau, M. LA-ICP-MS analyses of minor and trace elements and bulk Ge isotopes in zoned Ge-rich sphalerites from the Noailhac—Saint-Salvy deposit (France): Insights into incorporation mechanisms and ore deposition processes. *Geochim. Cosmochim. Acta* **2014**, *126*, 518–540. [CrossRef]

41. Liu, S.; Zhang, Y.; Ai, G.; Xue, X.; Li, H.; Shah, S.A.; Wang, N.; Chen, X. LA-ICP-MS trace element geochemistry of sphalerite: Metallogenic constraints on the Qingshuitang Pb–Zn deposit in the Qinhang Ore Belt, South China. *Ore Geol. Rev.* **2022**, *141*, 104659. [CrossRef]

42. Yang, Q.; Zhang, X.; Ulrich, T.; Zhang, J.; Wang, J. Trace element compositions of sulfides from Pb-Zn deposits in the Northeast Yunnan and northwest Guizhou Provinces, SW China: Insights from LA-ICP-MS analyses of sphalerite and pyrite. *Ore Geol. Rev.* **2022**, *141*, 104639. [CrossRef]

43. Oyebamiji, A.; Hu, R.; Zhao, C.; Zafar, T. Origin of the Triassic Qilinchang Pb-Zn deposit in the western Yangtze block, SW China: Insights from in-situ trace elemental compositions of base metal sulphides. *J. Asian Earth Sci.* **2020**, *192*, 104292. [CrossRef]

44. Zhou, J.; Yang, Z.; An, Y.; Luo, K.; Liu, C.; Ju, Y. An evolving MVT hydrothermal system: Insights from the Niujiaotang Cd-Zn ore field, SW China. *J. Asian Earth Sci.* **2022**, *237*, 105357. [CrossRef]

45. Luo, K.; Zhou, J.X.; Xu, C.; He, K.J.; Wang, Y.B.; Sun, G.T. The characteristics of the extraordinary germanium enrichment in the Wusihe large-scale Ge-Pb-Zn deposit, Sichuan Province, China and its geological significance. *Acta Petrol. Sin.* **2021**, *37*, 2761–2777. (In Chinese with English Abstract)

46. Bonnet, J.; Cauzid, J.; Testemale, D.; Kieffer, I.; Proux, O.; Lecomte, A.; Bailly, L. Characterization of Germanium Speciation in Sphalerite (ZnS) from Central and Eastern Tennessee, USA, by X-ray Absorption Spectroscopy. *Minerals* **2017**, *7*, 79. [CrossRef]

47. Johan, Z. Indium and germanium in the structure of sphalerite: An example of coupled substitution with copper. *Miner. Petrol.* **1988**, *39*, 211–229. [CrossRef]

48. Hu, Y.S.; Ye, L.; Huang, Z.L.; Li, Z.L.; Wei, C.; Danyushevskiy, L. Distribution and existing forms of trace elements from Maliping Pb-Zn deposit in northeastern Yunnan, China: A LA-ICPMS study. *Acta Petrol. Sin.* **2019**, *35*, 3477–3495. (In Chinese with English Abstract)

49. Wei, C.; Ye, L.; Hu, Y.; Huang, Z.; Danyushevsky, L.; Wang, H. LA-ICP-MS analyses of trace elements in base metal sulfides from carbonate-hosted Zn-Pb deposits, South China: A case study of the Maoping deposit. *Ore Geol. Rev.* **2021**, *130*, 103945. [CrossRef]

50. Yuan, B.; Zhang, C.; Yu, H.; Yang, Y.; Zhao, Y.; Zhu, C.; Ding, Q.; Zhou, Y.; Yang, J.; Xu, Y. Element enrichment characteristics: Insights from element geochemistry of sphalerite in Daliangzi Pb–Zn deposit, Sichuan, Southwest China. *J. Geochem. Explor.* **2018**, *186*, 187–201. [CrossRef]

51. Ye, L.; Li, Z.L.; Hu, Y.S.; Huang, Z.L.; Zhou, J.X.; Fan, H.F.; Danyushevskiy, L. Trace elements in sulfide from the Tianbaoshan Pb-Zn deposit, Sichuan Province, China: A LA-ICPMS study. *Acta Petrol. Sin.* **2016**, *32*, 3377–3393. (In Chinese with English Abstract)

52. Zhou, J.X.; Yang, D.Z.; Yu, J.; Luo, K.; Zhou, Z.H. Ge extremely enriched in the Zhulingou Zn deposit, Guiding City, Guizhou Province, China. *Geol. China* **2021**, *48*, 665–666. (In Chinese with English Abstract)

53. Hu, Y.; Wei, C.; Ye, L.; Huang, Z.; Danyushevsky, L.; Wang, H. LA-ICP-MS sphalerite and galena trace element chemistry and mineralization-style fingerprinting for carbonate-hosted Pb-Zn deposits: Perspective from early Devonian Huodehong deposit in Yunnan, South China. *Ore Geol. Rev.* **2021**, *136*, 104253. [CrossRef]

54. Cook, N.J.; Ciobanu, C.L.; Brugger, J.; Etschmann, B.; Howard, D.L.; Jonge, M.D.D.; Ryan, C.; Paterson, D. Determination of the oxidation state of Cu in substituted Cu-In-Fe-bearing sphalerite μ-XANES spectroscopy. *Am. Miner.* **2011**, *97*, 476–479. [CrossRef]

55. Belissont, R.; Muñoz, M.; Boiron, M.; Luais, B.; Mathon, O. Distribution and oxidation state of Ge, Cu and Fe in sphalerite by μ-XRF and K-edge μ-XANES: Insights into Ge incorporation, partitioning and isotopic fractionation. *Geochim. Cosmochim. Acta* **2016**, *177*, 298–314. [CrossRef]

56. Liu, W.; Mei, Y.; Etschmann, B.; Glenn, M.; Macrae, C.M.; Spinks, S.C.; Ryan, C.G.; Brugger, J.; Paterson, D.J. Germanium speciation in experimental and natural sphalerite: Implications for critical metal enrichment in hydrothermal Zn-Pb ores. *Geochim. Cosmochim. Acta* **2023**, *342*, 198–214. [CrossRef]

57. Fan, M.; Ni, P.; Pan, J.; Wang, G.; Ding, J.; Chu, S.; Li, W.; Huang, W.; Zhu, R.; Chi, Z. Rare disperse elements in epithermal deposit: Insights from LA–ICP–MS study of sphalerite at Dalingkou, South China. *J. Geochem. Explor.* **2023**, *244*, 107124. [CrossRef]

58. Pring, A.; Wade, B.; Mcfadden, A.; Lenehan, C.E.; Cook, N.J. Coupled Substitutions of Minor and Trace Elements in Co-Existing Sphalerite and Wurtzite. *Minerals* **2020**, *10*, 147. [CrossRef]

59. Guo, F.; Wang, Z.L.; Xu, D.R.; Yu, D.S.; Dong, G.J.; Ning, J.T.; Kang, B.; Peng, E.K. Trace element characteristics of sphalerite in the Lishan Pb-Zn-Cu polymetallic deposit in Hunan Province and the metallogenic implications. *Earth Sci. Front.* **2020**, *27*, 66–81. (In Chinese with English Abstract)

60. Ye, L.; Gao, W.; Yang, Y.L.; Liu, T.G.; Peng, S.S. Trace elements in sphalerite in Laochang Pb-Zn polymetallic deposit, Lancang, Yunnan Province. *Acta Petrol. Sin.* **2012**, *28*, 1362–1372. (In Chinese with English Abstract)

61. Oyebamiji, A.; Falae, P.; Zafar, T.; Rehman, H.U.; Oguntuase, M. Genesis of the Qilinchang Pb–Zn deposit, southwestern China: Evidence from mineralogy, trace elements systematics and S–Pb isotopic characteristics of sphalerite. *Appl. Geochem.* **2023**, *148*, 105545. [CrossRef]

62. Shen, H.; Zhang, Y.; Zuo, C.; Shao, Y.; Zhao, L.; Lei, J.; Shi, G.; Han, R.; Zheng, X. Ore-forming process revealed by sphalerite texture and geochemistry: A case study at the Kangjiawan Pb–Zn deposit in Qin-Hang Metallogenic Belt, South China. *Ore Geol. Rev.* **2022**, *150*, 105153. [CrossRef]
63. Knorsch, M.; Nadoll, P.; Klemd, R. Trace elements and textures of hydrothermal sphalerite and pyrite in Upper Permian (Zechstein) carbonates of the North German Basin. *J. Geochem. Explor.* **2020**, *209*, 106416. [CrossRef]
64. Li, X.; Zhang, Y.; Li, Z.; Zhao, X.; Zuo, R.; Xiao, F.; Zheng, Y. Discrimination of Pb-Zn deposit types using sphalerite geochemistry: New insights from machine learning algorithm. *Geosci. Front.* **2023**, *14*, 101580. [CrossRef]
65. Bauer, M.E.; Burisch, M.; Ostendorf, J.; Krause, J.; Frenzel, M.; Seifert, T.; Gutzmer, J. Trace element geochemistry of sphalerite in contrasting hydrothermal fluid systems of the Freiberg district, Germany: Insights from LA-ICP-MS analysis, near-infrared light microthermometry of sphalerite-hosted fluid inclusions, and sulfur isotope geochemistry. *Miner. Depos.* **2019**, *54*, 237–262.
66. Frenzel, M.; Cook, N.J.; Ciobanu, C.L.; Slattery, A.D.; Wade, B.P.; Gilbert, S.; Ehrig, K.; Burisch, M.; Verdugo-ihl, M.R.; Voudouris, P. Halogens in hydrothermal sphalerite record origin of ore-forming fluids. *Geology* **2020**, *48*, 766–770. [CrossRef]
67. Lee, J.H.; Yoo, B.C.; Yang, Y.; Lee, T.H.; Seo, J.H. Sphalerite geochemistry of the Zn-Pb orebodies in the Taebaeksan metallogenic province, Korea. *Ore Geol. Rev.* **2019**, *107*, 1046–1067. [CrossRef]
68. Fan, X.; Lü, X.; Wang, X. Textural, chemical, isotopic and microthermometric features of sphalerite from the Wunuer deposit, Inner Mongolia: Implications for two stages of mineralization from hydrothermal to epithermal. *Geol. J.* **2020**, *55*, 6936–6958. [CrossRef]
69. Sun, G.; Zeng, Q.; Zhou, J.; Zhou, L.; Chen, P. Genesis of the Xinling vein-type Ag-Pb-Zn deposit, Liaodong Peninsula, China: Evidence from texture, composition and in situ S-Pb isotopes. *Ore Geol. Rev.* **2021**, *133*, 104120. [CrossRef]
70. Benites, D.; Torró, L.; Vallance, J.; Laurent, O.; Valverde, P.E.; Kouzmanov, K.; Chelle-michou, C.; Fontboté, L. Distribution of indium, germanium, gallium and other minor and trace elements in polymetallic ores from a porphyry system: The Morococha district, Peru. *Ore Geol. Rev.* **2021**, *136*, 104236. [CrossRef]
71. Chu, X.; Li, B.; Shen, P.; Zha, Z.; Lei, Z.; Wang, X.; Tao, S.; Hu, Q. Trace elements in sulfide minerals from the Huangshaping copper-polymetallic deposit, Hunan, China: Ore genesis and element occurrence. *Ore Geol. Rev.* **2022**, *144*, 104867. [CrossRef]
72. Zhuang, L.; Song, Y.; Liu, Y.; Fard, M.; Hou, Z. Major and trace elements and sulfur isotopes in two stages of sphalerite from the world-class Angouran Zn–Pb deposit, Iran: Implications for mineralization conditions and type. *Ore Geol. Rev.* **2019**, *109*, 184–200. [CrossRef]
73. Liu, Y.; Qi, H.; Bi, X.; Hu, R.; Qi, L.; Yin, R.; Tang, Y. Two types of sediment-hosted Pb-Zn deposits in the northern margin of Lanping basin, SW China: Evidence from sphalerite trace elements, carbonate C-O isotopes and molybdenite Re-Os age. *Ore Geol. Rev.* **2021**, *131*, 104016. [CrossRef]
74. Wei, C.; Ye, L.; Huang, Z.; Hu, Y.; Wang, H. In situ trace elements and S isotope systematics for growth zoning in sphalerite from MVT deposits: A case study of Nayongzhi, South China. *Min. Mag.* **2021**, *85*, 364–378. [CrossRef]
75. Cugerone, A.; Cenki-tok, B.; Muñoz, M.; Kouzmanov, K.; Oliot, E.; Motto-ros, V.; Legoff, E. Behavior of critical metals in metamorphosed Pb-Zn ore deposits: Example from the Pyrenean Axial Zone. *Miner. Depos.* **2021**, *56*, 685–705. [CrossRef]
76. Cave, B.; Lilly, R.; Barovich, K. Textural and geochemical analysis of chalcopyrite, galena and sphalerite across the Mount Isa Cu to Pb-Zn transition: Implications for a zoned Cu-Pb-Zn system. *Ore Geol. Rev.* **2020**, *124*, 103647. [CrossRef]
77. Cave, B.; Lilly, R.; Hong, W. The Effect of Co-Crystallising Sulphides and Precipitation Mechanisms on Sphalerite Geochemistry: A Case Study from the Hilton Zn-Pb (Ag) Deposit. *Aust. Miner.* **2020**, *10*, 797. [CrossRef]
78. Cave, B.; Perkins, W.; Lilly, R. Linking uplift and mineralisation at the Mount Novit Zn-Pb-Ag Deposit, Northern Australia: Evidence from geology, U–Pb geochronology and sphalerite geochemistry. *Geosci. Front.* **2022**, *13*, 101347. [CrossRef]
79. Maurer, M.; Prelević, D.; Mertz-kraus, R.; Pačevski, A.; Kostić, B.; Malbašić, J. Genesis and metallogenetic setting of the polymetallic barite-sulphide deposit, Bobija, Western Serbia. *Int. J. Earth Sci.* **2019**, *108*, 1725–1740. [CrossRef]
80. Zhao, Y.; Chen, S.; Tian, H.; Zhao, J.; Tong, X.; Chen, X. Trace element and S isotope characterization of sulfides from skarn Cu ore in the Laochang Sn-Cu deposit, Gejiu district, Yunnan, China: Implications for the ore-forming process. *Ore Geol. Rev.* **2021**, *134*, 104155. [CrossRef]
81. Xiong, Y.; Zhou, T.; Fan, Y.; Chen, J.; Wang, B.; Liu, J.; Wang, F. Enrichment mechanisms and occurrence regularity of critical minerals resources in the Yaojialing Zn skarn polymetallic deposit, Tongling district, eastern China. *Ore Geol. Rev.* **2022**, *144*, 104822. [CrossRef]
82. Xing, B.; Mao, J.; Xiao, X.; Liu, H.; Jia, F.; Wang, S.; Huang, W.; Li, H. Genetic discrimination of the Dingjiashan Pb-Zn deposit, SE China, based on sphalerite chemistry. *Ore Geol. Rev.* **2021**, *135*, 104212. [CrossRef]
83. Lu, S.; Ren, Y.; Yang, Q.; Sun, Z.; Hao, Y.; Sun, X. Ore Genesis for Stratiform Ore Bodies of the Dongfengnanshan Copper Polymetallic Deposit in the Yanbian Area, NE China: Constraints from LA-ICP-MS in situ Trace Elements and Sulfide S–Pb Isotopes. *Acta Geol. Sin.* **2019**, *93*, 1591–1606. [CrossRef]
84. Torró, L.; Benites, D.; Vallance, J.; Laurent, O.; Ortiz-benavente, B.A.; Chelle-michou, C.; Proenza, J.A.; Fontboté, L. Trace element geochemistry of sphalerite and chalcopyrite in arc-hosted VMS deposits. *J. Geochem. Explor.* **2022**, *232*, 106882. [CrossRef]
85. Zhang, Q. Application of diagram of trace elements in the sphalerite and galena to the genetic type of lead and zinc deposit. *Geol. Geochem* **1987**, 64–66. (In Chinese with English Abstract)
86. Zhou, J.; Huang, Z.; Zhou, G.; Li, X.; Ding, W.; Bao, G. Trace Elements and Rare Earth Elements of Sulfide Minerals in the Tianqiao Pb-Zn Ore Deposit, Guizhou Province, China. *Acta Geol. Sin.* **2011**, *85*, 189–199.

87. Zhu, L.M.; Yuan, H.H.; Luan, S.W. Typomorphic characteristics and their significance of minor elements of sphalerite from Disu and Daliangzi Pb-Zn deposits, Sichuan. *Acta Geol. Sichuan* **1995**, *15*, 49–55. (In Chinese with English Abstract)
88. Li, Z.L.; Ye, L.; Huang, Z.L.; Nian, H.L.; Zhou, J.X. Primary research on trace elements in sphalerite from Tianqiao Pb-Zn deposit, northwestern Guizhou Province, China. *Acta Miner. Sin* **2016**, *36*, 183–188. (In Chinese with English Abstract)
89. Chen, C.H.; Song, Z.J.; Yang, Y.L.; Yang, D.P.; Gu, Y.; Chen, X.J. Characteristics of trace elements of sphalerite and genesis of the Dongzigou lead-zinc deposit in Xishui County, Guizhou Province, China. *Acta Miner. Sin.* **2019**, *39*, 485–493. (In Chinese with English Abstract)
90. Leach, D.L.; Bradley, D.; Lewchuk, M.T.; Symons, D.T.; Marsily, G.D.; Brannon, J. Mississippi Valley-type lead–zinc deposits through geological time: Implications from recent age-dating research. *Miner. Depos.* **2001**, *36*, 711–740. [CrossRef]
91. Leach, D.; Sangster, D.; Kelley, K.; Large, R.; Garven, G.; Allen, C.; Gutzmer, J.; Walters, S. Sediment-hosted lead-zinc deposits: A global perspective. *Econ. Geol.* **2005**, *100th Anniversary Volume*, 561–607.

 *minerals*

MDPI

Article

Genesis of the Giant Huoshaoyun Non-Sulfide Zinc–Lead Deposit in Karakoram, Xinjiang: Constraints from Mineralogy and Trace Element Geochemistry

Xiang Chen [1], Dengfei Duan [1], Yuhang Zhang [1], Fanyan Zhou [1], Xin Yuan [1,2] and Yue Wu [1,*]

[1] Hubei Key Laboratory of Petroleum Geochemistry and Environment, College of Resources and Environment, Yangtze University, Wuhan 430100, China; chenxiang199730@163.com (X.C.); cugddf@163.com (D.D.); zyhasw1@163.com (Y.Z.); zhoufanyan310@163.com (F.Z.); cpx56836@163.com (X.Y.)

[2] Geology Team No. 272 of Guangxi Zhuang Autonomous Region, Nanning 530031, China

* Correspondence: leadzinc@163.com

Abstract: The Huoshaoyun zinc–lead deposit, a giant non-sulfide deposit in Xinjiang, is one of the most significant discoveries of zinc–lead deposit in China and globally in recent years. The deposit is dominated by zinc–lead non-sulfides, with minor occurrences of sulfides such as sphalerite, galena, and pyrite. The non-sulfide minerals include smithsonite, cerussite, anglesite, and Fe-oxide. This study focuses on the mineralogical characteristics of sulfide and non-sulfide ores, as well as the trace element characteristics of sphalerite, smithsonite, and Fe-oxide. Mineralogical analysis reveals that smithsonite is derived from the oxidation of primary sulfide minerals and can be classified into three types that are generated during different stages of supergene oxidation. The three types of smithsonite are formed through replacing the sphalerite and host limestone, as well as directly precipitating in the fissures and vugs. Trace element analysis of sphalerite indicates that it is rich in Cd, Tl, and Ge, but poor in Fe and Mn. The ore-forming temperature, calculated using the GGIMFis geothermometer, is mostly within the range of 100~150 °C. Moreover, the trace element characteristics, ore-forming temperature, and S and Pb isotope compositions of the sulfide ores of the Huoshaoyun deposit are similar to those of the Jinding and Duocaima MVT lead–zinc deposits, which are also located in the Eastern Tethyan zinc–lead belt. This suggests that the sulfide orebody in the Huoshaoyun Zn-Pb deposit could also be the MVT deposit. Study of the trace element of the non-sulfide minerals shows that the Mn and Cd are relatively enriched in smithsonite, while Ga, Ge, and Pb are enriched in Fe-oxide. This can be attributed to distinct geochemical properties of the trace elements in the non-sulfide minerals of the Huoshaoyun deposit and is consistent with those of the other oxidized MVT deposits, thus indicating the supergene oxidation process of this deposit.

Keywords: sphalerite; smithsonite; trace elements; ore deposit genesis; supergene oxidation; Huoshaoyun deposit

Citation: Chen, X.; Duan, D.; Zhang, Y.; Zhou, F.; Yuan, X.; Wu, Y. Genesis of the Giant Huoshaoyun Non-Sulfide Zinc–Lead Deposit in Karakoram, Xinjiang: Constraints from Mineralogy and Trace Element Geochemistry. *Minerals* **2023**, *13*, 842. https://doi.org/10.3390/min13070842

Academic Editor: Ryan Mathur

Received: 6 May 2023
Revised: 19 June 2023
Accepted: 21 June 2023
Published: 22 June 2023

1. Introduction

Non-sulfide zinc–lead deposits are one of the important sources of lead, zinc, and other metals globally [1–3]. In the early 20th century, lead and zinc were primarily sourced from sulfide ores [4]. In recent years, the development of beneficiation technologies has enabled the reutilization of non-sulfide zinc–lead resources, particularly those dominated by smithsonite [5,6]. The economic viability of non-sulfide deposits has significantly improved, making them a major potential source of lead and zinc metals in the 21st century [1,6]. The Xinjiang Huoshaoyun deposit, located in the northwestern part of the Qinghai-Tibet Plateau, is one of the most important recent discoveries of zinc–lead deposits in China and globally. As of 2016, the Huoshaoyun deposit had a total lead and zinc metal resource of 18 million tons [7], making it the largest zinc–lead deposit in China. The deposit is mainly composed of non-sulfide zinc–lead ores (Zn-Pb carbonates: smithsonite and cerussite), which account for more than 95% of the zinc–lead reserves, with a small amount of sulfide ores. However, there is currently a hot debate regarding the genesis of both the zinc–lead carbonate and zinc–lead sulfide ore bodies at the Huoshaoyun deposit. Some scholars believe that both the non-sulfide and sulfide zinc–lead ores are of SEDEX (sedimentary exhalative) type related to the deep-seated magmatic-hydrothermal fluids [8–11], while others suggest that the Huoshaoyun deposit was formed through the oxidation of primary MVT (Mississippi Valley type) deposit [12]. Gao et al. (2020) proposed that the Huoshaoyun deposit was neither an oxidized MVT deposit nor a SEDEX deposit, but rather a type of hypogene hydrothermal deposit [13].

Extensive research has demonstrated that the trace element composition of sphalerite is a reliable indicator for distinguishing the genesis types of zinc–lead deposits and for revealing the physicochemical conditions of ore formation [14–18]. Furthermore, during the oxidation process of zinc–lead sulfides, certain trace elements (e.g., Cd, Ge, Ga, and Tl) show distinctive patterns of selective enrichment. For example, the decomposition of Ge-bearing sphalerite can lead to Ge enrichment in Fe-oxy-hydroxide minerals during the supergene process at relatively high pH values [19].

These findings provide evidence for elucidating the potential oxidation process and determining the genesis of the Huoshaoyun deposit.

In this study, the mineralogical research was conducted using an optical microscope, cathodoluminescence (CL), and TIMA (TESCAN Integrated Mineral Analyzer) analyses, to define the mineral paragenesis. Moreover, we performed laser ablation inductively coupled plasma mass spectrometry (LA-ICPMS) to reveal the trace element compositions of both sulfide (sphalerite) and non-sulfide minerals in the Huoshaoyun deposit. The results, combined with previous studies, provide evidence of the genesis of both the Zn-Pb sulfide and Zn-Pb carbonates orebodies of the Huoshaoyun deposit.

2. Regional Geology

The Huoshaoyun zinc–lead deposit is situated in the Karakorum area, Xinjiang, in the northwestern part of the Qinghai-Tibet Plateau. It is located at 79°00′00″ E, 35°00′00″ N in Hetian County, Xinjiang, approximately 300 km north of Hotan City. It is a giant zinc–lead deposit located in the western region of the eastern Tethyan zinc–lead metallogenic belt in China (Figure 1) [8,20–22]. The main zinc–lead deposits of this metallogenic belt are distributed within the Paleogene to Neogene basins in the Lanping-Simao and Tianshuihai North Qiangtang terranes. The Huoshaoyun Zn-Pb deposit occurs within the Linjitang Basin of the Tianshuaihai North Qiangtang Terrane, bounded by the Altyn fault to the southeast, the Jinsha River fault zone to the north, the Karakorum fault to the southwest, and the Qiaoertianshan fault passing through the basin (Figure 2) [23–27]. The exposed strata in the region consist mainly of Triassic and Jurassic formations, from bottom to top: the Middle Triassic Heweitan Formation, the Upper Triassic Keleqinghe Formation, and the Middle Jurassic Longshan Formation [8]. This area is characterized by well-developed fault structures, including several sets of faults trending NW, NE, and nearly EW. Among them, the NW-trending faults, represented by the Qiaoertianshan and Karakorum deep faults, are the largest in scale (Figure 2). The Qiaoertianshan Fault is not only a regional tectonic boundary but also could be the main ore-controlling structure for copper, zinc–lead, and other polymetallic deposits in the area [11,27]. Along with the Qiaoertianshan Fault and its secondary faults, several medium- to large-sized deposits have been discovered, including the Duobaoshan zinc–lead deposit, Baotashan zinc–lead deposit, and Tianshuihai zinc–lead deposit (Figure 2). The regional intrusive rocks are not developed and mostly consist of small-sized, intermediate-acidic intrusions. The volcanic activity in the region is weak [7,8,28].

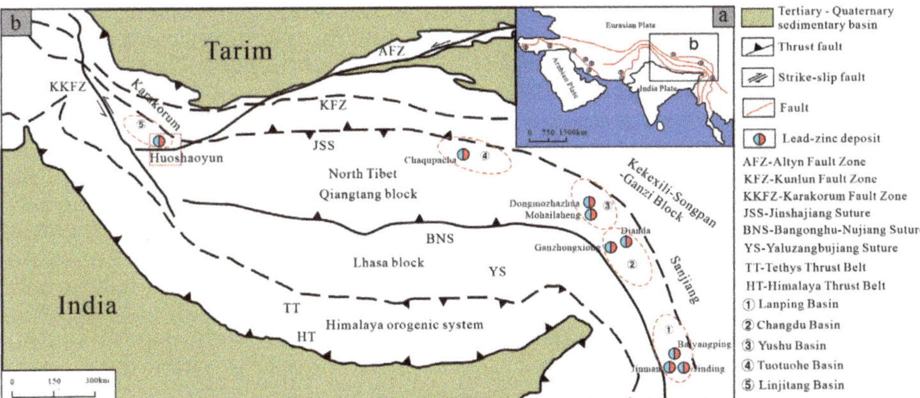

Figure 1. Distribution map of zinc–lead deposits in the Tethyan metallogenic belt [8]. (**a**). Simplified Diagram of the Tectonic Structure in the Western Tibetan Plateau. (**b**). Location of the Huoshaoyun deposit and the lead-zinc metallogenic basins within the Tibetan plateau. (**b**) is a detailed view of rectangle b in (**a**).

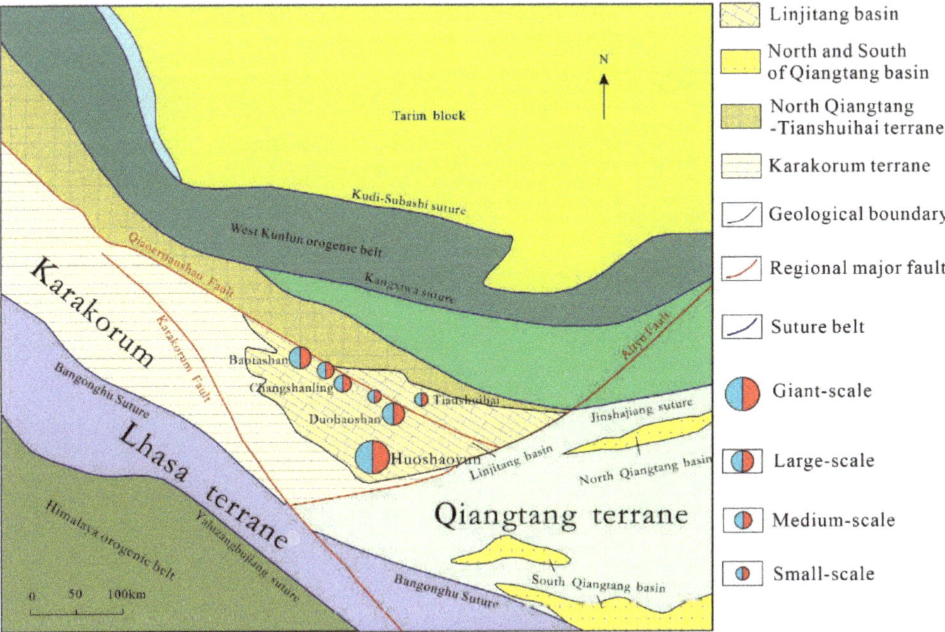

Figure 2. Geological map and distribution of the Huoshaoyun and other zinc–lead deposits in the Linjitang Basin [29].

3. Geology of the Huoshaoyun Deposit

The Huoshaoyun deposit is the largest zinc–lead deposit in China. It has over 18 Mt Zn-Pb metal reserves at an average grade of 23.58% zinc and 5.63% lead, as well as significant cadmium reserves [30]. It is noteworthy that the estimated resources of the studied deposit consist of 62 million tonnes of zinc–lead carbonate ore and 3 million tonnes of lead–zinc sulfide ore [22]. Based on these figures, the ratio of sulfidic to non-sulfidic ore in the deposit is relatively low, approximately 4.8%. This indicates that the majority of the ore deposit comprises zinc–lead carbonate ores, with a smaller proportion of lead–zinc sulfide ores. The mining area consists of the Upper Triassic Keleqinghe Formation, the Middle Jurassic Longshan Formation, and the Quaternary sedimentary group (Figure 3). A total of five Zn-Pb orebodies have been discovered in this deposit, with ore body I dominated by zinc–lead sulfide minerals and the others mainly composed of zinc–lead carbonate minerals. The orebodies are mainly stratiform, with dip angles ranging from 3° to 7° [26]. The ore-bearing host rocks of the deposit are mainly the Middle Jurassic Longshan Formation, with the second and fourth lithologic sections being the ore-bearing strata of the III and IV ore belts, and I and II ore belts, respectively. The second lithologic section of the Longshan Formation is composed mainly of argillaceous limestone, followed by bioclastic argillaceous limestone, oolitic limestone, and brecciated limestone, with occasional hematite mineralization. The fourth lithologic section is composed mainly of fine-grained limestone, with occasional hematite mineralization, and interbedded with bioclastic limestone, marly limestone, and mudstone. The underlying Keleqinghe Formation mainly consists of quartz sandstone and mudstone (Figures 3 and 4).

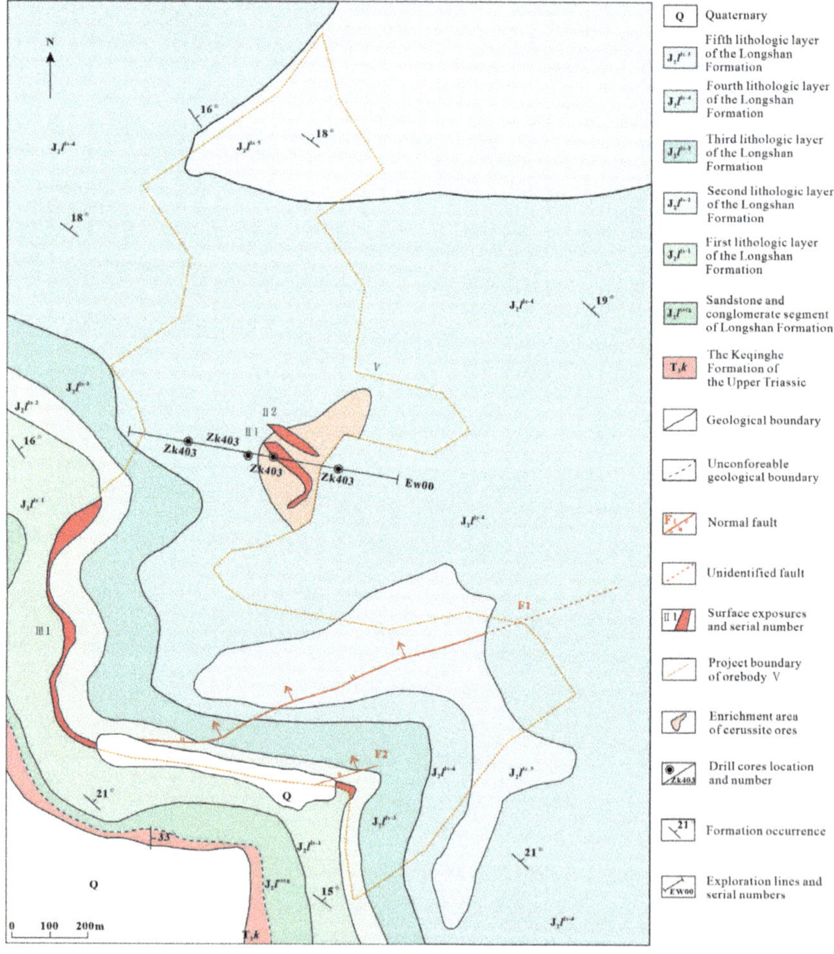

Figure 3. Geological structure, stratigraphy, and boundaries of the Huoshaoyun area [30].

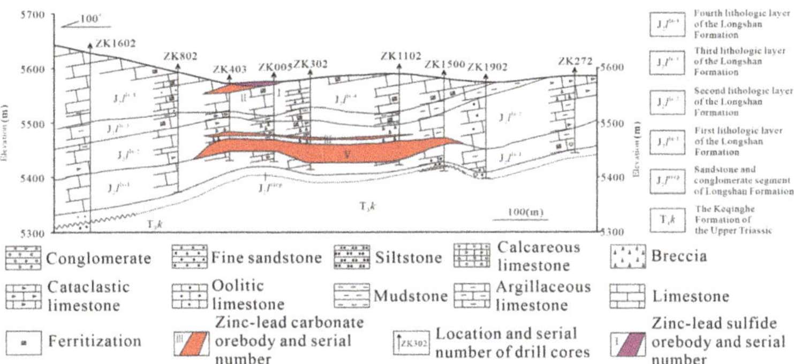

Figure 4. Distribution of ore bodies and their contact relationships with surrounding rocks in the Huoshaoyun zinc–lead deposit [30].

In the open pit of Huoshaoyun, mineralization can be classified into two parts: an upper sulfide orebody and a lower non-sulfide orebody (zinc–lead carbonates) (Figure 5a). The upper massive sulfide orebody is predominantly composed of galena, with small amounts of sphalerite and traces of pyrite (Figure 5b). The lower non-sulfide orebody is further divided into two parts: the upper part is composed of massive ore containing smithsonite as the primary mineral with rare cerussite (Figure 5d), while the lower part consists of a layered orebody composed of interbedded smithsonite and cerussite (Figure 5c,d). Near the fault, brecciated smithsonite ores are developed, containing cerussite, anglesite, and residual galena (Figure 5e). Moreover, gypsum, calcite, and other minerals can be observed locally within the smithsonite orebody (Figure 5f). The alteration of the wall rocks around the ore bodies is weak, mainly including calcification, pyritization, silicification, kaolinization, and hematite alteration [7,8,14,27].

Figure 5. The Zn-Pb mineralization in the open pit of the Huoshaoyun deposit. (**a**) Field photo of the upper Zn-Pb sulfide zone and lower non-sulfide mineralization zone in the Huoshaoyun open pit mine. (**b**) Galena-replaced limestone. (**c**) Interbedded smithsonite and cerussite. (**d**) Massive smithsonite orebody in the middle of the Huoshaoyun open-pit mine, underlying interbedded smithsonite and cerussite with minor galena remnant and anglesite. (**e**) Breccia of smithsonite ore near the fault with anglesite, cerussite, and galena remnant. (**f**) Gypsum developed in the cavity. Abbreviations: SZ—sulfide zone; SmtZ—smithsonite zone; SCZ—smithsonite and cerussite zone; Gn—galena; Smt—smithsonite; Cer—cerussite; Ang—anglesite; Gp—gypsum.

4. Sampling and Analytical Methods

The samples were collected from the open pit in the Huoshaoyun deposit. The collected samples were processed into thin sections, and representative thin sections and minerals were chosen for microscopic observation. Additionally, mineral automated quantitative analysis (TESCAN Integrated Mineral Analyzer—TIMA) and cathodoluminescence (CL) analysis were conducted, and LA-ICP-MS in situ trace element analysis was performed on selected minerals.

Microscopic observation was conducted under a Carl Zeiss Axio Scope.A1 microscope at the Hubei Key Laboratory of Petroleum Geochemistry and Environment, Yangtze University. Cathodoluminescence was performed using a CL8200MK5-2 cathodoluminescence instrument at the State Key Laboratory of Geological Processes and Mineral Resources at China University of Geosciences (Wuhan). The acceleration voltage of the cathodoluminescence instrument was set at 10–15 kV, with an electron beam current of 220–280 uA. The exposure time was set between 7 and 15 s, the gain was between 5 and 7, the saturation was 1.2, and the gamma value was between 1.3 and 1.8.

The TESCAN Integrated Mineral Analyzer (TIMA3 X GHM) system at the Xi'an Kuangpu Geological Exploration Technology Co., Ltd. (Xi'an, China) has obtained quantitative mineral abundances of these samples. The TIMA system comprises a TESCAN MIRA3 Schottky field emission SEM and nine detectors, including four high flux EDS detectors (EDAX Element 30) arranged at 90° intervals around the chamber. In this study, the dot mapping analysis mode was used with X-ray counts set to 1200, pixel spacing of BSE set to 3 μm, and dot spacing of EDS set to 9 μm. The measurements were conducted in a high vacuum environment, with an acceleration voltage of 25 kV, electricity of 9 nA, and a working distance of 15 mm. The electricity and BSE signals were calibrated by platinum Faraday cup and EDS signals by Mn standard. TIMA can automatically compare the measured BSE and EDS data of each different phase with the database and then distinguish their mineral phases and compute mineral abundances.

The LA-ICP-MS analysis was performed at Xi'an SampleSolution Analysis Technology Co., Ltd. (Wuhan, China). The laser beam spot size and frequency used in the analysis were 32 μm and 250 Hz, respectively. The single mineral trace element content was processed using multi-external standard without internal standard correction, with the use of NIST 610 and NIST 612 glass standard reference materials [31]. The USGS sulfide standard reference material MASS-1 was used as a monitoring standard to verify the reliability of the calibration method. Each time-resolved analysis data included approximately 20–30 s of blank signal and 50 s of sample signal. The offline processing of the analysis data, including the selection of sample and blank signals, correction for instrument sensitivity drift, and calculation of elemental content, was performed using the software ICPMSDataCal10.9 [31].

5. Results

5.1. Type and Texture of the Zn-Pb Ores

The Huoshaoyun deposit comprises three primary ore types: sulfide ore, mixed ore, and non-sulfide ore.

The sulfide ores are predominantly composed of massive fine-grained galena (Figure 6a) and oolitic sphalerite (Figure 6b). Galena commonly occurs as fine-grained in black massive ores (Figure 6a). It generally shows euhedral–subhedral granular texture and is intergrown with calcite (Figure 6e), locally replaced by anglesite when observed under the microscope (Figure 6c). Sphalerite appears as oolitic grains in hand specimens (Figure 6b) and frequently displays colloform textures in the microscope, with galena occurring frequently at both the edges and centers (Figure 6d). In addition, some samples also contain colloform galena and are enclosed by sphalerite (Figure 6f).

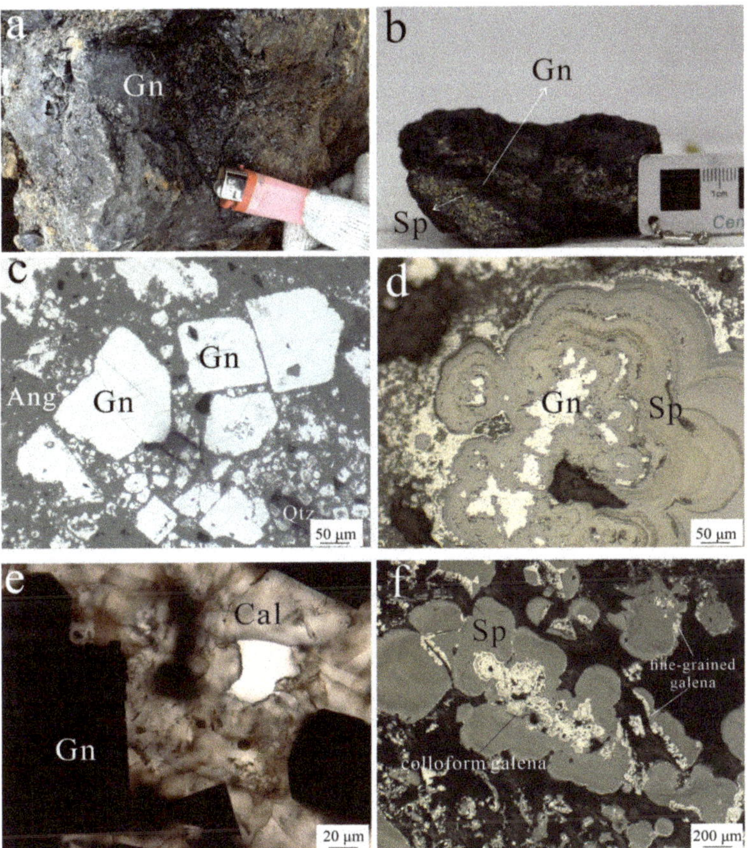

Figure 6. Textural features of sulfide ores in Huoshaoyun Zn-Pb deposit. (**a**) Massive fine-grained galena ore. (**b**) Oolitic sphalerite ore, with abundant oolitic sphalerite and galena. (**c**) Euhedral–subhedral fine-grained galena in the massive galena ore, which is partially replaced by anglesite. (**d**) Fine-grained galena enclosed by the colloform sphalerite. (**e**) Euhedral–subhedral galena intergrown with calcite. (**f**) Colloform galena is enveloped by colloform sphalerite. Abbreviations: Sp—sphalerite; Gn—galena; Ang—anglesite.

The mixed ore is predominantly composed of non-sulfide minerals, including smithsonite, cerussite, and Fe-oxide, with a minor amount of primary sulfides. The majority of these non-sulfide minerals are produced through direct replacement of sulfides or the host limestone (Figure 7a). Smithsonite, which exhibits gray oolitic textures under the microscope, is the most abundant non-sulfide mineral (Figure 7a). A small amount of extremely fine-grained smithsonite directly replaces sphalerite (Figure 7b). Galena and sphalerite occur together, with early-stage galena taking the form of dendritic–skeletal (Figure 7c) or massive blocks (Figures 6d and 7d) and being enveloped by colloform sphalerite (Figure 7d). A later shell of galena often forms in the outer part of the colloform sphalerite (Figures 6d,e and 7c,d). Anglesite appears relatively dull under the microscope and is usually distributed inside or on the edge of galena (Figure 7e,f). Cerussite is brighter and appears pale gray under the microscope, with a relatively complete crystal structure (Figure 7e,f). The replacement of galena by anglesite and cerussite is also observed (Figure 7e,f). Pyrite displays a euhedral–subhedral structure (Figure 7g,h) and is usually oxidized to form Fe-oxide, which is surrounded by cerussite (Figure 7e).

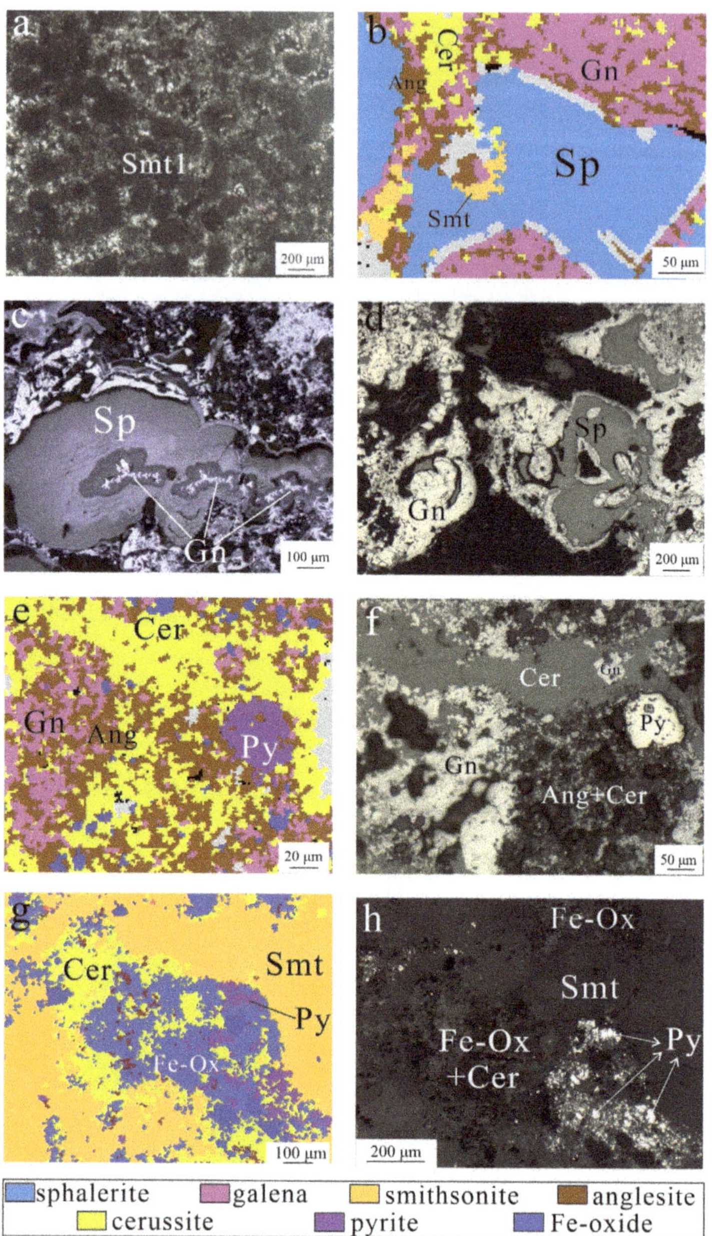

Figure 7. Textural features of the mixed ore. (**a**) Smithsonite replaces host limestone, forming the texture of dark cores with bright rims. (**b**) TIMA image shows smithsonite replacing sphalerite. (**c**) Dendritic–skeletal galena enclosed by colloform sphalerite. (**d**) Colloform sphalerite is intergrown with galena. (**e**) TIMA image shows anglesite replace galena directly, and both the anglesite and galena are replaced by later cerussite. (**f**) Microscopic image corresponding to (**e**). (**g**) TIMA image shows Fe-oxide and cerussite replacing pyrite. (**h**) Microscopic image corresponding to (**g**). Abbreviations: Smt1—grey smithsonite; Smt—smithsonite; Sp—sphalerite; Gn—galena; Py—pyrite; Ang—anglesite; Cer—cerussite; Fe-Ox—Fe-oxide.

Non-sulfide ore mainly consists of smithsonite, which occurs commonly as breccia, massive or banded ore. Under the microscope, smithsonite can be classified into three types: grey smithsonite (Smt1), yellow smithsonite (Smt2), and colorless smithsonite (Smt3).

Smt1 appears as light grey to dark grey on thin sections and appears grey under the microscope. It generally exhibits a bright rim with dark core (Figure 7a), with a small amount appearing in the form of small particles (Figure 8a). CL images show a dark red color (Figure 8c,f).

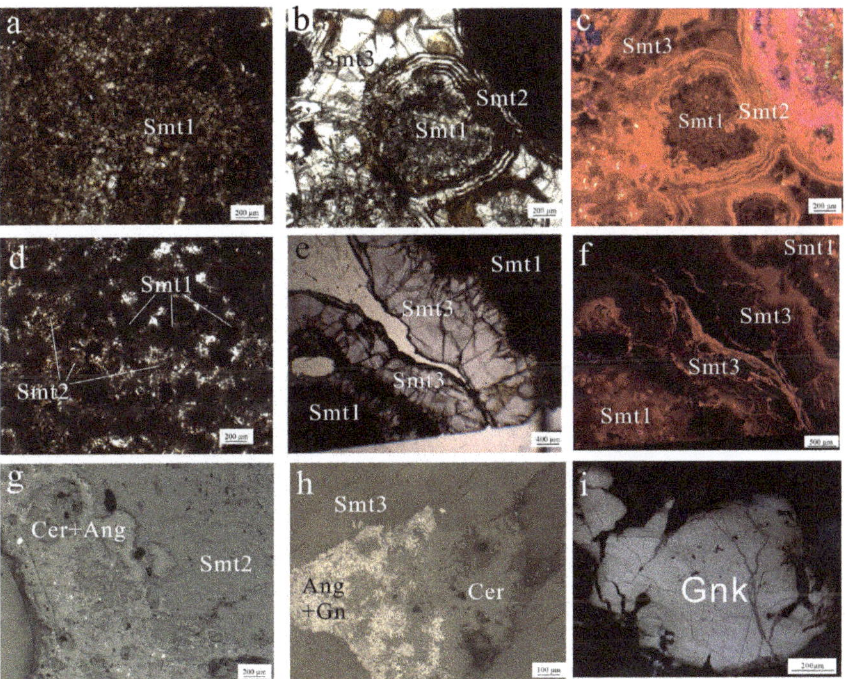

Figure 8. Textural features of the non-sulfide ore from the Huoshaoyun Zn-Pb deposit. (**a**) Small grains of Smt1. (**b**) Smt2 enclosing Smt1 and replaced by Smt3. (**c**) CL image corresponding to (**b**). (**d**) Smt2 enclosing Smt1. (**e**) Smt3 is growing symmetrically along the fractures from both sides toward the center. (**f**) CL image corresponding to (**e**). (**g**) Coexistence of smithsonite, anglesite, and cerussite. (**h**) Cerussite replacing anglesite and galena. (**i**) Microscopic image of greenockite. Abbreviations: Smt1—grey smithsonite; Smt2—yellow smithsonite; Smt3—colorless smithsonite; Gn—galena; Ang—anglesite; Cer—cerussite; Gnk—Greenockite.

Smt2 appears light to dark brown-yellow and shows bright red color in CL images (Figure 8c). It is generally dominated by a banding structure (Figure 8b) or small grains (Figure 8d), and has poor translucency. Grey smithsonite (Smt1) is often observed as an inclusion within it (Figure 8b,c).

Smt3 appears colorless on thin sections and mainly fills the fracture of the ores in a vein-like form (Figure 8b), growing symmetrically towards the center on both sides of the fractures (Figure 8e). The grey smithsonite (Smt1) and yellow smithsonite (Smt2) are often replaced by Smt3 (Figure 8b). CL images also show a dark red color (Figure 8f).

In addition to smithsonite, small amounts of Fe-oxide, cerussite, and anglesite (Figure 8g,h) are present. Occasionally, euhedral–subhedral greenockite also developed (Figure 8i).

Based on the above studies, the mineralization of the Huoshaoyun deposit underwent a primary sulfide period and a subsequent supergene oxide period (Table 1). According to the occurrence characteristics of the different types of smithsonite, which is the most important ore mineral of this deposit, the supergene oxide period can be further divided into three stages. The main mineral paragenesis sequence of the Huoshaoyun deposit is determined and listed in Table 1.

Table 1. Main mineral paragenesis sequence of the Huoshaoyun deposit.

Period	Sulfide	Supergene		
Stage	**Sulfide**	**Stage1**	**Stage2**	**Stage3**
Galena	+++			
Sphalerite	+++			
Pyrite	+			
Grey Smithsonite (Smt1)		++		
Yellow Smithsonite (Smt2)			+++	
Colorless Smithsonite (Smt3)				+
Anglesite		++	++	+
Cerussite		++	++	+
Fe-oxide		++		
Greenockite			+	
Calcite	++	+	+	+
Quartz	+	+	+	+
Gypsum		+		

Note: "+"s indicates their relative abundance.

5.2. The Results of LA-ICP-MS

The trace element content of the sphalerite, various types of smithsonite, and Fe-oxide was analyzed using LA-ICP-MS in this study. The results are presented in Tables 2 and 3.

Table 2. LA-ICP-MS testing results of sphalerite from the Huoshaoyun deposit, Xinjiang.

Simple	Fe	Mn	Cu	Cd	Ga	Ge	Tl	Pb	In	As	PC1	T (°C)
	52.3	12.8	0	3673	0.05	0.53	32.3	4318	0.02	12.4	1.05	151
	59.8	10.7	0.72	5940	0.15	0.45	28.4	5664	0	6.72	1.46	129
	62.1	12.1	0.53	4316	0.04	0	30.1	4772	0.01	8.4	-	-
HSY-10-1-7J	69.8	10.6	0.63	4995	0.86	1.12	28.0	5209	0.01	7.71	1.92	103
	61.5	10.9	0.66	6037	0.06	0	27.4	6056	0.01	6.45	-	-
	59.3	9.5	0.65	6282	0.02	0.06	27.7	6522	0.55	5.97	0.04	206
	59.4	11.0	0.02	6035	0	0.41	26.9	6265	0.01	5.24	-	-
HSY-17-1J	32.5	11.3	0.13	4383	0	28.4	52.5	4206	0.01	269	-	-
	10.7	12.4	0.17	3807	0.18	22.8	51.1	3954	0	262	-	-
	584	15.6	1.36	5416	0.38	12.8	17.2	4560	0.02	81.48	1.27	139
HSY-17-9J	1540	23.7	3.22	3130	1.65	9.45	17.3	9057	0.01	65.65	1.23	141
	387	18.2	0.48	5531	0.76	10.1	17.9	6772	0	67.37	1.71	115

Note: trace elements in ppm in the data table. "-" represents unable to calculate. For the meaning of "PC1" see below; "T" is the temperature calculated according to "PC1" parameters.

The results for the sphalerite are as follows:

(1) Sphalerite has low Fe and Mn element contents, with Mn content ranging from 10 ppm to 23.7 ppm and an average content of 13.2 ppm. The Fe content varies greatly, ranging from 10.7 ppm to 1540 ppm, with an average content of 248 ppm.

(2) Test on sphalerite reveals that Cd is the most enriched element among the rare dispersed elements, followed by Tl and Ge, while the contents of Ga and In are relatively low. Some measurement points of Ge, Ga, and In are below the detection limit. The Cd content in sphalerite is high and varies greatly, ranging from 3130 ppm to 6282 ppm with an average content of 4962 ppm. The Tl content ranges from 17.2 ppm to 52.5 ppm with an average content of 29.7ppm. The Ge content is unevenly distributed, with some samples having low content ranging from 0.06 ppm to 1.12 ppm, while some samples have higher content ranging from 9.45 ppm to 28.4 ppm. The Ga content

is generally low, ranging from 0.02 ppm to 1.65 ppm with an average content of 0.35 ppm. The In content is only detected in a few measurement points, ranging from 0.01 ppm to 0.55 ppm with an average content of 0.05 ppm.

(3) The sphalerite has relatively high Pb content and low As content. The Pb content ranges from 3954 ppm to 9057 ppm, with an average content of 5613 ppm. The distribution of As content is uneven, ranging from 5.24 ppm to 269 ppm, with an average content of 66.5 ppm.

Table 3. LA-ICP-MS results of smithsonite and Fe-oxide from the Huoshaoyun deposit, Xinjiang.

Simple	Mineral	Fe	Mn	Cd	Ga	Ge	Tl	Pb
	Smt1	12,189	3328	1010	0.14	0	0.16	8429
	Smt1	12,940	3267	1302	0.08	0.67	0.15	10,533
HSY-1-1-2J	Smt1	15,186	3487	994	0.04	1.38	0.12	15,343
	Smt1	17,221	3029	1165	0.07	0	0.29	48,890
	Smt1	15,722	3539	538	0.07	0	0.3	7407
HSY-1-1-2J	Smt2	27,649	5646	183	0.24	0	0.45	5535
	Smt2	5261	1598	1062	0.16	0.17	1.66	17,224
HSY-11-4J	Smt2	5512	1675	1067	0.31	0	1.23	16,044
	Smt2	11,084	3198	681	0.08	0.44	0.63	10,559
	Smt3	16,227	3941	270	0.04	1.03	0.88	9812
	Smt3	20,917	5512	83.9	0.19	3.75	0.45	4646
HSY-1-1-2J	Smt3	22,609	4827	199	0.03	0	0.78	6658
	Smt3	19,109	4507	159	0.01	3.68	0.43	5211
HSY-11-4J	Smt3	4849	2040	580	0.03	1.33	0.82	11,694
	Fe-Ox	711,876	0.10	33.1	5.74	60	0.94	51,301
	Fe-Ox	464,796	0.08	34.8	3.35	22.1	0.65	357,638
HSY-1-1-2J	Fe-Ox	473,549	0.02	33.7	5.2	26.2	3.06	272,532
	Fe-Ox	537,490	0.01	40.6	4.52	4.1	3.39	227,114
	Fe-Ox	460,070	0.02	31.8	4.4	26.7	3.17	300,398

Note: trace elements in ppm in the data table.

In summary, sphalerite from the Huoshaoyun deposit enriches rare dispersed elements Cd and Tl, and some samples also enrich Ge, while Fe and Mn are depleted in sphalerite.

Three types of smithsonite were analyzed. The trace element characteristics of the smithsonite samples are summarized as follows:

(1) The smithsonite enriches Fe and Mn elements, with Fe content ranging from 4849 ppm to 27,649 ppm and an average content of 14,748 ppm, and Mn content ranging from 1598 ppm to 5646 ppm with an average content of 3542 ppm.

(2) Cd is the most abundant rare dispersed element in smithsonite, while Tl, Ga, and Ge are of lower concentrations. The Cd content significantly varies among the different types of smithsonite. In Smt1, Cd content is relatively high, ranging from 994 ppm to 1302 ppm, with an average content of 1002 ppm. In Smt2, Cd content is lower, ranging from 183 ppm to 1067 ppm, with an average content of 748 ppm. In Smt3, Cd content is the lowest, ranging from 83.9 ppm to 580 ppm, with an average content of 258 ppm. Tl and Ga contents are low, ranging from 0.12 ppm to 1.66 ppm and 0.01 ppm to 0.31 ppm, respectively, while Ge content is the lowest, with only 8 points detected, ranging from 0.17 ppm to 3.75 ppm.

(3) Pb content is relatively high, ranging from 4646 ppm to 48,890 ppm, with an average content of 12,713 ppm.

In summary, smithsonite from the Huoshaoyun deposit enriches Cd, Fe, Mn, and Pb, and depletes Ge, Ga, and Tl.

Fe-oxide relatively enriches Cd, Ga, Ge, and Pb elements, while the contents of the Tl and Mn are low. Cd content ranges from 31.8 ppm to 40.6 ppm, Ga content ranges from 3.35 ppm to 5.74 ppm, and Ge content ranges from 4.1 ppm to 60 ppm.

6. Discussion

6.1. Genesis of the Sulfide Ores

6.1.1. Characteristics of Trace Elements in Sphalerite and Ore-Forming Temperature

Numerous previous studies have shown that trace elements (e.g., Fe, Mn, Cd, Ge, Ga, In, and Tl) are commonly present in the form of isomorphism in sphalerite via direct and coupled substitutions, as well as using crystal vacancy [15,16,32–34]. Consequently, the characteristics of the trace elements in sphalerite can record the signature of the ore-forming fluids and discriminate the different genetic types of Zn-Pb deposits [15,16,18,35].

The Fe and Mn contents in sphalerite of the Huoshaoyun deposit are significantly lower than those in magmatic/volcanic hydrothermal-related Zn-Pb deposits (e.g., skarn and VMS (volcanogenic massive sulfide) deposits) and SEDEX deposits (Figure 9a,b). Instead, they are consistent with those of MVT deposits, especially the Jinding and Duocaima (also named Chaqupacha deposit) MVT deposits, which are also located in the Eastern Tethyan zinc–lead belt (Figure 9a,b). However, compared to the Jinding and Duocaima deposits, as well as other MVT deposits, the Fe content in the Huoshaoyun deposit is slightly lower. This may be related to the exceptionally enriched Cd content in the sphalerite of the deposit (Figure 9c), since Fe^{2+} and Cd^{2+} may have competitive substitution of the Zn^{2+} in sphalerite, i.e., $(Fe^{2+}, Cd^{2+}) \leftrightarrow Zn^{2+}$ [15,16]. The Ge content of sphalerite in the Huoshaoyun deposit is unevenly distributed, with some points showing high concentrations, which is consistent with the findings of Yuan et al. (22.3 ppm on average) [14]. In contrast, the In content is generally low, typically below 0.1 ppm. The characteristic of high Ge and low In is distinct from magmatic/volcanic hydrothermal-related Zn-Pb deposits and SEDEX deposits, but similar to MVT deposits (Figure 9d,e). The sphalerite from the Huoshaoyun deposit enriched the rare dispersed element Tl (Figure 9f). Although some VMS and SEDEX deposits also have high Tl contents in the ores, the majority of Tl is enriched in pyrite and marcasite, while the enrichment of Tl in sphalerite is limited (Figure 9f) [36–39]. Additionally, Tl-enriched sphalerite is mainly found in MVT deposits (e.g., the Wiesloch deposit in Germany, the Upper Silesia ore field in Poland, and the Jinding and the Duocaima deposits in the Eastern Tethyan zinc–lead belt) [39–41].

In summary, the trace element characteristics of sphalerite in the Huoshaoyun deposit are similar to those of MVT deposits, especially the Jinding and Duocaima deposits, while they differ significantly from those of magmatic/volcanic hydrothermal-related Zn-Pb deposits (e.g., Skarn and VMS deposits) and SEDEX deposits.

Studies have revealed that sphalerite formed at high temperatures typically enriches Fe, Mn, and In, while sphalerite formed at medium–low temperatures commonly enriches Ge and Cd but depletes Fe, Mn, and In [43]. The sphalerite from the Huoshaoyun deposit is distinct from high-temperature sphalerite in that it depletes Fe and Mn but enriches Cd and Ge. Recently, Frenzel et al. (2016) conducted a comparative analysis of the trace element compositions of sphalerites from various types of Zn-Pb deposits against measured fluid homogenization temperatures. They systematically synthesized the data and proposed a series of calculation formulas (GGIMFis) that establish a relationship between the trace elements Ga, Ge, Fe, Mn, and In in sphalerite and ore-forming temperatures [44]:

$$PC1 = \ln\left(\frac{C_{Ga}^{0.22} \cdot C_{Ge}^{0.22}}{C_{Fe}^{0.37} \cdot C_{Mn}^{0.20} \cdot C_{In}^{0.11}} \right), \tag{1}$$

$$T = -54.4 \times PC1 + 208 \tag{2}$$

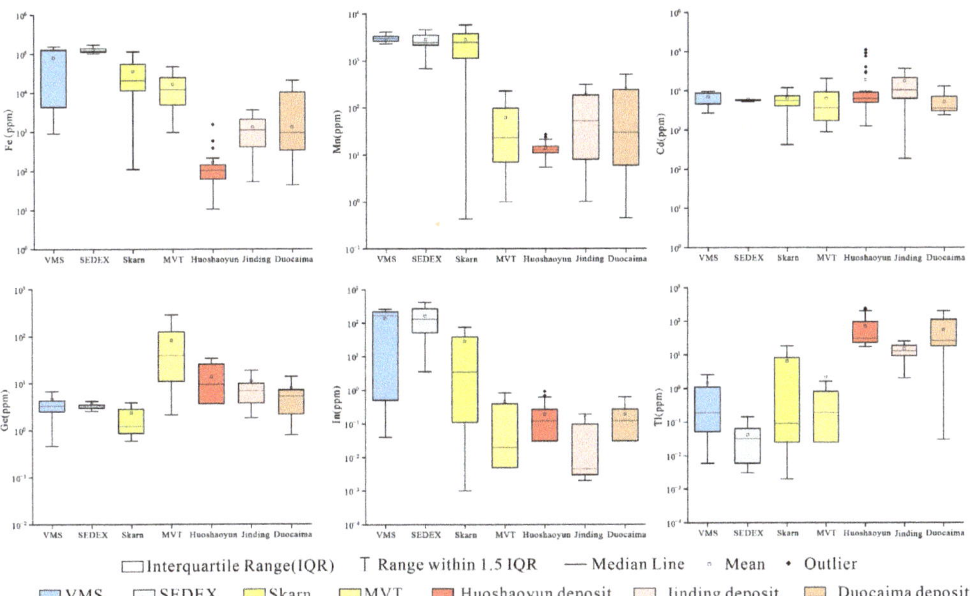

Figure 9. Box–whisker plots of the trace element compositions of sphalerite from Huoshaoyun, Jinding, Duocaima, MVT, VMS, Skarn, and SEDEX deposits. Note: the plots are based on data from this study, Yuan et al. [14], Ye et al. [15], Cook et al. [16], and Zhang [42].

In Equation (1), C_{Fe} denotes the weight percent concentration of Fe, C_{Ga} denotes the parts per million concentration of Ga, and the concentration of all other elements is equivalent to that of Ga.

The ore-forming temperature of the sphalerite (sulfide mineralization) in the Huoshaoyun deposit was calculated to be within the range of 103~206 °C using the GGIMFis (Table 2). The results indicate that the formation temperature of sphalerite in this deposit is comparatively low and significantly distinct from that of magmatic-hydrothermal deposits, but falls within the typical temperature range (70~180 °C) of MVT zinc–lead deposits [45].

6.1.2. Genesis of the Sulfide Ores in the Huoshaoyun Deposit

The binary plots of trace elements (or the rations of different elements) in sphalerite have been demonstrated to be the reliable discrimination of the genetic types of Zn-Pb deposits [46–49]. The sphalerite samples of the Huoshaoyun deposit all plot within the field of the MVT deposit in the In/Cd-Mn, Mn-In/Ge, Mn-Cd, Mn-Fe, Mn-In, and Mn-Ge discrimination diagrams, indicating the sulfide ores in this deposit could be of the MVT deposit (Figure 10). Moreover, the sphalerite samples of the Huoshaoyun deposit are overlapped with those of the Jinding and Duocaima MVT deposits in these diagrams (Figure 10), further suggesting that they could have similar ore-forming material/fluids source, metallogenic environment, and origins. In order to further elucidate the genetic type of the sulfide ore body in the Huoshaoyun deposit, this study compares the S-Pb isotopic compositions, metallogenic temperature, and the typical texture of the sulfide ores in the Huoshaoyun deposit with those of the Jinding and Duocaima MVT deposits.

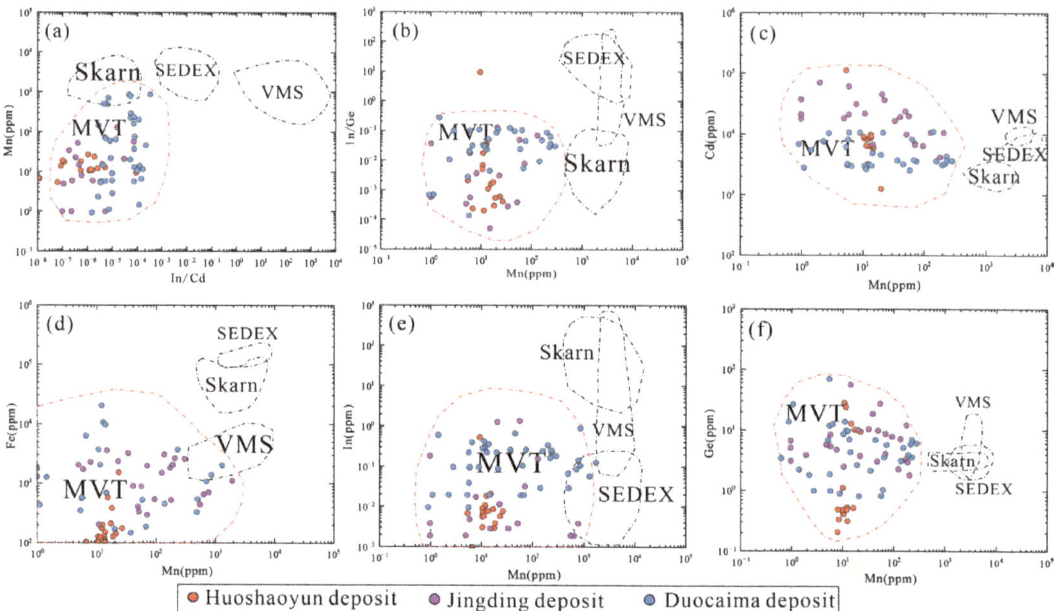

Figure 10. (a–f) Binary plots In/Cd vs. Mn, Mn vs. In/Ge, Mn vs. Cd, Mn vs. Fe, Mn vs. In, Mn vs. Ge in sphalerite from the Huoshaoyun deposit. Note: plotting fields were obtained from this study, Yuan et al. [14], Ye et al. [15], Cook et al. [16], and Zhang [42].

The sulfur isotopic compositions of sulfides from the Huoshaoyun, Jinding, and Duocaima deposits in the Eastern Tethyan zinc–lead belt show similarity (Figure 11). The sulfides in these deposits all have a wide range of sulfur isotopic compositions, with the $\delta^{34}S$ values ranging from −33.2‰ to 31.6‰ for the Huoshaoyun deposit [22,50,51], −48.6‰ to 7.7‰ for the Jinding deposit [52,53], and −26.34‰ to −3.52‰ for the Duocaima deposit [42] (Figure 11). This characteristic distinguishes them from Zn-Pb deposits associated with magmatic-hydrothermal fluids ($\delta^{34}S$ = −5‰~5‰) [54]. The $\delta^{34}S$ values of most of the sulfides from these deposits fall within the range of $\delta^{34}S$ values related to bacteriogenic sulfate reduction (BSR) (very low $\delta^{34}S$ values), indicating that the sulfur in sulfide minerals mainly originated from BSR of seawater sulfates. This is consistent with that of some MVT Zn-Pb deposits worldwide (Figure 11).

Moreover, the sulfide ores from the Huoshaoyun (Figures 6d and 7c), Duocaima, and Jinding deposits contain abundant colloform sphalerite and skeletal, fine-grained galena [42,52,53], suggesting the rapid precipitation of zinc–lead sulfides from supersaturated fluids under far-from-equilibrium conditions, while the mixing of the hot Zn-Pb bearing fluids with low-temperature H_2S-rich brine is an efficient way to lead the rapid precipitation of sulfides in the carbonate-hosted Zn-Pb deposits [39,55]. Therefore, these deposits could be formed through the mixing between a hot metal-bearing (Zn, Pb, etc.) fluid and a low-temperature brine containing BSR-derived H_2S, while some sulfides with positive sulfur isotope values in these deposits can be attributed to the involvement of the thermochemical sulfate reduction (TSR)-related reduced sulfur and/or the sulfate-limited conditions during the sulfide precipitation process [50,52].

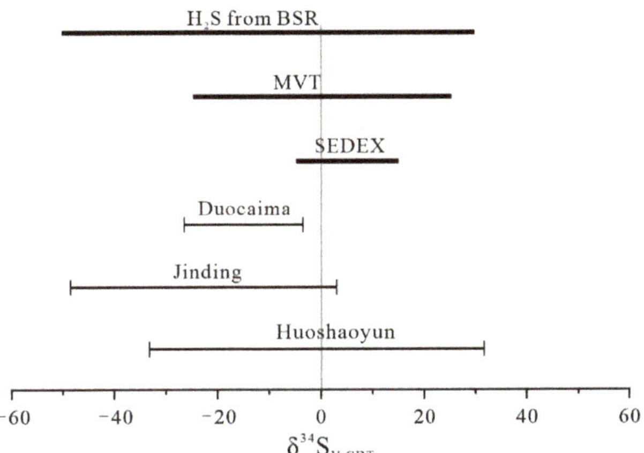

Figure 11. Sulfur isotope compositions of zinc–lead sulfides from the Huoshaoyun zinc–lead deposit. Note: δ34S values for H$_2$S from BSR from Chen et al. [56]; SEDEX and MVT zinc–lead deposits from Leach et al. [57]. Date for Huoshaoyun deposit from Yuan [50], Li et al. [22],Wu et al. [51]; dates for Jinding deposit from Tang et al. [52], Dai [53]; dates for Duocaima deposit from Zhang [42].

The ^{206}Pb/^{204}Pb ratios of sulfides in the Huoshaoyun Zn-Pb deposit range from 18.525 to 18.563, the ^{207}Pb/^{204}Pb ratios range from 15.673 to 15.710, and the ^{208}Pb/^{204}Pb ratios range from 38.879 to 39.005 [7,58]. These ratios are very similar to those of the Jinding Zn-Pb deposit (^{206}Pb/^{204}Pb ratios range from 18.410 to 18.523, ^{207}Pb/^{204}Pb ratios range from 15.620 to 15.662, and ^{208}Pb/^{204}Pb ratios range from 38.569 to 38.714) [59], and also close to the range of Pb isotopic composition of the Duocaima Zn-Pb deposit [42] (Figure 12). Additionally, the plot field of the sulfide date from these deposits in ^{207}Pb/^{204}Pb vs. ^{206}Pb/^{204}Pb diagrams indicates that the metal source of these deposits is derived from upper crustal rocks, which is consistent with the typical MVT deposits worldwide.

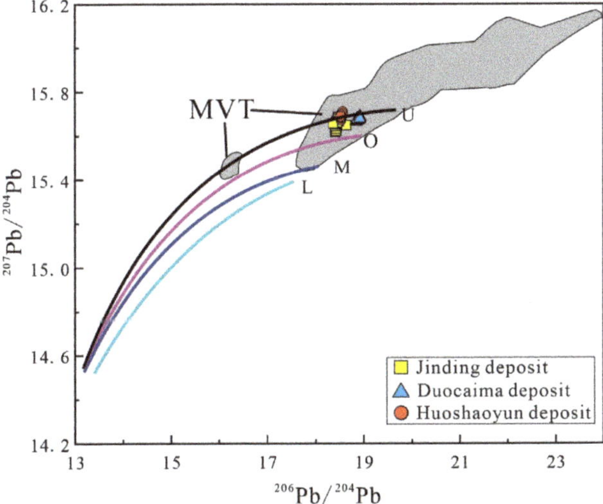

Figure 12. Correlative diagrams of ^{206}Pb/^{204}Pb vs. ^{207}Pb/^{204}Pb (modified from Xiong et al. [60]) for the Huoshaoyun, Jinding, and Duocaima deposits. Note: the plots are based on data from Gao et al. [7], Wu. [58], Zhang [59], Zhang [42]. U—upper crust; O—orogene; M—mantle; L—lower crust. The region shaded grey represents the extent of MVT deposits.

In addition, most of the ore-forming temperatures of the Huoshaoyun Zn-Pb deposit were determined to be about 100~150 °C using the GGIMFis geothermometer. These temperatures are consistent with those of the Jinding Zn-Pb deposit, which were mainly at 100~130 °C based on fluid inclusion measurements in sphalerite [61], and the Duocaima Zn-Pb deposit, which were mainly at 120~180 °C [62]. These results suggest that these deposits formed from the low-temperature hydrothermal fluids, which is consistent with the metallogenic temperature of the MVT deposits [57].

In summary, the trace-element signatures of sphalerite, ore-forming temperatures, S-Pb isotopes, and some special texture of the sulfide ores all indicate that the sulfide ores in the Huoshaoyun deposit are the MVT deposit and may have similar ore-forming processes to the Jinding and Duocaima deposits in the Eastern Tethyan Zn-Pb belt.

6.2. Trace Element Characteristics of the Non-Sulfide Minerals and Implications for Their Origins

During the supergene oxidation of zinc–lead sulfides, variations in the geochemical properties of the trace elements and/or physicochemical conditions (pH, temperature, etc.) can lead to the enrichment of distinct trace elements in different non-sulfide minerals [63].

The concentration of Fe and Mn in smithsonite from the Huoshaoyun Zn-Pb deposit is significantly greater than that in sphalerite (Figure 13a). This can be explained by the decomposition of both the Fe-bearing sphalerite and the pyrite during supergene oxidation (Figure 7), which released abundant Fe and Mn into the fluid. The capture of Fe and Mn by smithsonite occurred through isomorphic substitution.

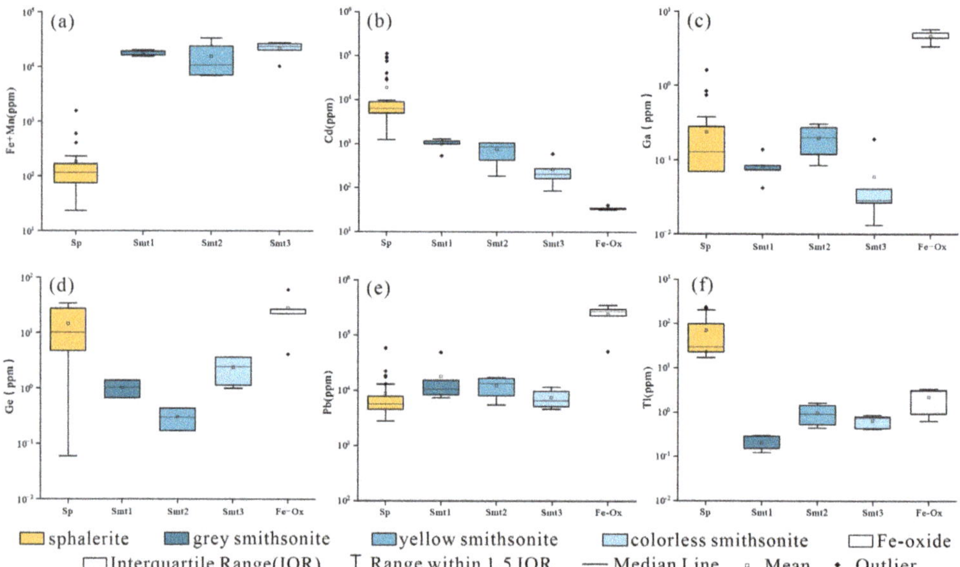

Figure 13. Trace element comparison of sphalerite, different stages of smithsonite, and Fe-oxide in the Huoshaoyun Zn-Pb deposit. (**a**)–(**f**) represent the comparison of Fe+Mn, Cd, Ga, Ge, Pb, and Tl elements in different minerals.

Although the concentration of Cd is relatively high in smithsonite, it is significantly lower compared to that in sphalerite. This indicates that some Cd may have been lost during the oxidation process. Moreover, the development of the greenockite (CdS) in the non-sulfide ores (Figure 8f) suggests that a part of the Cd^{2+} in the oxidizing fluids, resulting from oxidation of the Cd-bearing sphalerite, formed its own independent mineral (e.g., via $CdSO_4 + ZnS_2 \rightarrow ZnSO_4 + CdS$) [64] rather than entered the smithsonite structure. On the other hand, compared to Fe-oxide, Cd is enriched in smithsonite (Figure 13b). This shows that Cd^{2+} tends to enter smithsonite instead of being adsorbed by Fe-oxide during the supergene oxidation process, which could be attributed to the very similar geochemical properties of Cd^{2+} and Zn^{2+}.

Ga and Ge are enriched in Fe-oxide compared to smithsonite (Figure 13c,d). During supergene oxidation, the large amount of sulfuric acid generated from the alteration of pyrite could have caused the pH of the fluid to decrease and become acidic, leading to the incorporation of Ga into Fe-oxide when the pH of the metal-bearing solution is extremely low (pH < 4) [65]. Ge can not only octahedrally coordinate with oxygen and co-precipitation with Fe-oxide, but can also be adsorbed by Fe-oxide [19,66]. Therefore, Fe-oxide formed through the oxidization of the Ge- and Ga-bearing sulfide ores commonly has high contents of Ga and Ge (e.g., crystal non-sulfide Zn prospect, northern Peru) [19].

All type of smithsonites from the Huoshaoyun deposit have a relatively high content of Pb (Figure 13e), suggesting that during the oxidation process of zinc–lead sulfides, besides forming the cerussite and anglesite, Pb^{2+} in the oxidizing fluid can also enter into smithsonite [67]. In comparison, Fe-oxide has much higher Pb content than the smithsonite and sphalerite (Figure 13e). This could be because Pb^{2+} can not only coprecipitate with Fe-oxide but also can be continuously adsorbed by Fe-oxide [68].

Compared to sphalerite, both smithsonite and Fe-oxide exhibit significant depletion in Tl (Figure 13f), with the highest content being less than 4 ppm and the majority being less than 1 ppm. It is speculated that Tl was released into the surrounding environment during the oxidation process and underwent significant loss.

In summary, compared with sphalerite, the contents of Fe and Mn increase significantly in smithsonite. Ge and Ga are less concentrated in smithsonite but are more enriched in Fe-oxide. These features are consistent with the characteristics of the trace element migration and enrichment during the oxidizing process of the MVT deposits. For instance, Stavinga observed significantly higher Fe and Mn concentrations in oxidized smithsonite compared to primary sphalerite in the Prairie Creek MVT deposit [67], while Santoro et al. found that Ga and Ge were mainly enriched in Fe-oxide minerals during the supergene oxidation process of four MVT Zn-Pb deposits [65]. These findings indicate significant oxidation of sulfide ores in the Huoshaoyun Zn-Pb deposit.

6.3. From MVT Deposits to Non-Sulfide Ores

Zn-Pb-Fe sulfides not only occur in sulfide ores but also in the mixed ores of the Huoshaoyun deposit. These minerals, i.e., sphalerite, galena, and pyrite, are typically replaced by anglesite, cerussite, smithsonite, and Fe-oxide (Figure 7b,e–h), providing clear evidence of the oxidation process of the Zn-Pb-Fe sulfides. Therefore, combined with the other results of this study, we argued that the Huoshaoyun Pb-Zn deposit is a supergene MVT deposit and its formation process can be summarized as follows:

The sulfide ores are formed thought the mixing between a hot metal-bearing (Zn, Pb, etc.) fluid and a low-temperature brine containing BSR-derived H_2S.

The continuous periodic uplift of the Qinghai-Tibet Plateau since the Cenozoic provided long-term uplift, erosion, rapid exhumation of the sulfide ores, and exposure to oxygenated surface waters. Pyrite has quicker oxidation rates and contributes Fe^{3+}, SO_4^{2-}, and H^+ to the fluid, which led to acidification of the fluids and formation of Fe-oxide (Figure 7g,h).

Oxidation of sphalerite and galena also generated Zn^{2+}, Pb^{2+}, H^+, and SO_4^{2-}. The high sulfate and low pH condition, created by the oxidation of these sulfides, is conducive to the formation of anglesite from galena: $Pb^{2+} + SO_4^{2-} \rightarrow PbSO_4$ (Anglesite) (Figure 7e,f), while sulfates also react with the host rocks to form gypsum and liberate CO_3^{2-}: $SO_4^{2-} + CaCO_3 \rightarrow CaSO_4$ (Gypsum) $+ CO_3^{2-}$. The termination of the certain oxidation stage and/or depletion of the sulfides can cause the decreasing of SO_4^{2-} in the fluids and increasing in its pH. This can lead the anglesite to become unstable and alter to cerussite: $PbSO_4 + CO_3^{2-} \rightarrow PbCO_3$ (Cerussite) $+ SO_4^{2-}$ (Figure 7f). Occasionally, the galena could be directly replaced by cerussite, i.e., $PbS + H_2O + CO_2 + 2O_2 \rightarrow PbCO_3$ (Cerussite) $+ SO_4^{2-} + 2H^+$ (Figure 7b).

The formation of smithsonite can be divided into three main stages.

(1) Smt1: Sphalerite oxidation generates Zn^{2+} which migrates through fractures and reacts with host rocks, replacing the calcite and forming grey smithsonite: $Zn^{2+} + CaCO_3 \rightarrow ZnCO_3$ (Smithsonite) $+ Ca^{2+}$. The oolitic textures (Figure 7a) indicates its replacement of the host limestone, while the development of a small amount of fine-grained grey smithsonite (Figure 8a) suggests that a portion of Zn^{2+} in the fluid could have directly crystallized into smithsonite: $Zn^{2+} + H_2O + CO_2 \rightarrow ZnCO_3$ (Smithsonite) $+ 2H^+$.

(2) Smt2: The second-stage supergene oxidizing fluids formed this yellow smithsonite (Smt2) through replacing the host rock (Figure 8d), filling the cavities, and enclosing the grey smithsonite (Smt1) as banding-shaped or replacing Smt1 (Figure 8b,c).

(3) Smt3: The colorless smithsonite was formed by the third-stage fluid carrying Zn^{2+} filling along the fractures from both sides to the center, or replacing Smt1 and Smt2.

7. Conclusions

(1) The Huoshaoyun zinc–lead deposit is composed of sulfide and non-sulfide ores. The former is mainly composed of sphalerite, galena, and pyrite, whereas the latter is primarily composed of smithsonite, with minor cerussite, anglesite, and Fe-oxide. The non-sulfide minerals clearly replaced the sulfides, suggesting the oxidation of the primary sulfide ores.

(2) The trace element analysis of sphalerite indicates that it is rich in Cd, Tl, and Ge, but poor in Fe and Mn. The ore-forming temperature, calculated using the GGIMFis geothermometer, is most within the range of 100~150 °C. Moreover, the trace element characteristics, ore-forming temperature, and S and Pb isotope compositions of the sulfide ores of the Huoshaoyun deposit are similar to those of the Jinding and Duocaima MVT lead–zinc deposits, which are also located in the eastern Tethyan zinc–lead belt. This suggests that the sulfide orebody in the Huoshaoyun Zn-Pb deposit could be also the MVT deposit.

(3) The trace element of the non-sulfide minerals shows that the Mn and Cd are relatively enriched in smithsonite, while Ga, Ge, and Pb are enriched in Fe-oxide. This characteristic is consistent with that of the other oxidized MVT deposits worldwide, thus indicating the supergene oxidation process of the precursor MVT ores in the Huoshaoyun deposit.

Author Contributions: Conceptualization, X.C., D.D. and Y.W.; Data curation, X.C., Y.Z., F.Z. and X.Y.; Funding acquisition, D.D. and Y.W.; Investigation, X.C., D.D., Y.Z., F.Z. and Y.W.; Methodology, X.C., D.D. and Y.W.; Writing—original draft, X.C. and Y.Z.; Writing—review and editing, X.C., D.D., Y.Z. and Y.W. All authors have read and agreed to the published version of the manuscript.

Funding: This work was supported by the National Natural Science Foundation of China (42172088, 42203065) and the Natural Science Foundation of Hubei Province (2019CFB586).

Data Availability Statement: The authors declare that all analytical data supporting the findings of this study are available within the paper or cited in peer-review references.

Acknowledgments: We appreciate Sen Zhao for his help during fieldwork.

Conflicts of Interest: The authors declare no conflict of interest.

References

1. Hitzman, M.W.; Reynolds, N.A.; Sangster, D.F.; Allen, C.R.; Carmen, C.E. Classification, genesis, and exploration guides for nonsulfide zinc deposits. *Econ. Geol.* **2003**, *98*, 685–714. [CrossRef]

2. Fan, T.B.; Li, H.; Xu, X.W.; Dong, L.H. Research Status and Progress of Nonsulfide Zinc-Lead Deposit. *Northwestern Geol.* **2018**, *51*, 147–159. (In Chinese)

3. Harkins, S.A.; Appold, M.S.; Nelson, B.K.; Brewer, A.M.; Groves, L.M. Lead isotope constraints on the origin of nonsulfide zinc and sulfide zinc-lead deposits in the Flinders Ranges, South Australia. *Econ. Geol.* **2008**, *103*, 353–364. [CrossRef]

4. Rezaeian, A.; Rasa, I.; Amiri, A.; Jafari, M.R. Geochemistry of Oxygen and Carbon Stable Isotopes in Non-Sulfide Zn-Pb Deposits, Case Study: Chah-Talkh non-Sulfide Zn-PbDeposit (Sirjan-South of Iran). *World Appl. Sci. J.* **2013**, *24*, 1163–1171.

5. Boni, M.; Mondillo, N. The "Calamines" and the "Others": The great family of supergene nonsulfide zinc ores. *Ore Geol. Rev.* **2015**, *67*, 208–233. [CrossRef]
6. Large, D. The geology of non-sulphide zinc deposits: An overview. *Erzmetall* **2001**, *54*, 264–276.
7. Gao, Y.B.; Li, K.; Teng, J.X.; Zhao, X.M.; Zhao, X.J.; Yan, Z.Q.; Jin, M.S.; Zhao, H.B.; Li, X.T. Mineralogy, Geochemistry and Genesis of Giant Huoshaoyun Zn-Pb Deposit in Karakoram Area, Xinjiang, NW China. *Northwestern Geol.* **2019**, *52*, 152–169. (In Chinese)
8. Dong, L.H.; Xu, X.W.; Fan, T.B.; Qu, X.; Li, H.; Wan, J.L.; An, H.T.; Zhou, G.; Li, J.H.; Chen, G.; et al. Discovery of the Huoshaoyun Super-Large Exhalative-Sedimentary Carbonate Pb-Zn Deposit in the Western Kunlun Area and its Great Significance for Regional Metallogeny. *Xinjiang Geol.* **2015**, *33*, 41–50. (In Chinese)
9. Xu, Z.P.; Xu, Y. Genesis of the Huoshoaoyun deposit in the Karakorum district in Hotan, Xinjiang Province. *Resour. Inf. Eng.* **2017**, *32*, 26–27. (In Chinese)
10. Fan, T.B.; Yu, Y.J.; Xia, M.Y.; Jiang, G.P.; Wang, M. Geological Features and Prospect for the Huoshaoyun Pb-Zn Deposit in Hotan, Xinjiang. *Acta Geol. Sichuan* **2017**, *37*, 578–582. (In Chinese)
11. Jin, H.Z. Analysis of metallogenic condition and prospecting potential of lead-zinc deposit in Tianshuihai-Huoshaoyun area, Karakoram. *J. Geol.* **2018**, *42*, 17–22. (In Chinese)
12. Song, Y.C.; Hou, Z.Q.; Liu, Y.C.; Zhang, H.R. Mississippi Valley-Type (MVT) Pb-Zn deposits in the Tethyan domain: A review. *Geol. China* **2017**, *44*, 664–689. (In Chinese)
13. Gao, L.; Du, Y.; Song, L.S.; Ren, X.D.; Ding, J.H.; Wei, H.T.; Cui, N.; Zhang, Q. Geological characteristics and genesis of the Huoshaoyun hypogene smithsonite deposit, Karakoram Mountains, northwest China. *Geol. Rev.* **2020**, *66*, 365–379. (In Chinese)
14. Yuan, X.; Wu, Y.; Duan, D.F.; Zhu, J.; Ouyang, H.G.; Cao, L.; Zhou, B. Trace (Dispersed) Elements in Sphalerite from the Giant Huoshaoyun Lead-Zinc Deposit, Xinjiang and Their Geological Implications. *Geol. Explor.* **2022**, *58*, 545–560. (In Chinese)
15. Ye, L.; Cook, N.J.; Ciobanu, C.L.; Yuping, L.; Qian, Z.; Tiegeng, L.; Wei, G.; Yulong, Y.; Danyushevskiy, L. Trace and minor elements in sphalerite from base metal deposits in South China: A LA-ICPMS study. *Ore Geol. Rev.* **2011**, *39*, 188–217. [CrossRef]
16. Cook, N.J.; Ciobanu, C.L.; Pring, A.; Skinner, W.; Shimizu, M.; Danyushevsky, L.; Saini-Eidukat, B.; Melcher, F. Trace and minor elements in sphalerite: A LA-ICPMS study. *Geochim. Cosmochim. Acta* **2009**, *73*, 4761–4791. [CrossRef]
17. Qian, Z. Trace elements in galena and sphalerite and their geochemical significance in distinguishing the genetic types of Pb-Zn ore deposits. *Chin. J. Geochem.* **1987**, *6*, 177–190. [CrossRef]
18. Li, X.M.; Zhang, Y.X.; Li, Z.K.; Zhao, X.F.; Zuo, R.G.; Xiao, F.; Zheng, Y. Discrimination of Pb-Zn deposit types using sphalerite geochemistry: New insights from machine learning algorithm. *Geosci. Front.* **2023**, *14*, 101580. [CrossRef]
19. Mondillo, N.; Arfè, G.; Herrington, R.; Boni, M.; Wilkinson, C.; Mormone, A. Germanium enrichment in supergene settings: Evidence from the Cristal nonsulfide Zn prospect, Bongará district, northern Peru. *Miner. Depos.* **2018**, *53*, 155–169. [CrossRef]
20. Richards, J.P. Tectonic, magmatic, and metallogenic evolution of the Tethyan orogen: From subduction to collision. *Ore Geol. Rev.* **2015**, *70*, 323–345. [CrossRef]
21. Hou, Z.; Zhang, H. Geodynamics and metallogeny of the eastern Tethyan metallogenic domain. *Ore Geol. Rev.* **2015**, *70*, 346–384. [CrossRef]
22. Li, H.; Xu, X.W.; Borg, G.; Gilg, H.A.; Dong, L.H.; Fan, T.B.; Zhou, G.; Liu, R.L.; Hong, T.; Ke, Q.; et al. Geology and Geochemistry of the giant Huoshaoyun zinc-lead deposit, Karakorum Range, northwestern Tibet. *Ore Geol. Rev.* **2019**, *106*, 251–272. [CrossRef]
23. Deng, W.M. Geological Features of Ophiolite and Tectonic Significance in the Karakorum-West Kunlun Mts. *Acta Petrol. Sin.* **1995**, *11*, 98–111. (In Chinese)
24. Li, R.S.; Ji, W.H.; Yang, Y.C.; Pan, X.P. Geology of Kunlun Mountains and Adjacent Areas. Geology Publishing House: Beijing, China, 2008; pp. 1–400. (In Chinese)
25. Dong, L.H.; Feng, J.; Liu, D.Q.; Tang, Y.L.; Qu, X.; Wang, K.Z.; Yang, Z.F. Research for classification of metallogenic unit of Xinjiang. *Xinjiang Geol.* **2010**, *28*, 1–15. (In Chinese)
26. Zhang, Z.; Shen, N.; Peng, J.; Yang, X.; Feng, G.; Yu, F.; Zhou, L.; Li, Y.; Wu, C. Syndeposition and epigenetic modification of the strata-bound Pb–Zn–Cu deposits associated with carbonate rocks in western Kunlun, Xinjiang, China. *Ore Geol. Rev.* **2014**, *62*, 227–244. [CrossRef]
27. Fan, T.B.; Jin, H.Z.; Yu, Y.J.; Jiang, G.P.; Xia, M.Y. Metallogenic characteristics and prospecting progress of lead-zinc deposits in the Tianshuihai area of west Kunlun. *J. Geol.* **2019**, *43*, 184–197. (In Chinese)
28. Ren, G.L.; Fan, T.B.; Yu, Y.J.; Yang, M.; Liang, N.; Zhang, Z.; Li, J.Q.; Yang, J.L. Application of Multi-Source Remote Sensing Information to Metallogenic Prediction in the Huoshaoyun Region of Karakorum, Xinjiang. *Geol. Explor.* **2017**, *53*, 1164–1173. (In Chinese)
29. Tang, J.L.; Li, H.; Tan, K.B.; Ke, Q.; Dong, L.H.; Xu, X.W.; Ha, Y.R.T. Ore-Bearing Strata of Lead-Zinc Deposits in the Linjitang Basin of Karakorum in a Marine Sedimentary Environment: Constraints from Trace Elements in Limestone and Sulfur Isotope of Gypsum in Jurassic. *Geol. Explor.* **2020**, *56*, 1134–1144. (In Chinese)
30. Wan, J.L. *Exploration Report on the Huoshaoyun Lead-Zinc Deposit in Hetian County, Xinjiang*; The Eighth Geological Brigade of the Geological and Mineral Exploration and Development Bureau of Xinjiang Uygur Autonomous Region: Xinjiang, China, 2017. (In Chinese)
31. Liu, Y.; Hu, Z.; Gao, S.; Günther, D.; Xu, J.; Gao, C.; Chen, H. In situ analysis of major and trace elements of anhydrous minerals by LA-ICP-MS without applying an internal standard. *Chem. Geol.* **2008**, *257*, 34–43. [CrossRef]

32. Cook, N.J.; Etschmann, B.; Cristiana, C.L.; Geraki, K.; Howard, D.L.; Williams, T.; Rae, N.; Pring, A.; Chen, G.; Johannessen, B.; et al. Distribution and Substitution Mechanism of Ge in a Ge−(Fe)−Bearing Sphalerite. *Minerals* **2015**, *5*, 117–132. [CrossRef]

33. Fan, M.S.; Ni, P.; Pan, J.Y.; Wang, G.G.; Ding, J.Y.; Chu, S.W.; Li, W.S.; Huang, W.Q.; Zhu, R.Z.; Chi, Z. Rare disperse elements in epithermal deposit: Insights from LA–ICP–MS study of sphalerite at Dalingkou, South China. *J. Geochem. Explor.* **2023**, *244*, 107124. [CrossRef]

34. Ivashchenko, V. Geology, geochemistry and mineralogy of indium resources at Pitkäranta Mining District, Ladoga Karelia, Russia. *J. Geochem. Explor.* **2022**, *240*, 107046. [CrossRef]

35. Zhang, J.; Shao, Y.; Liu, Z.; Chen, K. Sphalerite as a record of metallogenic information using multivariate statistical analysis: Constraints from trace element geochemistry. *J. Geochem. Explor.* **2022**, *232*, 106883. [CrossRef]

36. Duan, H.Y.; Wang, C.M. Geochemistry and mineralization of a critical element: Thallium. *Acta Petrol. Sin.* **2022**, *38*, 1771–1794. (In Chinese)

37. Nestmeyer, M.; Keith, M.; Haase, K.M.; Klemd, R.; Voudouris, P.; Schwarz-Schampera, U.; Strauss, H.; Kati, M.; Magganas, A. Trace element signatures in pyrite and marcasite from shallow marine island arc-related hydrothermal vents, Calypso Vents, New Zealand, and Paleochori Bay, Greece. *Front. Earth Sci.* **2021**, *9*, 641654. [CrossRef]

38. Kelley, K.D.; Leach, D.L.; Johnson, C.A.; Clark, J.L.; Fayek, M.; Slack, J.F.; Anderson, V.M.; Ayuso, R.A.; Ridley, W.I. Textural, compositional, and sulfur isotope variations of sulfide minerals in the Red Dog Zn-Pb-Ag deposits, Brooks Range, Alaska: Implications for ore formation. *Econ. Geol.* **2004**, *99*, 1509–1532. [CrossRef]

39. Pfaff, K.; Koenig, A.; Wenzel, T.; Ridley, I.; Hildebrandt, L.H.; Leach, D.L.; Markl, G. Trace and minor element variations and sulfur isotopes in crystalline and colloform ZnS: Incorporation mechanisms and implications for their genesis. *Chem. Geol.* **2011**, *286*, 118–134. [CrossRef]

40. Heijlen, W.; Muchez, P.; Banks, D.A.; Schneider, J.; Kucha, H.; Keppens, E. Carbonate-hosted Zn-Pb deposits in Upper Silesia, Poland: Origin and evolution of mineralizing fluids and constraints on genetic models. *Econ. Geol.* **2003**, *98*, 911–932. [CrossRef]

41. Wang, C.; Yang, L.; Bagas, L.; Evans, N.J.; Chen, J.; Du, B. Mineralization processes at the giant Jinding Zn–Pb deposit, Lanping Basin, Sanjiang Tethys Orogen: Evidence from in situ trace element analysis of pyrite and marcasite. *Geol. J.* **2018**, *53*, 1279–1294. [CrossRef]

42. Zhang, H.S. Mineralization of Sediment-hosted Lead-Zinc Deposits in the Middle-East Segment of the Neo-Tethys Tectonic Domain: Examples from the Duocaima Deposit in Qinghai, China and the Duddar Deposit in Pakistan. Doctoral Dissertation, University of Science and Technology of China, Beijing, China, 2021. (In Chinese).

43. Shen, H.; Zhang, Y.; Zuo, C.; Shao, Y.; Zhao, L.; Lei, J.; Shi, G.; Han, R.; Zheng, X. Ore-forming process revealed by sphalerite texture and geochemistry: A case study at the Kangjiawan Pb–Zn deposit in Qin-Hang Metallogenic Belt, South China. *Ore Geol. Rev.* **2022**, *150*, 105153. [CrossRef]

44. Frenzel, M.; Hirsch, T.; Gutzmer, J. Gallium, germanium, indium, and other trace and minor elements in sphalerite as a function of deposit type—A meta-analysis. *Ore Geol. Rev.* **2016**, *76*, 52–78. [CrossRef]

45. Basuki, N.I.; Spooner, E.T.C. A review of fluid inclusion temperatures and salinities in Mississippi Valley-type Zn-Pb deposits: Identifying thresholds for metal transport. *Explor. Min. Geol.* **2002**, *11*, 1–17. [CrossRef]

46. Xingyu, L.; Bo, L.; Xinyue, Z.; Chengnan, Z.; Huaikun, Q.; Gao, L. Trace element composition and genesis mechanism of the Fuli Pb-Zn deposit in Yunnan: LA-ICP-MS and in situ S-Pb isotopic constraints. *Front. Earth Sci.* **2023**, *11*, 352.

47. Liu, S.; Zhang, Y.; Ai, G.; Xue, X.; Li, H.; Shah, S.A.; Wang, N.; Chen, X. LA-ICP-MS trace element geochemistry of sphalerite: Metallogenic constraints on the Qingshuitang Pb–Zn deposit in the Qinhang Ore Belt, South China. *Ore Geol. Rev.* **2022**, *141*, 104659. [CrossRef]

48. Xing, B.; Mao, J.; Xiao, X.; Liu, H.; Zhang, C.; Guo, S.; Li, H.; Huang, W.; Lai, C. Metallogenic discrimination by sphalerite trace element geochemistry: An example from the Fengyan Zn-Pb deposit in central Fujian, SE China. *Ore Geol. Rev.* **2022**, *141*, 104651. [CrossRef]

49. Hu, Y.; Ye, L.; Wei, C.; Li, Z.; Huang, Z.; Wang, H. Trace Elements in Sphalerite from the Dadongla Zn-Pb Deposit, Western Hunan–Eastern Guizhou Zn-Pb Metallogenic Belt, South China. *Acta Geol. Sin.-Engl. Ed.* **2020**, *94*, 2152–2164. [CrossRef]

50. Yuan, X. Study on Genesis of Primary Sulfide Ore Body from the Giant Huoshaoyun Lead-Zinc Deposit, Xinjiang. Master's Dissertation, Yangtze University, Wuhan, China, 2022. (In Chinese).

51. Wu, Z.Y.; Song, Y.C.; Hou, Z.Q.; Liu, Y.C.; Zhuang, L.L. The World-Class Huoshaoyun Nonsulfide Zinc-Lead Deposit, Xinjiang, NW China: Formation by Supergene Oxidization of a Mississippi Valley-Type Deposit. *Earth Sci.* **2019**, *44*, 1987–1997. (In Chinese)

52. Tang, Y.Y.; Bi, X.W.; Fayek, M.; Hu, R.Z.; Wu, L.Y.; Zou, Z.C.; Feng, C.X.; Wang, X.S. Microscale sulfur isotopic compositions of sulfide minerals from the Jinding Zn–Pb deposit, Yunnan Province, Southwest China. *Gondwana Res.* **2014**, *26*, 594–607. [CrossRef]

53. Dai, Z.J. Bacteriogenic Textures and Bacterial Fossils in Sulfide Ores and Their Metallogenic Implication in the Giant Jinding Pb-Zn Deposit, Yunnan, China. Master's Dissertation, China University of Geosciences, Beijing, China, 2016. (In Chinese).

54. Ohmoto, H. Stable isotope geochemistry of ore deposits. In *Stable Isotopes in High Temperature Geological Processes*; De Gruyter: Berlin, Germany, 2018; pp. 491–560.

55. Gagnevin, D.; Menuge, J.F.; Kronz, A.; Barrie, C.; Boyce, A.J. Minor elements in layered sphalerite as a record of fluid origin, mixing, and crystallization in the Navan Zn-Pb ore deposit, Ireland. *Econ. Geol.* **2014**, *109*, 1513–1528. [CrossRef]

56. Chen, X.; Xue, C.J. Origin of H2S in Uragen large-scale Zn-Pb mineralization, western Tien Shan: Bacteriogenic structure and S-isotopic constraints. *Acta Petrol. Sin.* **2016**, *32*, 1301–1314. (In Chinese)

57. Leach, D.L.; Sangster, D.F.; Kelley, K.D.; Large, R.R.; Garven, G.; Allen, C.R.; Gutzmer, J.; Walters, S. Sediment-Hosted Lead-Zinc Deposits: A Global Perspective. *Econ. Geol.* **2005**, *100*, 561–607.
58. Wu, Z.Y. The World-Class Huoshaoyun Nonsulfide Zinc-Lead Deposit, Xinjiang, NW China: Formation by Supergene Oxidization of a Mississippi Valley-Type Deposit. Master's Dissertation, China University of Geosciences (Beijing), Beijing, China, 2019. (In Chinese).
59. Zhang, R.J. Metallogenic Metal Source of Jinding Giant Lead-Zinc Deposit, Yunnan Province: Constraints of Trace Elements of Sphalerite and Lead, Zinc and Strontium Isotopic Compositions. Master's Dissertation, China University of Geosciences (Beijing), Beijing, China, 2020. (In Chinese).
60. Xiong, S.F.; Jiang, S.Y.; Chen, Z.H.; Zhou, J.X.; Ma, Y.; Zhang, D.; Duan, Z.P.; Niu, P.P.; Xu, Y.M. A Mississippi Valley–type Zn-Pb mineralizing system in South China constrained by in situ U-Pb dating of carbonates and barite and in situ S-Sr-Pb isotopes. *Bulletin* **2022**, *134*, 2880–2890. [CrossRef]
61. Mu, L.; Hu, R.; Bi, X.; Tang, Y.; Lan, T.; Lan, Q.; Zhu, J.; Peng, J.; Oyebamiji, A. New insights into the origin of the world-class Jinding sediment-hosted Zn-Pb deposit, Southwestern China: Evidence from LA-ICP-MS analysis of individual fluid inclusions. *Econ. Geol.* **2021**, *116*, 883–907. [CrossRef]
62. Liu, C.Z.; Li, S.J.; Chen, Y.L.; Li, D.P.; Gao, Y.W.; Guo, H.M.; Li, L.S.; Wang, Y.K. Characteristics and Genetic Type of Fluid Inclusion of the Duocaima Pb-Zn Deposit in Qiangtang Area. *Geotecton. Metallog.* **2015**, *39*, 658–669. (In Chinese)
63. Mondillo, N.; Herrington, R.; Boyce, A.J.; Wilkinson, C.; Santoro, L.; Rumsey, M. Critical elements in non-sulfide Zn deposits: A reanalysis of the Kabwe Zn-Pb ores (central Zambia). *Mineral. Mag.* **2018**, *82*, S89–S114. [CrossRef]
64. Ye, L.; Pan, Z.P.; Li, C.Y.; Liu, T.G.; Xia, B. The present situation and prospects of geochemical researches on cadmium. *Acta Petrol. Mineral.* **2005**, *24*, 339–348. (In Chinese)
65. Santoro, L.; Putzolu, F.; Mondillo, N.; Boni, M.; Herrington, R. Influence of genetic processes on geochemistry of fe-oxy-hydroxides in supergene Zn non-sulfide deposits. *Minerals* **2020**, *10*, 602. [CrossRef]
66. Chirico, R.; Mondillo, N.; Boni, M.; Joachimski, M.M.; Ambrosino, M.; Buret, Y.; Mormone, A.; Leigh, L.E.N.B.; Flores, W.H.; Balassone, G. Genesis of the Florida Canyon Nonsulfide Zn Ores (Northern Peru): New Insights Into the Supergene Mineralizing Events of the Bongará District. *Econ. Geol.* **2022**, *117*, 1339–1366. [CrossRef]
67. Stavinga, D. Trace Element Geochemistry and Metal Mobility of Oxide Mineralization at the Prairie Creek Zinc-Lead-Silver Deposit. Doctoral Dissertation, Queen's University, Kingston, ON, Canada, 2014.
68. Martínez, C.E.; McBride, M.B. Cd, Cu, Pb, and Zn coprecipitates in Fe oxide formed at different pH: Aging effects on metal solubility and extractability by citrate. *Environ. Toxicol. Chem. Int. J.* **2001**, *20*, 122–126. [CrossRef]

 *minerals*

Article

Integration of Electrical Resistivity Tomography and Induced Polarization for Characterization and Mapping of (Pb-Zn-Ag) Sulfide Deposits

Mosaad Ali Hussein Ali [1,*], Farag M. Mewafy [2], Wei Qian [3], Fahad Alshehri [4], Mohamed S. Ahmed [4,*] and Hussein A. Saleem [1,5]

[1] Department of Mining and Metallurgy Engineering, Assiut University, Assiut 71515, Egypt; hasmohamad@kau.edu.sa
[2] Boone Pickens School of Geology, Oklahoma State University, Stillwater, OK 74078, USA; farag.mewafy@okstate.edu
[3] School of Earth Sciences and Engineering, Hohai University, Nanjing 211100, China; wei.geoserve@gmail.com
[4] Abdullah Alrushaid Chair for Earth Science Remote Sensing Research, Geology and Geophysics Department, College of Science, King Saud University, Riyadh 11451, Saudi Arabia; falshehria@ksu.edu.sa
[5] Mining Engineering Department, King Abdulaziz University, Jeddah 21589, Saudi Arabia
[*] Correspondence: mossad_ali2000@aun.edu.eg (M.A.H.A.); mohahmed@ksu.edu.sa (M.S.A.)

Abstract: The accurate characterization and mapping of low-grade ore deposits necessitate the utilization of a robust exploration technique. Induced polarization (IP) tomography is a powerful geophysical method for mineral exploration. An integrated survey using electrical resistivity tomography (ERT) and IP was employed in this study to characterize and map (Zn-Pb-Ag) ore deposits in NE New Brunswick, Canada. The survey encompassed twelve parallel lines across the study area. The 2D and 3D inversion of the results provided a detailed image of the resistivity and chargeability ranges of subsurface formations. The boundaries of sulfide mineralization were determined based on resistivity values of (700–2000 Ohm.m) and chargeability values of (3.5 mV/V) and were found to be located at an approximate depth of 80–150 m from the surface. The findings were validated through a comparison with data from borehole logs and mineralogy data analysis. The size and shape of sulfide deposits were successfully characterized and mapped in the study area using this cost-effective mapping approach.

Keywords: resistivity; chargeability; mineral exploration; 2D and 3D inversion

Citation: Ali, M.A.H.; Mewafy, F.M.; Qian, W.; Alshehri, F.; Ahmed, M.S.; Saleem, H.A. Integration of Electrical Resistivity Tomography and Induced Polarization for Characterization and Mapping of (Pb-Zn-Ag) Sulfide Deposits. *Minerals* **2023**, *13*, 986. https://doi.org/10.3390/min13070986

Academic Editors: Jia-Xi Zhou, Changqing Zhang, Tao Ren, Yue Wu and Amin Beiranvand Pour

Received: 15 June 2023
Revised: 12 July 2023
Accepted: 22 July 2023
Published: 24 July 2023

1. Introduction

Subsurface mineral prospecting presents significant challenges, especially within geologically complex formations. These challenges become even more challenging when we explore low-grade ore deposits. Metal sulfides are crucial ore minerals for global non-ferrous metal supplies [1]. Volcanic-related massive sulfide deposits, referred to as "Volcanogenic Massive Sulfide" (VMS) deposits, are widely spread and represent the most frequently occurring type of such deposits. These deposits are influential sources of various valuable metals, including zinc (Zn), lead (Pb), copper (Cu), silver (Ag), and gold (Au), while also serving as notable sources for cobalt (Co), selenium (Se), manganese (Mn), cadmium (Cd), indium (In), bismuth (Bi), tin (Sn), tellurium (Te), gallium (Ga), and germanium (Ge). Canada stands out with a substantial number of more than 350 VMS deposits, which contribute significantly to the production of different metals. Specifically, in Canada, VMS deposits account for 27% of copper production, 49% of zinc production, 20% of lead production, 40% of silver production, and 3% of gold production [1–3]. The prevalence of copper–zinc and zinc–copper VMS deposits in Canada is attributed to the abundance of primitive oceanic arc settings in the Precambrian era [1].

In recent decades, geophysical techniques have played a crucial role in providing valuable insights into subsurface mineralization [4–8]. These techniques gained paramount importance in mineral exploration for several reasons. Firstly, each technique relies on a unique physical property such as resistivity, conductivity, chargeability, gravity, magnetic, and seismic properties. This diversity allows for a comprehensive understanding of the subsurface by integrating multiple data sets for comparison. Secondly, geophysical approaches are cost-effective, non-invasive, and easy to deploy. Thirdly, they have the capability to cover both small and large areas, making them suitable for exploration purposes. Furthermore, data acquired through these techniques can be interpreted instantly in the field, providing initial information about the explored targets.

Induced polarization (IP) tomography has proven to be particularly effective in identifying and delineating sulfide deposits [9–12], being the sole geophysical technique with the capability to distinguish conductive or semi-conductive minerals dispersed within a background of high electrical resistivity (host rocks) [13–15]. By utilizing resistivity and chargeability measurements, we can effectively differentiate the mineral deposit content within rocks [16].

This study focuses on the characterization and mapping of a low-grade (Pb-Zn-Ag) sulfide deposit using electrical resistivity tomography (ERT) and IP (ERT-IP) in Nash Creek (NC), NE New Brunswick, Canada. The ore deposit of interest is located along the western edge of the Jacquet River Graben and is known for its low-grade mineralization [3,17]. The exploration history of the NC deposits has been extensively described by [18]. Mineral exploration in the NC region dates back to at least the 1930s, with a staking rush starting in the early 1950s following the discovery of the Heath Steele Mine in the Bathurst–Miramichi region. Several regions in the vicinity, including NC, Knowles Vein, Mitchell Settlement, Falls Brook, Jack Burns Lake, and McNeil Brook, have been periodically explored. Previous exploration programs relied on selected geophysical, geochemical, and shallow drilling data, with no DCIP exploration conducted in the past 15 years.

To better understand the distribution and extent of sulfide mineralization, a 2D ERT-IP survey was conducted in the study area. The ERT-IP survey comprised twelve parallel lines, covering a significant portion of the target area. The acquired data were subjected to 2D and 3D inversion processes to generate resistivity and chargeability models. The resulting models provided valuable information about the subsurface formation resistivity, ranging from 4 to 5000 Ωm, and chargeability values ranging from 0 to 12 mV/V. Based on the resistivity and chargeability models, the boundaries of the sulfide mineralization were identified. The mineralized zones exhibited specific resistivity values in the range of 700–2000 Ωm and chargeability values of approximately greater than or equal to 3.5 mV/V. These mineralization zones were found at depths of approximately 80–150 m below the surface. The reliability of these findings was validated by comparing them with borehole logging data and mineralogical analysis.

This study aims to contribute to the understanding of (Pb-Zn-Ag) sulfide deposits and their geological characteristics through the application of ERT-IP tomography. The results obtained from the 2D ERT-IP survey and subsequent inversion techniques provide valuable insights into the spatial distribution and characteristics of the mineralization. The findings from this study have the potential to enhance exploration strategies and resource estimation in similar geological settings.

2. Geology and Mineralization

The study area is situated approximately 5.6 km away from NC, with a latitude of 47.88° and a longitude of −66.11°. The geological map (Figure 1) encompasses various types of volcanic and sedimentary rocks, including rhyolites, volcanic mafic flows, pillow lavas, tuffs, breccias, siltstones, and limestones [18,19]. The NC sulfide mineralization is formed within a bi-modal volcanic–sedimentary sequence located in the half-graben. Generally, there are three main lithologic units in the NC area: mafic rock, felsic rock, and sedimentary rocks [19]. These volcanic and sedimentary rocks were formed within a half-graben structure that is bounded by faults on the western side [20]. The prospective "Dalhousie Group" is locally overlain by carboniferous rocks. Previous studies show that the formations within the Dalhousie Group have been the main target for mineral exploration at NC. These formations are the Mitchell Settlement Fm, Jacquet River Fm, Archibald Fm, Sunnyside Fm and Big Hole Brook Fm. Most of the discovered mineralization exists within the more felsic Archibald Settlement Fm of the Hayes Zone and in the Sunnyside Fm at the Hickey Zone.

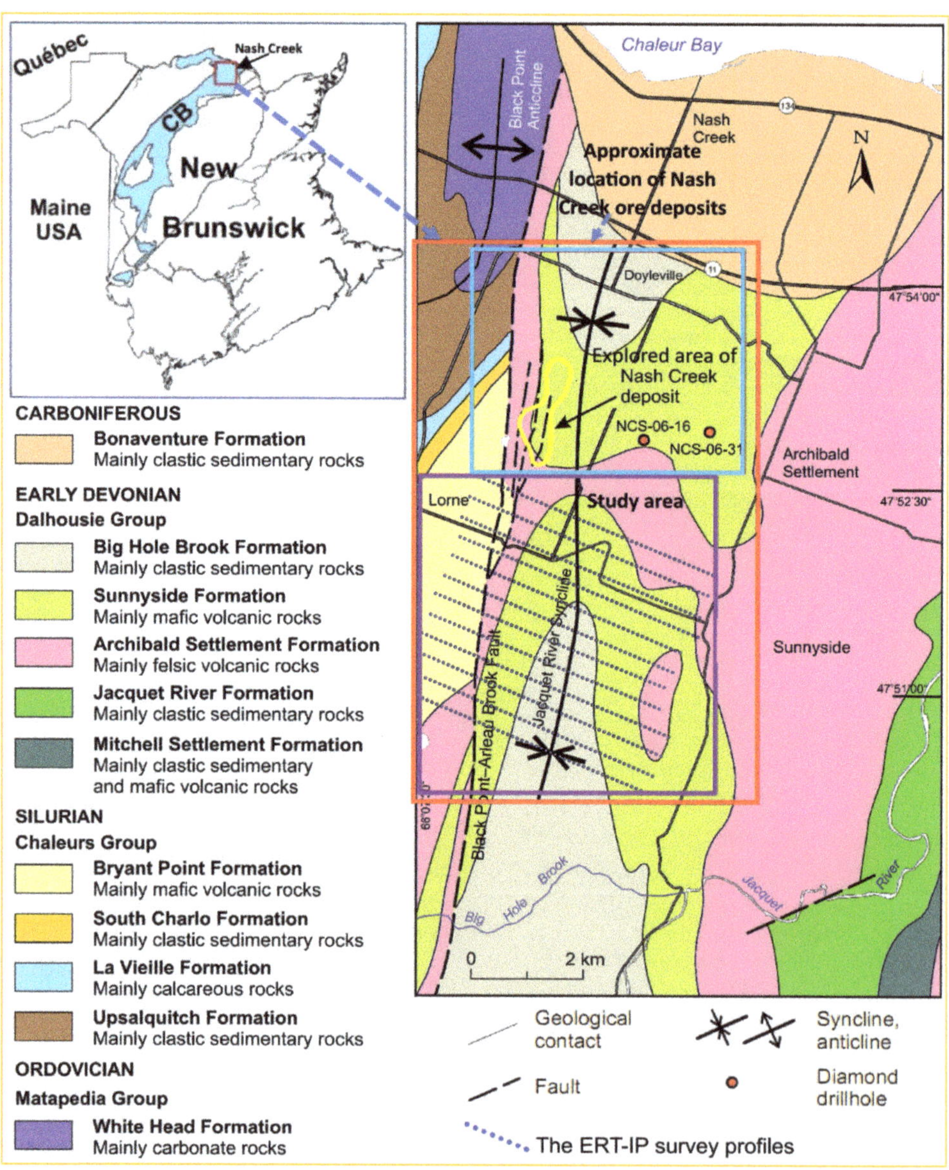

Figure 1. The location of the study area and geological features for Nash Creek deposit [21]. The light blue square indicates the previously explored area, while the purple square represents the current area of study.

The previous works such as the drilling program conducted in NC has intersected a sulfide mineralization deposit containing sphalerite, galena, pyrite, and occasionally chalcopyrite. The silver (Ag) grades show a moderate correlation with the (Zn-Pb) sulfides. Overall, the distribution of the sulfides ore reveals an increasing trend in assay results from the northern portion of the study area. These drilling programs have revealed the presence of the mineralization deposit at approximately 180 m depth [19].

3. Techniques

3.1. Electrical Resistivity

The ERT method uses the measurement of electrical resistivity to produce images of subsurface structures. The principle of ERT is based on the fact that different materials have different electrical resistivities. By injecting electrical current into the ground through two current electrodes (A, B) and tracking the potential difference between two potential electrodes (M, N), the electrical resistivity distribution can be mapped within the subsurface. Due to the electrical homogeneous and isotropic medium of the subsurface, the data collected during resistivity surveys are called apparent resistivity ρ_a. The apparent resistivity can be calculated using the equation below.

$$\rho_a = G\,U/I, \tag{1}$$

where, U = potential difference; I = Applied current; ρ_a = Apparent resistivity of the medium; G = Geometrical factor that depends on electrode array.

3.2. Induced Polarization

Induced polarization (IP) is observed when a direct current passing through two electrodes is cut off: the voltage decays slowly and takes time to reach zero. This is an indication that charge was stored within the rocks. Four nonpolarizable electrodes are used during IP surveys: two electrodes to inject the electrical current and two additional electrodes to measure the resulting difference of the electrical potential. This phenomenon can be quantified in either the time domain (TDIP) by observing the rate of decay of voltage, or in the frequency domain induced polarization (FDIP) or spectral induced polarization (SIP). In the TDIP, the chargeability value in the IP measurements is derived from the integration of an IP decay curve. TDIP imaging measures the time-dependent response of the subsurface to electrical current to identify variations in chargeability or polarization properties. On the other hand, FDIP or SIP are determined by measuring phase shifts between sinusoidal currents and voltages. Chargeability can be expressed in terms of primary voltage (existing in steady-state conditions during the injection of the electrical current) divided by the secondary voltage (see Figure 2) and Equation (2).

$$M = U_{IP}/U_{DC}, \tag{2}$$

where, M = Chargeability; U_{IP} = Secondary potential; U_{DC} = Initial potential.

3.3. The Integration of ERT-IP and Tomography Measurements

The integration of ERT-IP tomography methods involves combining the measurements obtained from both techniques to gain a more comprehensive understanding of subsurface properties. ERT measures the electrical resistivity distribution of the subsurface, providing information about the lithology, moisture content, and presence of geological structures. IP, on the other hand, measures the polarization or chargeability of subsurface materials, which is related to the presence of mineralization, clay content, and fluid conductivity. By integrating ERT and IP data, it is possible to distinguish between resistive and polarizable materials, allowing for more accurate characterization of subsurface conditions and improved detection of geological features such as mineral deposits. This integration can enhance the interpretation of geophysical data and aid in various applications like groundwater exploration, mineral prospecting, and environmental studies.

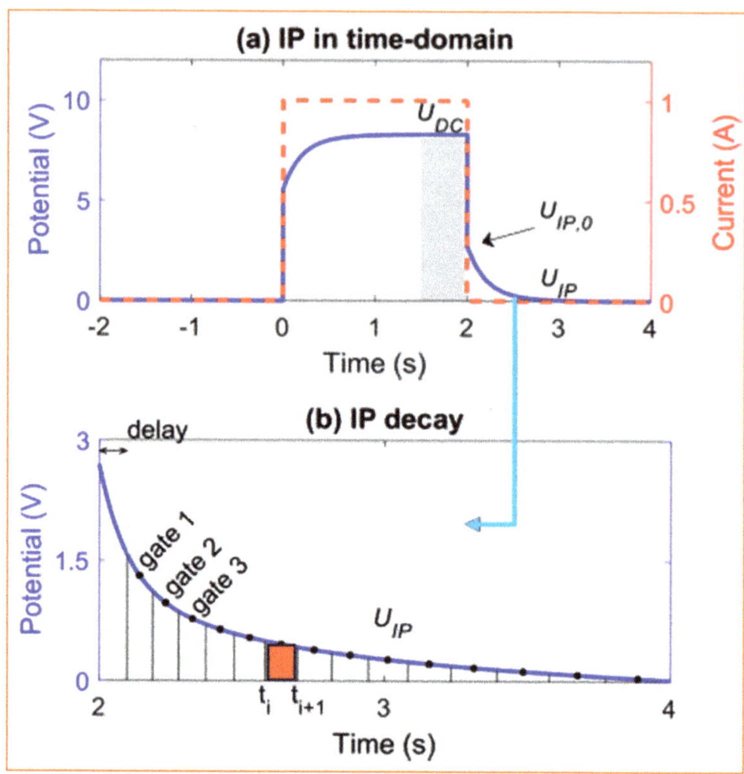

Figure 2. Induced polarization measured in the time domain as the potential response of 2 s current on-time. (**a**) the red dotted line represents the relationship between the applied current and time, while the blue line displays the behavior of the measured potential. (**b**) For full-decay IP modelling, the decay is divided into time gates and an apparent chargeability value is defined in each gate [22].

In a 2D survey using the ERT-IP method, the sequence of data measurement typically involves the following steps:

- Electrode Placement: Electrodes are placed on the ground surface in a specific configuration. In ERT, this usually involves the use of a pair of current electrodes (current injection) and a set of potential electrodes (voltage measurement) that are arranged in a linear or rectangular array. In IP, additional electrodes may be used to apply a voltage waveform and measure the resulting electrical response;
- Current Injection: A known electric current is injected into the ground through the current electrodes. The injected current flows through the subsurface, and its distribution is influenced by the electrical properties of the materials encountered;
- Voltage Measurement: The potential electrodes measure the voltage distribution in the ground resulting from the injected current. These measurements are recorded and used to determine the electrical resistivity distribution in ERT and the polarization or chargeability distribution in IP;
- Data Acquisition: A series of measurements are taken by varying the electrode positions along the survey line. This involves moving the electrode array and repeating steps 2 and 3 at different locations along the line. The electrode spacing may be kept constant or adjusted depending on the desired level of resolution and the subsurface conditions;
- Data Interpretation: Once the data acquisition is complete, the collected voltage and current data are processed and analyzed. In ERT, inverse modeling techniques are

used to create a resistivity model of the subsurface, representing the distribution of different geological units or features. In IP, the chargeability or polarization data are analyzed to identify zones of interest, such as mineralized areas;

- Data Integration: Finally, the ERT and IP data are integrated and correlated to obtain a more comprehensive understanding of the subsurface. This may involve overlaying the resistivity and chargeability models, identifying areas of anomalous responses, and interpreting the geological implications of the integrated results.

The sequence of data measurement may vary depending on the specific survey designs and equipment used, but the general steps outlined above provide a framework for conducting a 2D ERT-IP survey and analyzing the collected data.

4. Data Acquisition and Processing

4.1. Data Acquisition

TDIP data were collected using the IRIS Syscal Pro system. Syscal Pro is a resistivity and IP measurement system that can perform electrical resistivity tomography (ERT) and IP imaging for various near-surface applications. It can be used to measure the primary voltage and the voltage decay curve and thus provides resistivity and chargeability (IP) data. Notably, the Syscal Pro system features 20 chargeability slices, enhancing its capability to accurately capture and measure discharge phenomena. It also has a 1 μV resolution on the primary voltage. This is helpful for geophysical surveys that require a shallow investigation of the subsurface structures and the characterization of the electrical properties of the subsurface materials.

The resistivity and chargeability data sets were collected simultaneously. The study area consists of 12 surface lines, each about 2 km long, 100 m apart and covering a total area of about 9.5 km^2 (Figure 3). In ERT-IP surveys, electrode spacing refers to the distance between the current injection electrodes (A, B, source electrodes) and the potential measurement electrodes (M, N, receiver electrodes). The choice of electrode spacing depends on several factors, including the survey's objectives, the subsurface's geological conditions, and the desired depth of investigation. Generally, different electrode spacings are used to target different depths and resolutions. In surveys targeting shallow depths, such as environmental or engineering applications, the electrode spacing is usually smaller, ranging from a few meters upwards. This helps to provide high-resolution data and to capture subtle changes in subsurface resistivity or induced polarization. For surveys targeting greater depths, such as mineral exploration or groundwater studies, larger electrode spacings are employed. These can range from several meters upwards, allowing for deeper penetration into the subsurface but at the cost of reduced resolution. It is important to note that the choice of electrode spacing is a trade-off between the depth of the investigation, resolution, and practical considerations such as the size of the survey area and the available equipment. A pole–dipole configuration with an initial electrode spacing of 25 m was applied for the ERT-IP tomography survey. The pole–dipole array is commonly used in mineral exploration due to its ability to provide detailed information about subsurface targets [23,24]. It offers good depth penetration and excellent lateral resolution [24]. This array configuration is cost-effective, and is efficient in data acquisition and processing, making it a preferred choice for this study, which requires the accurate characterization of subsurface mineral deposits. In the ERT-IP data acquisition using the pole–dipole array, the sequence of moving electrodes for data acquisition follows a specific pattern. With this initial electrode spacing, the survey progressed through a series of 10 levels. At each level, a current injection electrode (source electrode) is selected, and potential measurement electrodes (receiver electrodes) are positioned at varying distances along the survey line. The first receiver electrode was placed 25 m away from the source electrode, creating the initial 25 m electrode spacing. Subsequently, the other receiver electrodes were positioned at increasing distances from the source electrode, maintaining the same spacing interval. The electrode sequence continues with the second receiver electrode being positioned 50 m away from the source, the third receiver electrode at 75 m, and so on, until all 10 levels have been completed. This sequential movement of the electrodes allows for systematic data acquisition at different distances and provides measurements at increasing depths as the survey progresses. To account for the topographic effect on the data resolution, the GPS data of each electrode along the survey lines were collected using handheld Garmin GPS devices. The GPS measurements had an accuracy of approximately ±5 m for horizontal coordinates and ±10 m for elevation data. The specific parameters used for ERT-IP measurements can be summarized as follows:

- Number of survey lines: 12 lines;
- Geophysical instrument: Syscal Pro;
- Array types: Pole–dipole;
- Electrode spacing: Electrode spacing refers to the distance between the current injection electrode (A, source) and the potential measurement electrode (M, receiver). In this survey, a pole–dipole configuration with an initial electrode spacing of 25 m was utilized;
- Number of levels: 10 levels;
- IP domain: time domain;
- Current injection duration (on-time): 240 s;
- Voltage measurement window (off-time): 20 time windows (gates), semi-logarithmic;
- Time interval between measurements: 80 s.

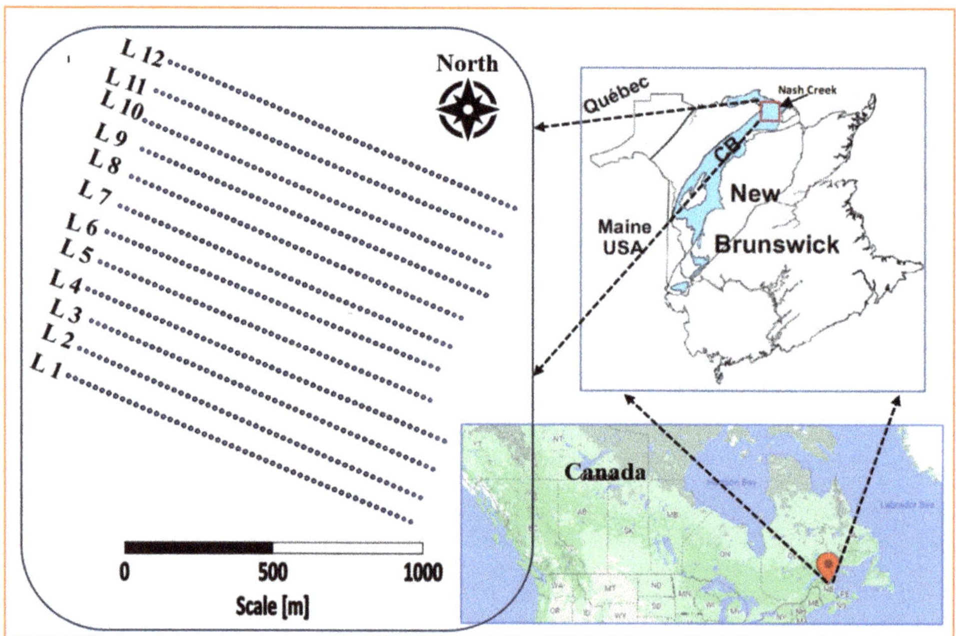

Figure 3. Map of the geophysical survey profiles (from L_1 to L_{12}) in the study area.

4.2. Data Inversion

RES2DINV (V4.08, Geotomo Software, Penang, Malaysia) and RES3DINV (V3.14, Geotomo Software, Penang, Malaysia) [25,26] RES2DINV and RES3DINV are popular commercial inversion software packages used for the interpretation and modelling of electrical resistivity and IP data obtained from geophysical surveys. RES2DINV is designed for 2D ERT-IP surveys, while RES3DINV is specifically developed for 3D and quasi-3D (2.5D) surveys. RES2DINV allows the interpretation of ERT-IP data collected along 2D survey lines and provides a 2D model of the subsurface resistivity distribution. The software utilizes the finite element method to perform the inversion process, which involves determining the best-fit resistivity model that can reproduce the observed resistivity data. RES2DINV incorporates various regularization techniques to stabilize the inversion process and minimize artifacts in the resulting resistivity models. On the other hand, RES3DINV is tailored for the interpretation of 3D and 2.5D ERT-IP data acquired from surveys using different electrode configurations, such as dipole–dipole or pole–dipole arrays. RES3DINV employs the voxel-based inversion method, dividing the subsurface into a three-dimensional grid of small cells and determining the resistivity values of each cell to create a 3D resistivity model. The software incorporates sophisticated algorithms to handle the complexities of 3D data inversion, including the handling of varying electrode spacing and geometry. Both RES2DINV and RES3DINV provide user-friendly interfaces with robust visualization capabilities, allowing users to view the input data, the resulting models, and the associated inversion parameters. They also offer tools for data quality control, sensitivity analysis, and model validation. The software packages are widely used in various geophysical applications, including environmental studies, groundwater exploration, mineral exploration, and engineering investigations.

These two software packages were utilized for the ERT-IP data processing and inversion in this study through employing the finite element method for discretizing the subsurface. The topography of the study area was incorporated into the inversion models seamlessly. The inversion process divided the subsurface into rectangular cells and aimed to produce a model that closely matched the measured data through iterative updates using the Gauss–Newton optimization technique. Also, instead of the conventional least square method, the robust L1-norm inversion method, which considers absolute errors, was employed due to the presence of sharp boundaries expected in the subsurface mineralized features. By minimizing the absolute difference between the measured and estimated data values, the impact of outlier data points was reduced to an acceptable range of error (i.e., reaching minimal further changes in RMS after a certain number of iterations) [25,27]. As per Loke [25], it is important to note that the model with the lowest RMS error may exhibit unrealistic variations and may not necessarily be the most accurate geological representation. Instead, Loke suggests selecting a model where the RMS error remains stable and does not undergo significant changes [25]. Taking this into consideration, the work models presented here are derived from the fifth iteration of the inversion process. The appendix contains the error statistics, including absolute and RMS errors, for the resistivity and IP inverse models. Please refer to Figures A1 and A2 in the appendix for the respective figures.

5. Results and Discussion

5.1. 2D Resistivity and Chargeability Models

In order to avoid redundancy when presenting the 2D inversion results (resistivity and chargeability models) for each individual line, we have chosen to display the results for only two lines located in the middle. Specifically, lines 5 and 6 were selected to showcase the inversion results, as shown in Figure 4. The complete figure featuring the 2D inversion results of all the survey lines can be found in the appendix (Figure A3). The survey sequence of these lines was performed in the northwest-trending zone (Figure 3) The inverted models were represented in the west–east direction (Figure 4). The resistivity model exhibited a range of formation resistivity values from 4 to 5000 Ωm, indicating significant variations in the subsurface lithology and mineralization. In general, lower resistivity values are associated with the presence of conductive zones, representing sulfide mineralization, while higher resistivity values represent the surrounding non-conductive host rocks (Mafic Lithic Tuff) [28]. However, it is important to note that in certain cases, the zones of sulfide mineralization exhibited moderate to high resistivity values, as influenced by the broader geology of the study area [9]. This inconsistency in the interrelationship between resistivity values and mineralization zones makes it difficult to directly isolate mineralization zones from other features that could represent low resistivity values. Moreover, it has become hard to identify the boundaries of mineralization zones solely based on resistivity anomalies. The chargeability models, on the other hand, revealed chargeability values ranging from 0 to 12 mV/V. The number of chargeability anomalies is lower and shows well-identified boundaries compared to the resistivity anomalies. Higher chargeability values were observed in zones with increased sulfide mineralization, indicating the presence of conductive minerals. The chargeability response was particularly useful in delineating the boundaries and extent of the mineralized zones [28]. While high chargeability anomalies can be an indication of the presence of either clay minerals or sulfide mineral deposits, previous geological and petrophysical studies on the area did not report any presence of clay minerals. For example, Figure 5a shows that the lithology of the area is fully dominated by igneous rocks. Thus, we assume that the high chargeability anomalies in the study area are mainly driven by the presence of sulfide ore deposits.

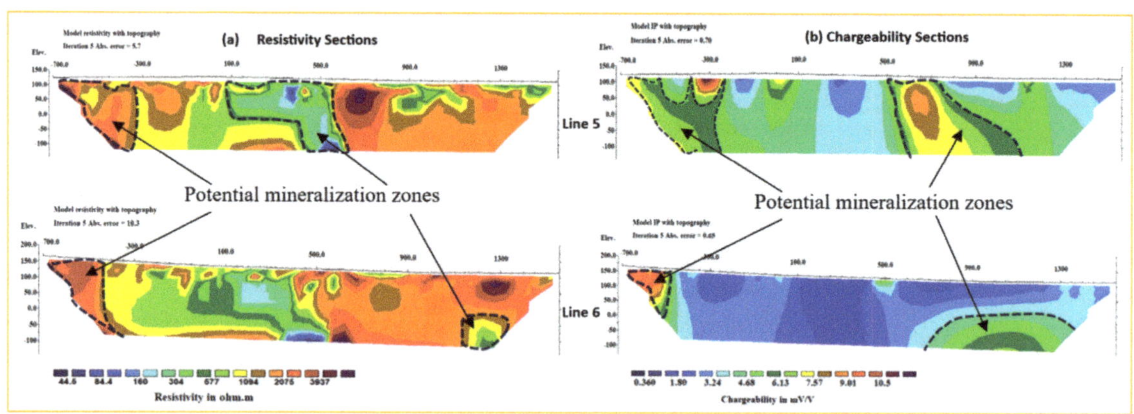

Figure 4. The 2D inversion models of resistivity and chargeability for survey lines 5 and 6. The dashed contours indicate potential mineralization zones. (**a**) The resistivity sections. (**b**) The chargeability sections.

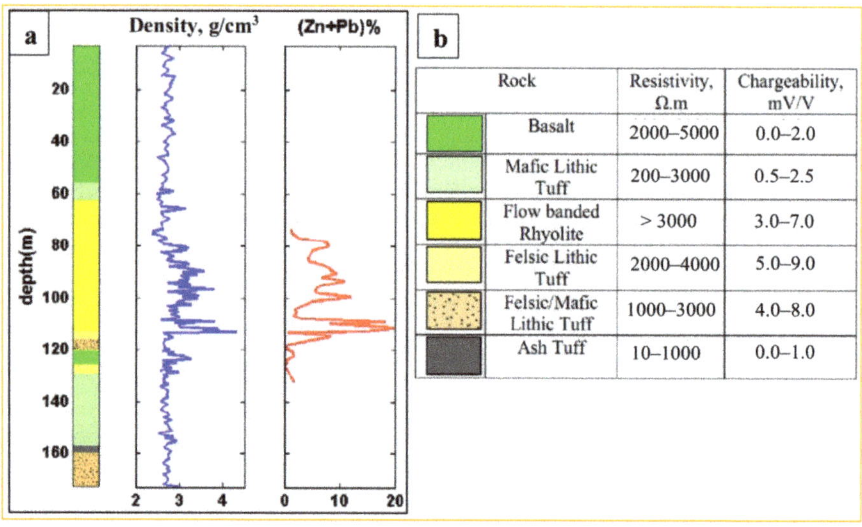

	Rock	Resistivity, Ω.m	Chargeability, mV/V
	Basalt	2000–5000	0.0–2.0
	Mafic Lithic Tuff	200–3000	0.5–2.5
	Flow banded Rhyolite	> 3000	3.0–7.0
	Felsic Lithic Tuff	2000–4000	5.0–9.0
	Felsic/Mafic Lithic Tuff	1000–3000	4.0–8.0
	Ash Tuff	10–1000	0.0–1.0

Figure 5. (**a**) An example of the borehole logs showing the density and assay (Zn + Pb%) analysis [3,20,29]. The location of this borehole is depicted in Figure A4, which is included in the appendix. (**b**) The resistivity and chargeability quantities of the rock units.

As the study area is fully dominated by mafic volcanic rocks (Figure 1), the mineralization zones were abundant in the inverted models, with relatively low chargeability and moderate to high resistivity values. Previous petrophysical surveys conducted in the study area have indicated the presence of high-grade sulfide mineralization in Pyroclastic units composed of Felsic Lithic Tuff, while low-grade mineralization is embedded in Mafic Lithic Tuff and medium-grade mineralization is found in flow-banded Rhyolite units (Figure 5a) [3,20,29]. So, the mineralization zones appear with relatively low to moderate chargeability and moderate to high resistivity values. The inversion models for both resistivity and chargeability are presented in Figure 4. These data were also correlated to the borehole logging data (Figure 5b). The resistivity sections show various anomalies relatively with low resistivity values along the middle of most of the resistivity lines. Additionally, some other low resistivity anomalies are dispersed across some lines. This makes it difficult to pinpoint specific zones for potential mineralization.

On the other hand, the chargeability lines show a smaller number of anomalies compared to the resistivity lines. Moreover, the anomalies along the chargeability lines are narrower and well-defined. The presence of a low number of chargeability anomalies could indicate that the majority of the sulfide deposit within the exploration area is of medium to low grade. Furthermore, the limited extent of the mineralization zone suggests a weak occurrence of the deposit.

The mineralized zone in the 2D inversion models (Figure A3) appears at a depth of around 80 m to 140 m from the surface. From line 1 to line 4 (Figure A3), the mineralization zone seems to be a narrow ore body with a shallower depth and a low-grade ore. In contrast, line 5 and 6 (Figure 4) display a deeper mineralization zone of a higher grade and extent compared to the previous lines. Lines 7 to 9 (Figure A3) exhibit relatively shallow mineralization zones with a low grade and a medium extent. Finally, for lines 10 to 12, the mineralization zone is deeper and has a high grade and extent.

5.2. 3D Resistivity and Chargeability Models

Performing a 3D inversion of a 2D ERT-IP survey using RES3DINV involves several steps to ensure accurate and reliable results. The following is a detailed description of the process:

I.   Data Acquisition and Preprocessing: The first step was acquiring the 2D ERT-IP survey data using appropriate measurement equipment, as mentioned earlier. This involved positioning electrodes along the survey lines and measuring the voltage and current data. Once the data sets were collected, they underwent preprocessing, which included removing any noise or interference, correcting for electrode offsets, and applying geometric corrections to ensure accurate spatial positioning of the data;

II.   3D Data model: The 3D model was constructed through combining the 2D lines [30,31]. These data were merged using the GPS coordinates of survey lines. This merged 3D view serves as a valuable resource for further interpretation, modelling, and exploration planning, facilitating a comprehensive assessment of the mineral deposits in the study area;

III.   Inversion: The 3D inversion process was performed using RES3DINV. The 3D model is helpful for visualizing the lateral distribution of the ore deposits.

The utilization of 3D inversion techniques for resistivity and chargeability modeling, based on 2D surveys, provides significant advantages over traditional 2D inversion approaches [32,33]. The incorporation of the third dimension allows for improved accuracy in representing lateral variations, enhanced depth resolution, realistic geometry representation, and improved visualization capabilities [34–36]. These advantages contribute to a more comprehensive characterization and mapping of sulfide deposits, assisting in the development of effective exploration strategies and resource evaluation.

The results obtained from the 3D inversion method provide a significant improvement over 2D inversion in terms of accurately characterizing the resistivity and chargeability distribution of the mineralized zones. Moreover, the 3D model reveals an N–S trend on both resistivity and chargeability models that align with the N–S fault that crosses the study area (Figure 6). Detailed analysis of the presented horizontal and vertical sections (Figures A5–A8 in the appendix) reveals key insights. To facilitate effective comparison, particular attention is given to the fifth slice from the horizontal sections (Figure 6) and the 18th slice from the vertical sections (X-Z direction), spanning a distance from 9500 to 11500 m in the Y direction (Figure 7). These slices depict the resistivity and chargeability distribution, along with the accompanying RMS errors of the inverted models. The resistivity model derived from the 3D inversion showcases a broad range of formation resistivity values, spanning from 4 to 5000 Ωm. Similarly, the chargeability values range from 0 to 12 mV/V. Notably, when comparing the mineralized zones, the results obtained from the 3D inversion demonstrate a notable improvement over the 2D inversion. Specifically, the resistivity values within the mineralization zones, as shown in Figures 6a and 7a, exhibit a moderate to high range, ranging from 800 to 1500 Ωm. This finding aligns closely with the interpretations derived from the 2D inversion models. Furthermore, the chargeability values associated with the mineralization zones, depicted in Figures 6b and 7b, fall within a relatively low to moderate range. The chargeability signature zones >3.5 mV/V were more broadly clearly developed than the resistivity signature zones. These characteristics were relatively harmonious with those in the 2D models. Accordingly, the continuity of the chargeability signature zones was effectively visualized in the 3D model (Figure 8). The 3D model visualizations were more useful for characterizing the shape and size of sulfide deposits than the 2D models. Figure 9 shows the 3D model of the occurrence of sulfide signature zones at a 3.5 mV/V cut-off chargeability value. The size of the sulfide deposit ore (potential geological reserve) was estimated to be 393,107 m^3 based on the inversion distance method (in the visualization software used (Voxler 4), there exists an option to calculate the volume of the iso-surface). While the data show an N–S trend that aligns with the N–S fault, it is not clear if the mineralization in the study area is structurally controlled by the fault and this point may require further work.

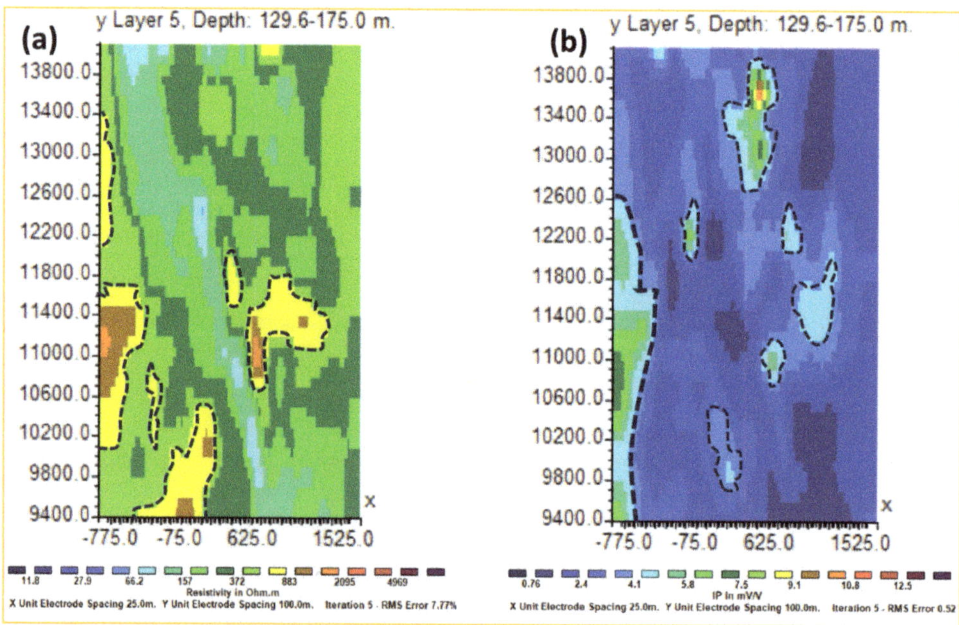

Figure 6. Horizontal sections of the 3D inversion result. (**a**) The resistivity inversion model. (**b**) The chargeability inversion model. The dashed contours indicate potential mineralization zones.

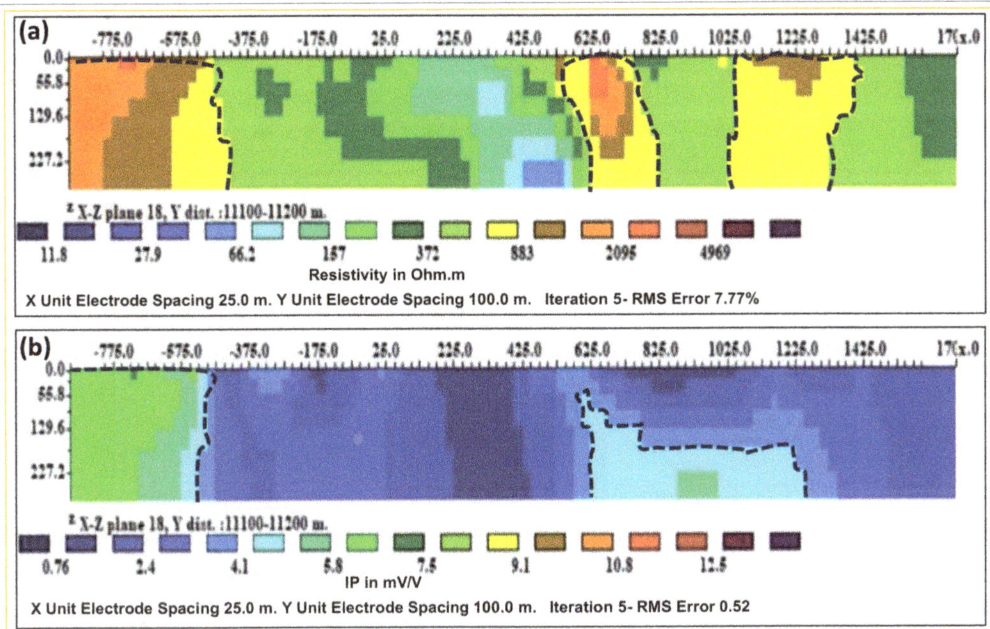

Figure 7. Vertical sections (X−Z direction) of the 3D inversion result. (**a**) The resistivity inversion model. (**b**) The chargeability inversion model. The dashed contours indicate potential mineralization zones.

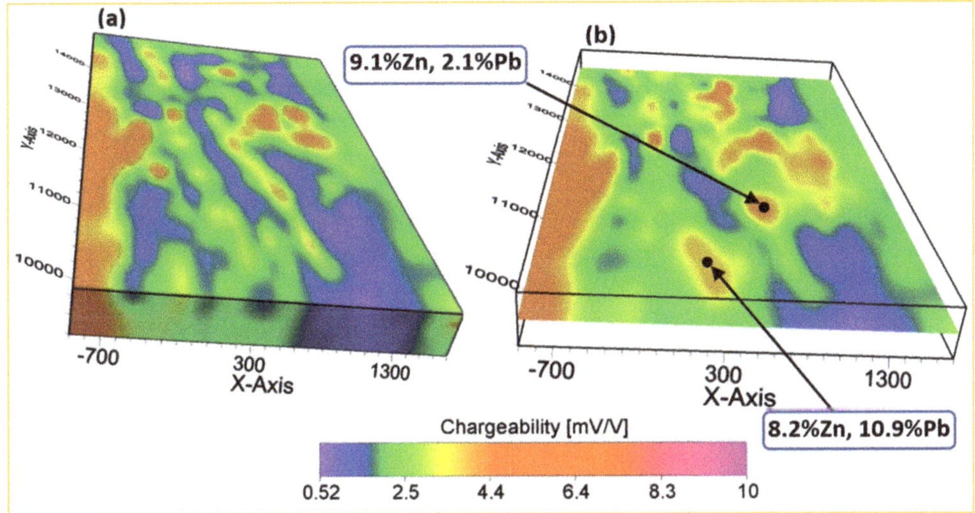

Figure 8. (**a**) The full 3D chargeability inversion model. (**b**) A slice at 150 m depth of the 3D chargeability inversion model with the inclusion of two core sample locations as an example.

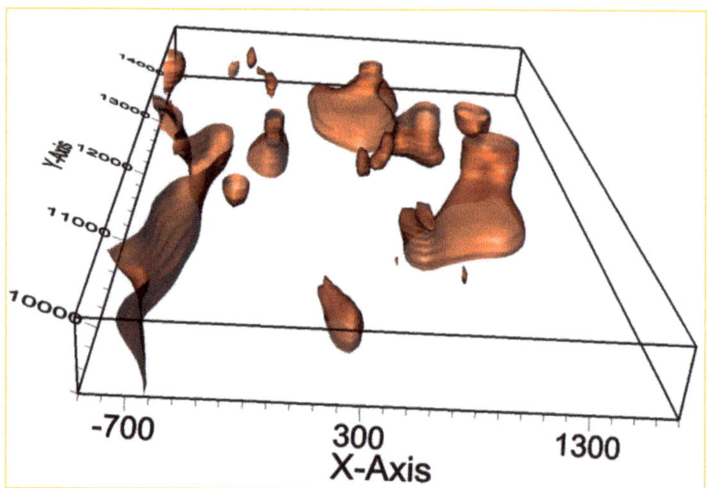

Figure 9. 3D iso-surface model of the occurrence of sulfide deposit zones at 3.5 mV/V cut-off chargeability value.

Overall, the 3D inversion method offers a significant advancement over the 2D inversion by providing a more accurate characterization of the resistivity and chargeability distribution within the mineralized zones. These improved results contribute to a better understanding of subsurface mineralization and further enhance the reliability and confidence of the study's findings.

5.3. Validation with Borehole Logging and Mineralogy Data

Two candidate hotspot locations were identified on the chargeability 3D model (Figure 8b), and two test boreholes were drilled to validate our findings. Core samples were extracted from these boreholes for mineralogical analysis. The analysis confirmed the presence of sulfide minerals in the core samples (8.2% Zn, 10% Pb and 9.1% Zn, 2.1% Pb), which is consistent with the mineralization zones identified on the chargeability model (Figure 8b). The agreement between the ERT-IP survey results and the mineralogical analysis supports our interpretation of the characterization and mapping of this low-grade sulfide deposit using ERT-IP tomography.

The boundaries of the sulfide mineralization zone, as determined by ERT-IP criteria, coincided with the geological and petrophysical features of the area, such as faults and lithological contacts. This alignment provided confidence in the accuracy and reliability of the ERT-IP survey results. Consequently, it is considered that 2D/3D ERT-IP tomography is useful for the determination of the boundaries of sulfide mineralization alteration zones.

6. Conclusions

This study focused on the integrated analysis of electrical resistivity tomography and induced polarization (ERT-IP) for the characterization and mapping of low-grade (Pb-Zn-Ag) sulfide deposits at Nash Creek in NE New Brunswick, Canada. Both 2D and 3D inversion models were conducted and compared, providing a comprehensive understanding of the subsurface characteristics and distribution of the sulfide deposit. The ERT-IP inversion yielded detailed 2D and 3D resistivity and chargeability models, revealing the range of resistivity (4 to 5000 Ohm.m) and chargeability (0–12 mV/V) values for subsurface formations. Based on these models and considering existing geological and petrophysical studies, potential sulfide mineralization zones were identified using resistivity values (700–2000 Ohm.m) and chargeability values ($\geq$3.5 mV/V) at depths of approximately 80–150 m. A 3D iso-surface model was constructed to visualize the distribution of these potential mineralization zones, allowing for a 3D geological model of the study area. The estimated size of the sulfide deposit ore (potential geological reserve) was determined to be 393,107 m^3 using the inversion distance approach. The accuracy of the findings was confirmed through comparison with borehole logs and mineralogy data analysis. The study successfully characterized and mapped the size and shape of sulfide deposits in the study area, showcasing the effectiveness of this cost-effective mapping approach. The findings contribute to the understanding of sulfide deposits and their geological characteristics in the study area. The application of ERT-IP tomography demonstrated its efficacy in delineating mineralized zones and can enhance exploration strategies and resource estimation in similar geological settings. Further investigations and integration with additional geoscientific data can lead to a more comprehensive understanding of the deposit and optimize resource extraction techniques.

Author Contributions: Conceptualization, M.A.H.A. and F.M.M.; methodology, M.A.H.A. and F.M.M.; software, M.A.H.A. and F.M.M.; validation, M.A.H.A. and W.Q.; formal analysis, M.A.H.A. and F.A.; investigation, M.A.H.A. and W.Q.; resources, M.A.H.A. and M.S.A.; data curation, W.Q.; writing—original draft preparation, M.A.H.A. and W.Q.; writing—review and editing, M.A.H.A., F.M.M., and H.A.S.; visualization, M.A.H.A., F.M.M. and W.Q.; supervision, W.Q. and H.A.S.; project administration, W.Q.; funding acquisition, F.A. and M.S.A. All authors have read and agreed to the published version of the manuscript.

Funding: Abdullah Alrushaid Chair for Earth Science Remote Sensing Research at King Saud University, Riyadh, Saudi Arabia.

Data Availability Statement: Data sharing not applicable.

Acknowledgments: The authors extend their appreciation to Abdullah Alrushaid Chair for Earth Science Remote Sensing Research for funding.

Conflicts of Interest: The authors declare no conflict of interest.

Appendix A

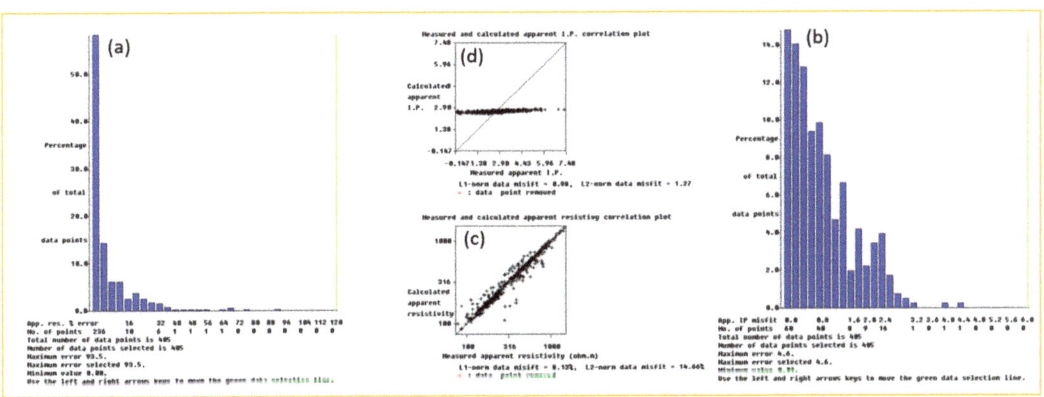

Figure A1. Showcases the outcomes of the 2D inversion results for Line 1, illustrating the absolute and RMS error statistics without the exclusion of data outliers. The figure is divided into four parts: (**a**) a histogram depicting the misfit between the measured and calculated apparent resistivity values, (**b**) a histogram illustrating the misfit between the measured and calculated apparent IP values, (**c**) a scatter plot demonstrating the misfit between the measured and calculated apparent resistivity values, and (**d**) a scatter plot exhibiting the misfit between the measured and calculated apparent IP values.

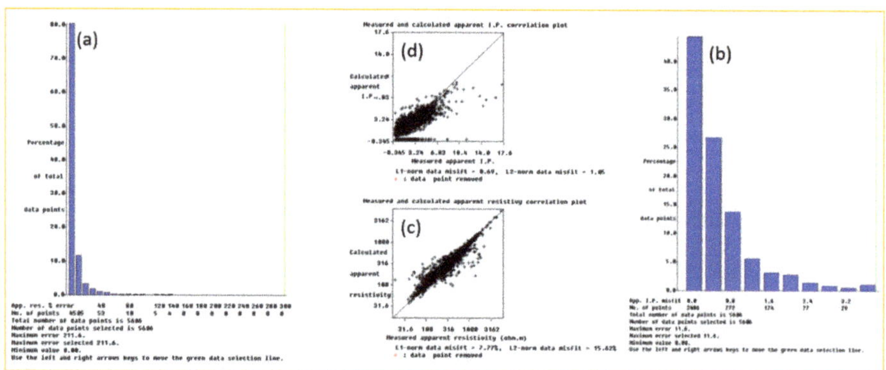

Figure A2. Showcases the outcomes of the 3D inversion results, illustrating the absolute and RMS error statistics without the exclusion of data outliers. The figure is divided into four parts: (**a**) a histogram depicting the misfit between the measured and calculated apparent resistivity values, (**b**) a histogram illustrating the misfit between the measured and calculated apparent IP values, (**c**) a scatter plot demonstrating the misfit between the measured and calculated apparent resistivity values, and (**d**) a scatter plot exhibiting the misfit between the measured and calculated apparent IP values.

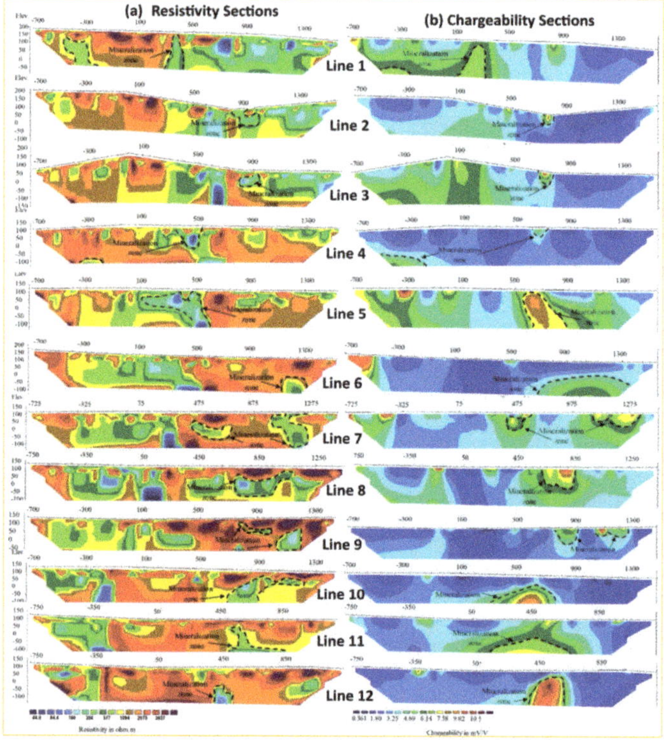

Figure A3. The 2D inversion models of resistivity and chargeability for the survey lines. The dashed contour zones in the figure indicate potential mineralized zones. (**a**) The resistivity sections. (**b**) The chargeability sections.

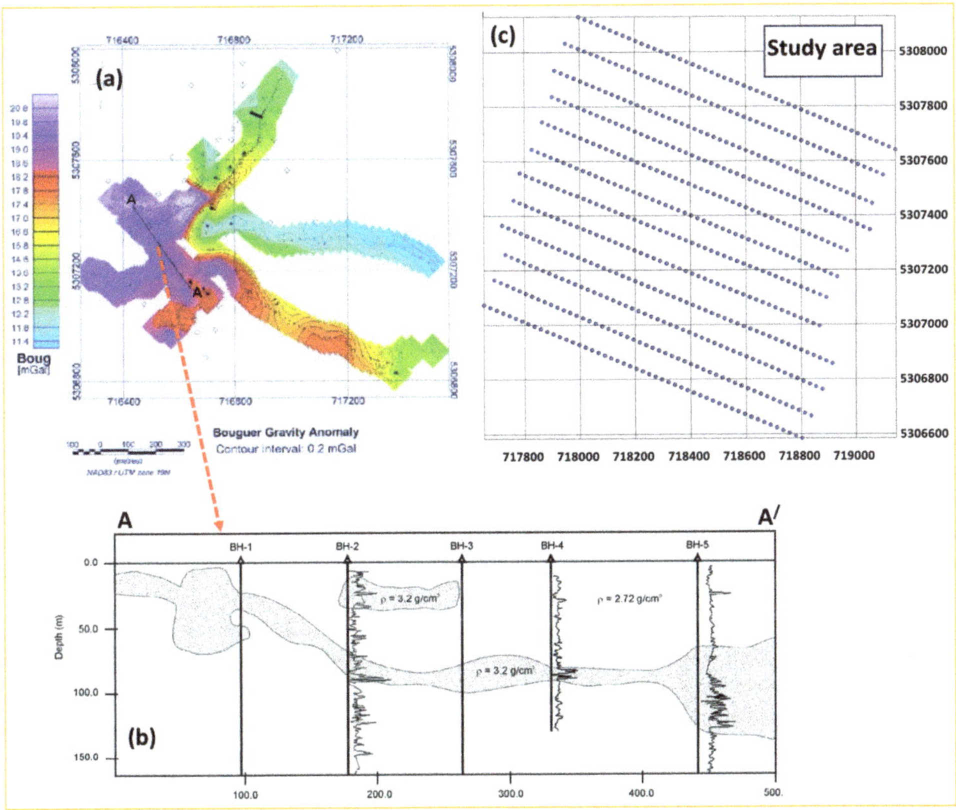

Figure A4. Illustrates the location of the borehole logs, which is approximately 1.2 km west of the study area: (**a**) presents a map depicting the Bouguer gravity and the position of the borehole logs along the A-A' line [3], (**b**) represents the vertical cross-section along the A-A' line, indicating the specific location of the borehole logs [3], and (**c**) shows the study area.

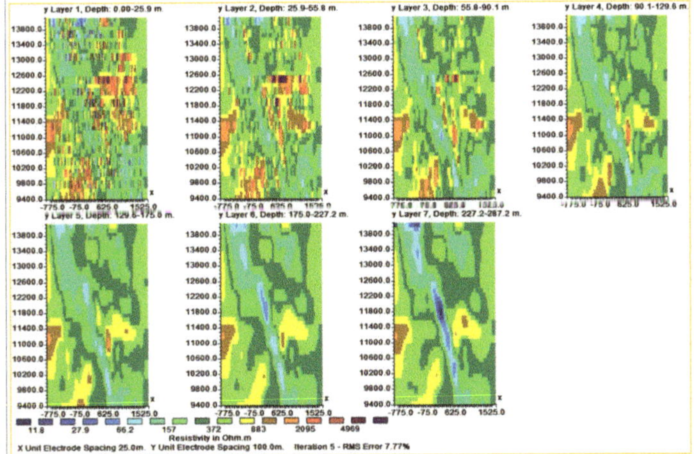

Figure A5. Horizontal sections of the 3D resistivity inversion model.

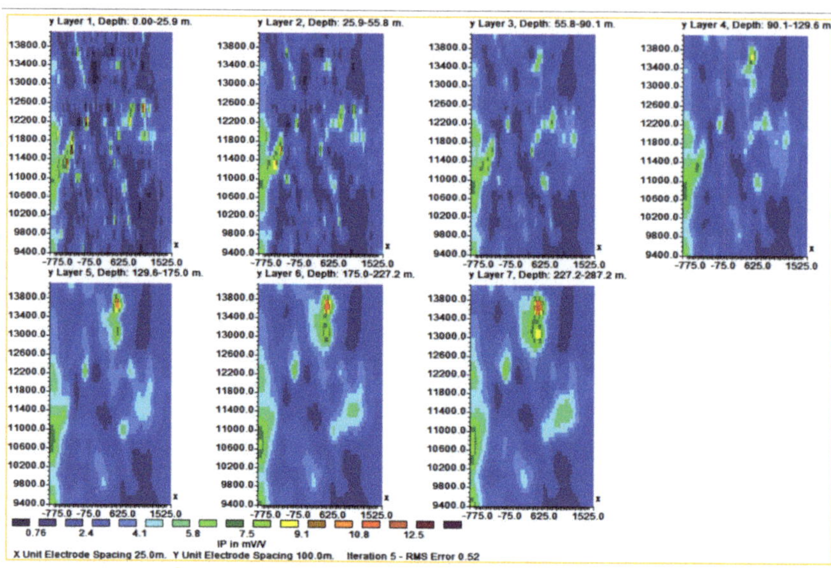

Figure A6. Horizontal sections of the 3D chargeability inversion model.

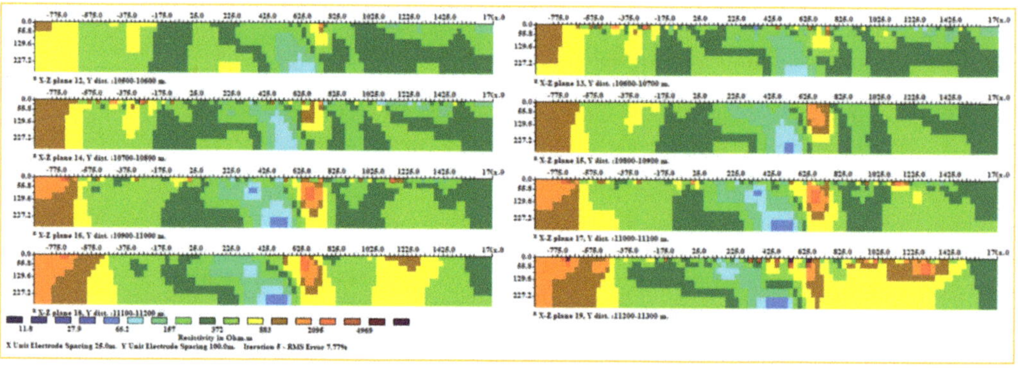

Figure A7. Vertical sections (X-Z direction) of the 3D resistivity inversion model.

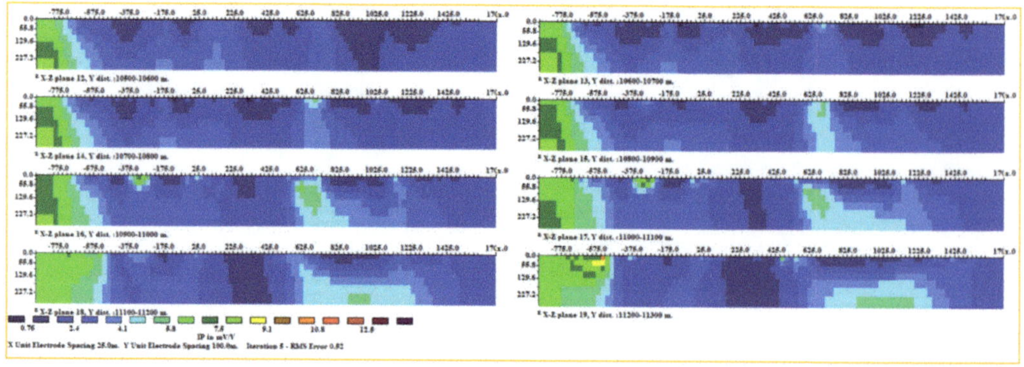

Figure A8. Vertical sections (X-Z direction) of the 3D chargeability inversion model.

References

1. Galley, A.G.; Hannington, M.D.; Jonasson, I.R. Volcanogenic Massive Sulphide Deposits. In *Mineral Deposits of Canada: A Synthesis of Major Deposit-Types, District Metallogeny, the Evolution of Geological Provinces, and Exploration Methods*; Geological Association of Canada, Mineral Deposits Division: Sudbury, ON, Canada, 2007; Volume 5, pp. 141–161.
2. Lydon, J.W. Ore deposit models-8. Volcanogenic massive sulphide deposits Part I: A descriptive model. *Geosci. Canada* **1984**, *11*, 4.
3. Ugalde, H.; L'Heureux, E.; Milkereit, B. An integrated geophysical study for orebody delineation, Nash Creek, New Brunswick. In Proceedings of the Exploration 07: Fifth Decennial International Conference on Mineral Exploration 2007, Toronto, ON, Canada, 9–12 September 2007; pp. 1055–1058.
4. Eldougdoug, A.; Abdelazeem, M.; Gobashy, M.; Abdelwahed, M.; Abd El-Rahman, Y.; Abdelhalim, A.; Said, S. Exploring gold mineralization in altered ultramafic rocks in south Abu Marawat, Eastern Desert, Egypt. *Sci. Rep.* **2023**, *13*, 7293. [CrossRef]
5. Bhadra, B.K.; Jain, A.K.; Karunakar, G.; Meena, H.; Rehpade, S.B.; Rao, S.S. Integrated remote sensing and geophysical techniques for shallow base metal deposits (Zn, Pb, Cu) below the gossan zone at Kalabar, Western Aravalli Belt, India. *J. Appl. Geophys.* **2021**, *191*, 104365. [CrossRef]
6. Gandhi, S.M.; Sarkar, B.C. Chapter 5—Geophysical Exploration. In *Essentials of Mineral Exploration and Evaluation*; Gandhi, S.M., Sarkar, B.C., Eds.; Elsevier: Amsterdam, The Netherlands, 2016; pp. 97–123.
7. Reedman, J.H. *Techniques in Mineral Exploration*; Springer Science & Business Media: Berlin, Germany, 2012.
8. Faruwa, A.R.; Qian, W.; Obafunmilayo, O.S.; Daramola, B.B.; Ali, M.A.H.; Dusabemariya, C.; Markus, U.I. Airborne magnetic and radiometric mapping for litho-structural settings and its significance for bitumen mineralization over Agbabu bitumen-belt southwestern Nigeria. *J. Afr. Earth Sci.* **2021**, *180*, 104222. [CrossRef]
9. Dusabemariya, C.; Qian, W.; Bagaragaza, R.; Faruwa, A.R.; Ali, M. Some Experiences of Resistivity and Induced Polarization Methods on the Exploration of Sulfide: A Review. *J. Geosci. Environ. Prot.* **2020**, *8*, 68–92. [CrossRef]
10. Ali, M.; Sun, S.; Qian, W.; Bohari, A.D.; Claire, D.; Zhang, Y. Geoelectrical tomography data processing and interpretation for Pb-Zn-Ag mineral exploration in Nash Creek, Canada. *E3S Web Conf.* **2020**, *168*, 00003. [CrossRef]
11. Mashhadi, S.R. Detecting resistivity and induced polarization anomalies of galena veins in the presence of highly chargeable and conductive geological units at Daryan barite deposit in Iran. *J. Asian Earth Sci. X* **2022**, *7*, 100086. [CrossRef]
12. Ali, M.; Sun, S.; Qian, W.; Bohari, A.D.; Claire, D.; Faruwa, A.R.; Zhang, Y. Borehole resistivity and induced polarization tomography at the Canadian Shield for Mineral Exploration in north-western Sudbury. *E3S Web Conf.* **2020**, *168*, 00002. [CrossRef]
13. Nelson, P.H.; Van Voorhis, G.D. Estimation of sulfide content from induced polarization data. *Geophysics* **1983**, *48*, 62–75. [CrossRef]
14. Gurin, G.; Titov, K.; Ilyin, Y. Induced Polarization of Rocks Containing Metallic Particles: Evidence of Passivation Effect. *Geophys. Res. Lett.* **2019**, *46*, 670–677. [CrossRef]
15. Pelton, W.H.; Ward, S.H.; Hallof, P.G.; Sill, W.R.; Nelson, P.H. Mineral discrimination and removal of inductive coupling with multifrequency IP. *Geophysics* **1978**, *43*, 588–609. [CrossRef]
16. Heritiana, A.R.; Riva, R.; Ralay, R.; Boni, R. Evaluation of flake graphite ore using self-potential (SP), electrical resistivity tomography (ERT) and induced polarization (IP) methods in east coast of Madagascar. *J. Appl. Geophys.* **2019**, *169*, 134–141. [CrossRef]
17. Goodfellow, W.D.; McCutcheon, S.R.; Peter, J.M. Massive Sulfide Deposits of the Bathurst Mining Camp, New Brunswick, and Northern Maine: Introduction and Summary of Findings. *Econ. Geol. Monogr.* **2003**, *11*, 1–16.
18. Brown, D.F. *Technical Report on Mineral Resource Estimate*; Nash Creek Project; Government of Canada Publications: Restigouche County, NB, Canada, 2007; pp. 1–190.
19. James, P.; Barr, F. *Technical Report and Updated Mineral Resource Estimate on the Nash Creek Project*; Government of Canada Publications: Restigouche County, NB, Canada, 2018.
20. Milkereit, B.; Qian, W.; Ugalde, H.; Bongajum, E.; Gräber, M. Geophysical Imaging of a "Blind" Zn-Pb-Ag Deposit. In Proceedings of the EAGE Expand, Rome, Italy, 9–12 June 2008.
21. Walker, J.A. Stratigraphy and lithogeochemistry of Early Devonian volcano-sedimentary rocks hosting the Nash Creek Zn-Pb-Ag Deposit, northern New Brunswick. In *Geological Investigations in New Brunswick for 2009*; Martin, G.L., Ed.; Mineral Resource Report 2010-1; New Brunswick Department of Natural Resources, Lands Minerals and Petroleum Division. Government of Canada Publications: Restigouche County, NB, Canada, 2010; pp. 52–97.
22. Madsen, L.M. Monte Carlo Analysis and 3D Spectral Inversion of Full-Decay Induced Polarization Data. Ph.D. Thesis, Aarhus University, Aarhus, Denmark, 2019.
23. Tavakoli, S.; Bauer, T.E.; Rasmussen, T.M.; Weihed, P.; Elming, S.-Å. Deep massive sulphide exploration using 2D and 3D geoelectrical and induced polarization data in Skellefte mining district, northern Sweden. *Geophys. Prospect.* **2016**, *64*, 1602–1619. [CrossRef]
24. Zhang, G.; Lü, Q.-T.; Lin, P.-R.; Zhang, G.-B. Electrode array and data density effects in 3D induced polarization tomography and applications for mineral exploration. *Arab. J. Geosci.* **2019**, *12*, 221. [CrossRef]
25. Loke, M.H. Rapid 2D Resistivity & IP Inversion Using the Least-Squares Method. 2018. Available online: https://www.geotomosoft.com (accessed on 2 February 2023).
26. Loke, M.H. Rapid 3-D Resistivity & IP Inversion Using the Least-Squares Method. 2018. Available online: https://www.geotomosoft.com (accessed on 27 March 2023).

27. Han, M.H.; Shin, S.W.; Park, S.; Cho, S.J.; Kim, J.H. Induced polarization imaging applied to exploration for low-sulfidation epithermal Au-Ag deposits, Seongsan mineralized district, South Korea. *J. Geophys. Eng.* **2016**, *13*, 817–823. [CrossRef]
28. Moreira, C.A.; Lopes, S.M.; Schweig, C.; da Rosa Seixas, A. Geoelectrical prospection of disseminated sulfide mineral occurrences in Camaquã sedimentary basin, Rio Grande do Sul state, Brazil. *Braz. J. Geophys.* **2012**, *30*, 169–179. [CrossRef]
29. Bongajum, E.; White, I.; Milkereit, B.; Qian, W.; Morris, W.A.; Nasseri, M.H.B.; Collins, D.S. *Multiparameter Petrophysical Characterization of an Orebody: An Exploration Case History*; ROCKENG09: Toronto, ON, Canada, 2009; pp. 1–9.
30. Loke, M. 2-D and 3-D Electrical Imaging Surveys. *Tutorial* **2004**, *29*, 29–31.
31. Dahlin, T.; Loke, M.H. Quasi-3D resistivity imaging-mapping of three dimensional structures using two dimensional DC resistivity techniques. In Proceedings of the 3rd EEGS Meeting, Aarhus, Denmark, 9–11 August 1997.
32. Embeng, S.B.N.; Meying, A.; Ndougsa-Mbarga, T.; Moreira, C.A.; Amougou, O.U.O. Delineation and Quasi-3D Modeling of Gold Mineralization Using Self-Potential (SP), Electrical Resistivity Tomography (ERT), and Induced Polarization (IP) Methods in Yassa Village, Adamawa, Cameroon: A Case Study. *Pure Appl. Geophys.* **2022**, *179*, 795–815. [CrossRef]
33. Kim, J.-H.; Tsourlos, P.; Karmis, P.; Vargemezis, G.; Yi, M.-J. 3D inversion of irregular gridded 2D electrical resistivity tomography lines: Application to sinkhole mapping at the Island of Corfu (West Greece). *Near Surf. Geophys.* **2015**, *14*, 275–285. [CrossRef]
34. Yi, M.-J.; Kim, J.; Song, Y.; Cho, S.-J.; Chung, S.-H.; Suh, J.H. Three-dimensional imaging of subsurface structures using resistivity data. *Geophys. Prospect.* **2001**, *49*, 483–497. [CrossRef]
35. Loke, M.H.; Barker, R.D. Practical techniques for 3D resistivity surveys and data inversion1. *Geophys. Prospect.* **1996**, *44*, 499–523. [CrossRef]
36. Thomas, G.; Carsten, R.; Spitzer, K. Three-dimensional modelling and inversion of dc resistivity data incorporating topography—II. Inversion. *Geophys. J. Int.* **2006**, *166*, 506–517.

MDPI
St. Alban-Anlage 66
4052 Basel
Switzerland
Tel. +41 61 683 77 34
Fax +41 61 302 89 18
www.mdpi.com

Minerals Editorial Office
E-mail: minerals@mdpi.com
www.mdpi.com/journal/minerals